Hunger, Poverty and Voice

Entitlement and Empowerment

The Authors

Dr. Anannya Chakraborty is currently working as an Assistant Professor in Lovely Professional University in Jalandhar, Punjab. She has completed her Doctoral Research as a UGC NET Junior Research Fellow under the guidance of Prof. S. K. Acharya, in the Department of Agricultural Extension, BCKV, Mohanpur WB. She has obtained her B.Sc. Agriculture degree and completed her M.Sc. In Agricultural Extension from the same University. Both UGC NET with JRF and ICAR NET have been qualified by her. So far, she has authored 5 books and 18 of her papers (2 international publications) have successfully published in peer reviewed journals and some are in progress. She has taken part actively in a number of national and international seminars and conferences. She was honoured with best paper awards twice.

Prof. Sankar Kr Acharya, former Head, Dept. of Agril Extension and Director, Extension Education, Bidhan Chandra Krishi Viswavidyalaya, Mohanpur, WB, born on 6th October, 1960, started his career as Assistant Professor at BCKV in 1988 and has been in teaching, research and extension over 30 years. Research Publication: 217 in National and International Journals. Book Publication: 97 books authored/ co-authored Membership of Distinction: Chancellor Nominee on selection committee for professors' positions; Member of Agriculture Commission; Marketing and Extension Sub Committee, Govt. of West Bengal; Expert member WWF projects in BTR (Buxa Tiger Reserve Project); Expert member; DFID project on Primary Education (DPEP); REIA, Visited Italy, France, Germany, China, Sri Lanka and Bangladesh. Delivered 18 Key note addresses in International Congress, Chaired 19 Sessions in International conferences, Editors of a score of national and international journals. He has so far been awarded with 13 best paper awards in National and International Conferences Organizing Secretary of two International Conferences, He has also been honoured to be selected as the Convener of Panel (PE-32) entitled The hunger , poverty and silence of IUAES, University of Manchester, UK, 2013. He is the fellow, ISEE, IARI, New Delhi.

Dr. Amit Baran Sharangi is a Professor of BCKV (Agricultural University) attached with teaching, research and extension for about 23 years. He completed his Post-Doctorate from the University of Melbourne, Australia. He is specially trained from IISR, IIHR & ARO (Tel Aviv, Israel); experienced in herbs, spices & MAPs. He worked at laboratories of Australia, USA and UK for continuing research on those crops as well as teaching Post Graduate students in the Universities abroad. Prof. Sharangi has several international awards including FULBRIGHT (USA), INSA-RSE Visiting Scientist (UK), Endeavour Award (Australia), Marquis Who'sWho (USA), Young Achievers Award (SADHNA), Higher Education Leadership Award, Outstanding Scientist Award, Bharat Ratna Mother Teresa Gold Medal Award, etc. He is Fellow of eight prestigious academies/societies. He has published about 86 peer-reviewed papers, 25 books from Springer Nature, Taylor & Francis, Nova Publisher, CRC Press etc and numerous book chapters within his subject area. He is also associated with 50 journals worldwide as editor-in-chief, regional & technical editor, editorial board member and active reviewer. He has administrative experience for four years as HOD, chaired several sessions of national and international conferences, run six projects so far as PI and Co-PI, developed as well as restructured PG courses as BSMA committee member (DG, ICAR, Govt of India) at national level. New Delhi; World Scientist Ranking, AD Scientific Index, 2021.

Hunger, Poverty and Voice

Entitlement and Empowerment

Anannya Chakraborty
Sankar Kr Acharya
Amit Baran Sharangi

2022
Scholars World
A Division of
Astral International Pvt. Ltd.
New Delhi – 110 002

ISBN: 978-93-5461-362-3 (HB)

Published by : **Scholars World**
A Division of
Astral International Pvt. Ltd.
– ISO 9001:2015 Certified Company –
4736/23, Ansari Road, Darya Ganj
New Delhi-110 002
Ph. 011-4354 9197, 2327 8134
E-mail: info@astralint.com
Website: www.astralint.com

Digitally Printed at : **Replika Press Pvt. Ltd.**

Foreword

Prof. J. P. Srivastava
Former Prof. Emeritus
Dept. of Agriculture Extension
& Communication SHUATS
Naini Prayagraj, U.P

The correlation between hunger and poverty stands so linear that it is well discernible and explicable; people are hungry because they are poor, or poor people are destined to go hungry. But its non-linear passage to voice in dovetailing huger-poverty-voice is both empirically and theoretically has been a complex proposition. The issues of voice related to hunger and poverty have seldom been dealt with. There are several books on hunger and poverty. But voice as a determinant in resolving hunger has been studied that much. Albeit, the World Bank's' voices of the poor' project was a huge undertaking which focused attention on both the issue of voice as self-expression in terms of people speaking for themselves about their own circumstances and, what they feel about poverty-related issues, and, on the need for participatory processes in wider conceptualizations and definitions of poverty. So, the proposed book is going to be a unique creation, which, based on empirical evidences has made an attempt to find the missing links between hunger and poverty. It raises and resolves the question as to people remains poor because they are mostly voiceless and because they are voiceless and poor, they continue to remain hungry. A neo pedagogy of oppressed has taken care of lethal combination amongst and between hunger-poverty-aphonics. The methodological innovations and analytical processes has uniquely been applied to reinterpret and re-examine the lethality of interaction among and between hunger, poverty and voice have not only been unique, but possesses are huge theoretical and policy dimension. I strongly believe, this book is going to be a unique genre of social science intellectuality for this sub-continent and for the global audience as well.

Prof. J.P Srivastava

Preface

History of human civilization initiated with the struggle against hunger, it flourishes with the fight against poverty and turns oblivious with absence of voices being raised against these two bio-social lethals. If hunger is a pain, poverty is an agony, then, keeping silent against this carcinoma of civilization is no doubt a crime. Is it that poor people are put to an insurmountable silence by the ruthless hegemonies of political economy, or, they are silent, that's why they remain poor? Silence has been the lethal enzyme in transforming poverty into hunger, and, offers to a socio-chemical bond, poverty-hunger-silence; the bond that remains non-breakable as on date. Economists lament for poverty, nutritionists regret hunger, but the issues of silence and its cohesion to both hunger and poverty has seldom been scholastically delved into or dealt with. We can hope, a desperation of wishful thinking as it may be, this book shall drench the researchers who think and feel to explore a missing link between hunger and poverty, the silence or the social anemia of voices.

Anannya Chakraborty

Sankar Kr Acharya

Amit Baran Sharangi

Contents

1 CHAPTER

INTRODUCTION

Hunger is the sauce of life that drives us to the golden harvest; it's the fire either, that destroys socio-ecological balances to invite ever increasing entropy. The UNO in its recent press release express concerns that more millions will die not due to COVID pandemic only, it is because of hunger and poverty. The humongous up surging of joblessness, economic disaster and social disorder due to this pandemic, the poor, hungry and voiceless people are going to die, a repetition of social Titanic episode can be as worst as an analogy!

In India only close to 10 million migrant workers have become re-migrant for the worse destiny, a close to five million have lost their job or extremely vulnerable to a sudden joblessness due to economic apocalypse as reflected in a appalling -23 per cent growth revealed by the ministry of finance, India. The need of the hour is to elicit the factual data on the damaged and vulnerability status with an immediate restart of economical repairing and social healing process. We have no choice but to make people, those who are reeling under dire marginalization, free from fear psychosis by restoration of both economy and morals.

Silence is the moth lethal inputs that make poor the poorer, deprived with a stubborn silence and the hungry with an unacceptable negotiation with social hegemony and extortion inflicted on the poor by the mighty richer community. So, a comprehensive research offers an inevitability of a study that would delineate and decipher the lethality of a venomous combination amongst and between hunger-poverty-silence, the trio that causes deadening consequences of deprived people and the civilization at length.

Even with the swashbuckling claim on growth and prosperity on the present civilization, the other side of this prosperity is so brick and disastrous that have no match for the past centuries even. Out of around 7 billion population of the world, 1.5 billion are hungry. They don't have adequate access to food, if it is there, the quality doesn't stand anywhere near to fulfil their calorie requirement. In India, 350 million people are living below the poverty line and of them, 200 million people have become victim to moderate to extreme hunger indexes. 42 per cent of the new born babies are under weight. 60 per cent of the children are suffering from moderate to high level of anaemia experiencing stunted growth.

Beyond the curtain of hunger, there is another problem that is chronic hunger. Based on hunger index we the nation is occupying 100th position in the world (IFPRI Report, 2017). The scenario of chronic hunger is even worse and astoundingly it is worse than African nations as well.

Quoting back of Gunner Myrdal, a German analyst (Wrote' Asian Drama '), the reasons for sluggish growth of Indian Agriculture is her Agricultural work force, who themselves are suffering from malnutrition, hunger and social impoverishment.

Indian agriculture is the largest but unorganised economic sector of the world. The farm entrepreneurs, are suffering from both the vagaries of nature and market. The decelerating agricultural economy has thrown the growers into a vicious cycle of hunger, poverty and silence. Hunger here has been denoted by the level of food, calorie and nutrition intake by human bodies within the framework of minimum requirement set by World Health Organisation(WHO) .Poverty here has been measured in terms of per capita income per month from a unit of holding, which is 121 million in India(NSSO data, 2014-2015).

The silence here is measured in terms of ability and permeability to communicate with others or availing communication from others as to their need for survival and growth. The unique character of studying silence 'uninformed diaspora' in India is that for farmers huge pile of information are there, but only a minuscule proportion is being accessed by the farmers. The voice for Minimum Support Price, crop insurance, health and nutrition, security and sustainability, is either plan fully unheard or they are refrained from uttering their voices to defend their stakes. The silent killer is the lethargy of uninformed life, when an individual is restrained either from accessing information or uttering opinion.

So, these three lethality are both technically and operationally combined together. Simply enough, it is to withstand the corollary that states Indian farmers are poor and that is why they are hungry; they are both poor and hungry because they are silent.

OBJECTIVES

1. To organize selected base line survey for delineating the present scenario of hunger, poverty and voice.
2. To organize participatory analysis of poverty, hunger and voice to generate primary and qualitative information on the issues mentioned.
3. To analyse and estimate the intra and inter level relation and interaction by taking poverty hunger and voice as dependent variables to be predicted and elucidated from a score of socio-ecological and techno-economic variables selected for the study.
4. To generate a micro level policy based on empirical study and thereafter to create experimental paradigm for its further replication.

CHAPTER 2

THEORIZING HUNGER, POVERTY AND VOICE

HUNGER: A DEVIL TO DEFEAT

What is Hunger?

Hunger is an everyday phenomena. Most people are able to satisfy this need. Even if not immediately, they can count on having a meal or snack within hours. This is not the type of hunger that Bread is concerned with.

Inadequate supply of food while an individual is hungry leads to chronic hunger. They do not get enough calories, essential nutrients, or both to sustain . People who are hungry have a continuous problem with getting food to eat. They have a primary need — how to feed themselves and their children today and tomorrow. Anything else comes later.

In politics, humanitarian aid, and social science," **hunger** is a condition in which a person, for a sustained period, is unable to eat sufficient food to meet basic nutritional needs".

The world population has experienced a sustained period of hunger time and again throughout the history of our civilization. In many cases, this food supply disruptions were caused by war, plagues, adverse weather or sometimes even man made. For the first few decades after World War II, technological progress and enhanced political cooperation has made the struggle against hunger and famine quiet easier. While progress was uneven, by 2000 the threat of extreme hunger subsided for many of the world's people. According to the WFP (World Food Policy) statistics show that, "Some 795 million people in the world do not have enough food to lead a healthy active life. That's about one in nine people on earth. The vast majority of the world's hungry people live in developing countries, where 12.9 per cent of the population is undernourished."

The average international price of food had been predominantly stable for several decades until 2006. In the closing months of 2006, however, prices began to rise at a galloping pace. By 2008, rice had tripled in price in some regions, and this severely affected the developing countries. Food prices fell in early 2009, but rose to another record margin in 2011, and have since decreased slightly. The 2008 worldwide financial crisis further increased the number of people suffering from hunger, including dramatic increases even in advanced economies such as Great Britain, the Eurozone and the United States.

The Millennium Development Goals was implemented with a commitment to reduce 50% of the world's population suffering from extreme hunger by 2015. Though as of 2012, this target appeared difficult to achieve, due to inflation in food prices. However, in late 2012 the UN's Food and Agriculture Organization (FAO) stated it is still possible to hit the target with sufficient effort. In 2013, the FAO estimated that 842 million people are undernourished (i.e.12% of the global population). Malnutrition is a cause of death for more than 3.1 million children under 5 every year. UNICEF estimates 300 million children go to bed hungry each night; and that 8000 c

Hunger: The physiological definition

The physiological urge to eat food represents hunger. The physical sensation of hunger is related to contractions of the stomach muscles. These contractions—sometimes called hunger pangs once they become severe—are believed to be triggered by high concentrations of the ghrelin hormone. The hormones Peptide YY and Leptin can have an opposite effect on the appetite, causing the sensation of being full. Ghrelin can be released if blood sugar levels get low—a condition that can result from long periods without eating. Stomach contractions from hunger can be especially severe and painful in children and young adults.

The irregular consumption of meals makes the hunger pangs even worse. People who can't afford to eat more than once a day sometimes refuse one-off additional meals, because if they don't eat at around the same time on the next days, they may suffer extra severe hunger pangs. Older people may feel less violent stomach contractions when they get hungry, but still suffer the secondary effects resulting from low food intake: these include weakness, irritability and decreased concentration. Prolonged lack of adequate nutrition also causes increased susceptibility to disease and reduced ability for the body to self heal.

HUNGER: ITS ASPECTS AND PARADIGM

Hunger and Gender

In both developing and advanced countries, feeding children is a priority that the parents fulfill at the cost of their own stomach. Women, however, seem more likely to make this sacrifice than men. World Bank studies consistently find that about 60%

of those who are hungry are female. The apparent explanation for this imbalance is that, compared to men, women more often forgo meals in order to feed their children.

The greater sexual inequality has made this phenomena unique to the developing countries since a very long period of time. More recent findings suggested that mothers often miss meals in advanced economies too. For example, a 2012 study undertaken by Netmums in the UK found that one in five mothers sometimes misses out on food to save their children from hunger.

With regard to several periods and regions, gender has also played as a crucial factor in determining whether or not victims of hunger would make suitable examples for generating enthusiasm for hunger relief efforts. James Vernon, in his *Hunger: A Modern History*, wrote that in Britain before the 20th century, it was generally only women and children suffering from hunger who could arouse compassion. Men who failed to provide for themselves and their families were often regarded with contempt.

This changed after World War I, where securing employment was a struggle for the thousands of men who had proved their manliness in combat. Similarly, female gender could be advantageous for those wishing to advocate for hunger relief, with Vernon writing that being a woman helped Emily Hobhouse draw the plight of hungry people to wider attention during the Second Boer War.

Source-*Migrant Mother* **by Dorothea Lange(1936)**

Hunger and politics

"when I die, don't built a monument on me. Don't bestow me degrees from great universities. Just clothe the naked. Say that I tried to house the homeless. Let people say that I tried to feed the hungry"...Martin Luther King.

World War II gave birth to a new international politico-economic order, which was later described as Embedded liberalism. For at least the first decade after the war, the United States, by far the period's most dominant national actor, was strongly supportive of efforts to tackle world hunger and to promote international development. It heavily funded the United Nation's development programmes, and later the efforts of other multilateral organizations like the International Monetary Fund (IMF) and the World Bank (WB).The newly established United Nations became a leading player in co-ordinating the global fight against hunger. The UN has three agencies that work to promote food security and agricultural development: the Food and Agriculture Organization (FAO), the World Food Programme (WFP) and the International Fund for Agricultural Development (IFAD). FAO is the world›s agricultural knowledge agency, providing policy and technical assistance to developing countries to promote food security, nutrition and sustainable agricultural production, particularly in rural areas.WFP›s key mission is to deliver food into the hands of the hungry poor. The agency steps in during emergencies and uses food to aid recovery after emergencies. Its longer term approaches to hunger helps the transition from recovery to development. IFAD, with its knowledge of rural poverty and exclusive focus on poor rural people, designs and implements programmes to help those people access the assets, services and opportunities they need to overcome poverty. Following successful post WWII reconstruction of Germany and Japan, the IMF and WB began to turn their attention to the developing world. A great many civil society actors were also active in trying to combat hunger, especially after the late 1970s when global media began to bring the plight of starving people in places like Ethiopia to wider attention. Most significant of all, especially in the late 1960s and 70s, the Green revolution helped improved agricultural technology propagate throughout the world. The United States began to change its approach to the problem of world hunger from about the mid 1950s. Influential members of the administration became less enthusiastic about methods they saw as promoting an over reliance on the state, as they feared that might assist the spread of communism. Despite this view, during the 1960s postwar era hunger within the United States was overshadowed by hunger in Europe and Asia. John F. Kennedy did in fact use Executive Order to double the amount of commodities available from the surplus commodity program as well as initiated the pilot Food Stamp Program which later became permanent in 1964. By the 1980s, the previous consensus in favour of moderate government intervention had been displaced across the western world. The IMF and World Bank in particular began to promote market-based solutions. In cases where countries became dependent on the IMF, they sometimes forced national governments to prioritize debt repayments and sharply cut public services. This sometimes had a negative effect on efforts to combat hunger.

Increased use of irrigation played a major role in the Green Revolution. (Source-internet)

The issue of food sovereignty was brought forward by Organizations such as Food First, it further claimed that every country on earth (with the possible minor exceptions of some city-states) has sufficient agricultural capacity to feed its own people, but that the "free trade" economic order, which from the late 1970s to about 2008 had been associated with such institutions as the IMF and World Bank, had prevented this from happening.

The World Bank claimed to be a part of the solution to hunger, asserting that the best way for countries to break the cycle of poverty and hunger was to build export-led economies that provide the financial means to buy foodstuffs on the world market. However, in the early 21st century the World Bank and IMF became less dogmatic about promoting free market reforms. They increasingly returned to the view that government intervention does have a role to play, and that it can be advisable for governments to support food security with policies favourable to domestic agriculture, even for countries that do not have a Comparative advantage in that area. As of 2012, the World Bank remains active in helping governments to intervene against hunger.

Until at least the 1980s—and, to an extent, the 1990s—the problem of demand exceeding supply was a dominant academic outlook with regard to world hunger. Proposed solutions often focused on boosting food production, and sometimes on birth control. There were exceptions to this, even as early as the 1940s, Lord Boyd-Orr, the first head of the UN's FAO, had perceived hunger as largely a problem of distribution, and drew up comprehensive plans to correct this. Few agreed with him at the time, however, and he resigned after failing to secure support for his plans from the US and Great Britain. In 1998, Amartya Sen won a Nobel Prize in part for demonstrating that hunger in modern times is not typically the product of a lack of food. Rather, hunger usually arises from food distribution problems, or from governmental policies in the developed and developing world. It has since been broadly accepted that world hunger results from issues with the distribution as well as the production of food. Sen›s 1981 essay *Poverty and Famines: An Essay on Entitlement and Deprivation* played a prominent part in forging the new consensus. In 2007 and 2008, rapidly increasing food prices caused a *global food crisis*, increasing the numbers suffering from hunger by over a hundred million. Food riots erupted in several dozen countries; in at least two cases, Haiti and Madagascar, this led to the toppling of governments. A second *global food crisis* unfolded due to the spike in food prices of late 2010 and early 2011. Fewer food riots occurred, due in part to greater availability of food stock piles for relief. However, several analysts argue the food crisis was one of the causes of the Arab Spring.

In the Modernizing Agriculture The pathway and process book of Prof. S K Acharya and Prof. Hemanga Kalita it was seen that one of the major way outs of modern and costumed agricultural practice is to increase the farmers' B C ratio and to improve the farmers' livelihood.

Table 1.1: Difference in between hunger status of India and USA

Hunger in India	Hunger in USA
India has approximately 400 million people who face absolute ($1.25/day poverty), and is home to about one-quarter of the people facing hunger globally. More than 40% of Indian children experience malnutrition. These are awful facts. Yet, India is taking steps to change; small in comparison to rich-country budgets, but big for a developing country.	In the US, more than 45 million people face some kind of food insecurity and qualify for food assistance. It's not a lot by US standards, about $4 a day. But, politicians in Washington are looking for ways to cut them off.
The new law in India seeks to provide affordable food to 800 million Indians.	The bill in the House of Representatives will likely deny food assistance to 4 million Americans.
India is willing to commit approximately $20 billion annually to provide this food assistance, about 2% of national GDP.	The House of Representatives is trying to cut approximately $4 billion annually from a $75 billion food assistance program, which is about half of one-percent of US GDP.
A separate law in India, the National Rural Employment Guarantee Act enacted in 2005, guarantees the poorest people 100 days of employment to provide for basic income.	The bill in the House of Representatives would impose new work requirements on poor people to qualify for food assistance. It does not offer any help in meeting these requirement, much less guaranteeing employment.
The India Food Security Law was unopposed by any major political party. The political class in India is united by the goal to end hunger.	The House Bill was passed with an exclusively Republican majority. Only 15 Members of Congress crossed party lines for the vote. In the US, hunger has become a partisan issue.

Hunger and India

The inadequacy in dealing with the grave hunger problems makes India persistently secure a lower position of 100th in 118 developing countries as far as the Global Hunger Index (GHI) is concerned (IFPRI Report,2017).Countries worse than India include extremely poor African countries such as Niger, Chad, Ethiopia and Sierra Leone besides two of India's neighbours: Afghanistan and Pakistan. Whereas other neighbours Sri Lanka, Bangladesh, Nepal and China are all ranked above India. The GHI is calculated by taking into consideration four key parameters: share of undernourished population, wasted and stunted children aged under 5, and infant mortality rate of the same age group. Among the 131 countries studied, data was available for 118 countries. This year, for the first time, two measures of child hunger that are wasting and stunting have been used to give a more complete picture. Wasting refers to low weight in relation to a child's height, reflecting acute under nourishment and stunting refers to the less height in relation to age, reflecting the same. The International Food Policy Research Institute (IFPRI) makes the annual calculations of GHI. Based on its readings according to the most recent data, the 2016 GHI for India was derived from the fact that an estimated 15% population is undernourished and are lacking in adequate food intake, both in quantity and quality .The share of under 5 children who are `wasted› is about 15%, while the share of children who are `stunted› is have a staggering percentage of 39. This reflects widespread and chronic deficiency of balanced food. The under-5 mortality rate is 4.8% in India, partially reflecting the fatal synergy of inadequate nutrition and unhealthy environments.

WORLD HUNGER INDEX - HOW INDIA COMPARES

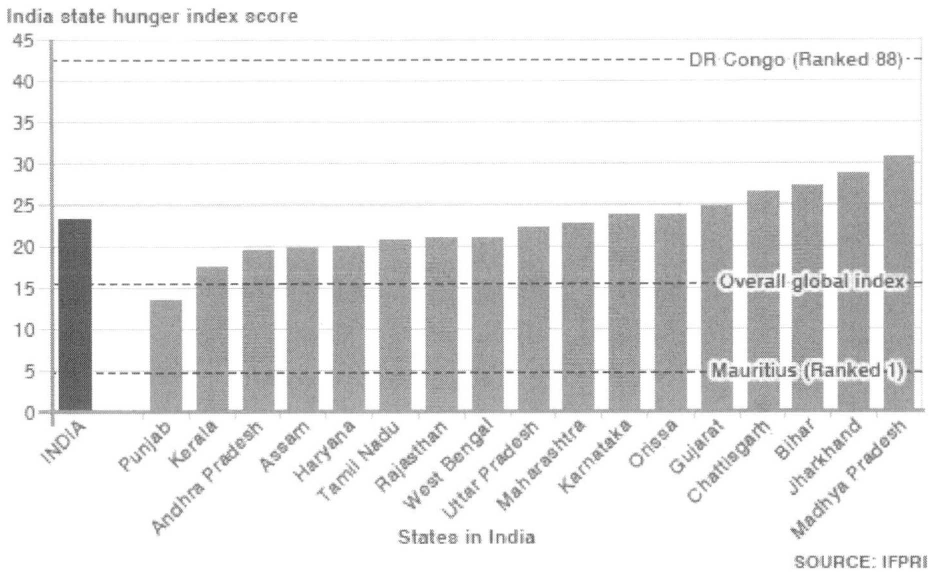

SOURCE: IFPRI

Poverty....the definition

There has umpteen definitions of poverty where the most widely held, accepted and understood definition of absolute poverty measures poverty strictly in economic terms — earning less than $1.90 a day. But the World Bank goes beyond the economic definition of poverty, instead of this they expand the definition of poverty into different aspects...

"Poverty is hunger. Poverty is lack of shelter. Poverty is being sick and not being able to see a doctor. Poverty is not having access to school and not knowing how to read. Poverty is not having a job, is fear for the future, living one day at a time. Poverty is losing a child to illness brought about by unclean water. Poverty is powerlessness, lack of representation and freedom."

This poverty definition specifically encircles living conditions, inability to meet basic needs because food, clean drinking water, proper sanitation, education, health care and other social services are inaccessible.

This poverty threshold breeds on the fear for the future and broadens to engulf dependence, oppression and even exploitation.

This larger measure of poverty expands the contributors and causes of poverty, the World Bank developed indicators to assess the non-income dimensions of poverty as well. The indicators include education, health, access to social services, vulnerability, social exclusion, and access to social capital.

Poverty described

How do the poor experience poverty?

The child and women are indubitably the most vulnerable section of the society as far as hunger is concerned in most of the under developed and developing countries. As they are dependent on others, their rights and freedoms is restricted. They live without support. Their dignity is assaulted and humiliated every day and their lives are abundant...in scarcity.

In a 2002 survey, the poor in Niger reported a standard of living characterized by:

- suffering relationships
- always having to seek out others, or to work for somebody else
- being alone, unsupported, uninvolved and "never consulted"
- having nothing to eat, lacking the means to meet clothing and financial needs and having nothing to sell
- restricted rights and freedoms
- incapacity to make decisions, to feed or clothe oneself, or to act on one's own initiative

Similar descriptions were found in a major World Bank study, *Voices of the poor: Can anyone hear us?*

But the humiliating, silencing, life-destroying violence of poverty is not inescapable.

Poverty can be overcome. And child sponsorship can help.

We work with thousands of local church partners in some of the world's poorest countries to identify the poorest of the poor. Our approach to helping the poor starts when they are children and when they're most vulnerable.

The poverty wheel and the world's poor

The problem of world poverty can be illustrated through a good visualization device i.e. a bicycle wheel. The wheel's hub represents a life in which survival is a daily accomplishment; the hub is a life of absolute poverty.

The wheel's spokes represent basic human needs, and the wheel's rim represents a life fulfilled.

Dimensions of Poverty

Social poverty

The "social spoke" of poverty encloses undervalued peoples group, people with few or no rights and people whose voices have been silenced. A culture or government that devalues people — especially children — cultivates social poverty.

We work with governments and cultural systems to encourage the idea that children are valuable, should be cherished, and should be given every opportunity to flourish.

Educational poverty

Education is an unaffordable luxury for hundreds of millions of children around the world. Lack of education (educational poverty) creates a lack of options. It makes finding future work difficult, and it makes children vulnerable to exploitation. An education offers the knowledge, skills, and training to create a better future. The chance to receive a quality education changes everything. An education offers hope of a life beyond poverty.

In order to encourage children to complete and go beyond a primary education we provide resources. We help students stay in school and receive an education that enables them to succeed and thrive in life. We also expand the circle of caring adults actively participating in a child's life as friends, teachers and mentors. We provide a consistent and dependable source of learning and support for children and their families, including literacy training for caregivers and parents.

Health poverty

Health poverty may seem absurd, but there are many people around the world who are still unaware of the importance of brushing their teeth or making sure the water they drink is clean.

Physical and emotional health is the basis for our ability to work, play and develop sound relationships. That's why we teach children how to monitor their own health for common diseases and provide hygiene training. We also offer medical assistance when more serious needs arise. We teach preventative care and help with emergency surgeries, organ transplants, casts, wheelchairs and canes, cancer treatment, hearing and eye care, and dental needs.

Spiritual poverty

Without an understanding of God's love, it is difficult to resist despair. Children are particularly vulnerable to the emotional and spiritual messages of worthlessness and hopelessness poverty delivers. But their lives undergo a revolution when they realize God loves and values them. They understand they were placed on the earth with a divine purpose in mind.

We devote ourselves to helping children of all faiths, cultures, backgrounds and races experience unbiased Almighty's love and embrace who they were created to be. Through a holistic approach to child development, we carefully blend physical, social, economic and spiritual care together. This includes teaching God's Word.

We demonstrate our love and live out our faith by extending care to others irrespective of caste, creed, class or religion, the help we offer is so pure without any ulterior motive and solely for the love of God.. We do not coerce anyone into becoming Christian, and we do not impose conversion requirements on those participating in our programs.

Environmental poverty

The circumstances of absolute poverty is not indifferent to extreme environmental risk, our well-being is a conglomeration of our physical surroundings — climate, water supply, housing and land . From insect and water-borne illnesses to extreme weather conditions, such as drought and flooding, the spirits, hope and health of the poor are depressed by the environment they live in. We work to ensure environmental conditions don't make extreme poverty inescapable.

Economic poverty

Almost half the world lives with a household income below $2.50 a day. This level of poverty is the equal of slavery. People need an income level which allows them to purchase what they cannot make or grow.

In order to help children create a life filled with economic opportunity we provide vocational training and offer job-skills development. Our income-generation training includes teaching business and entrepreneurship, financing start-ups, providing seed capital and micro-loans. We also offer financial training for a child's caregivers and parents.

Poverty: a battle to win

Poverty is the scarcity or the lack of a certain (variant) amount of material possessions or money. It is a multifaceted concept, which may include social, economic, and political elements. Absolute poverty, extreme poverty, or destitution refers to the complete lack of the money necessary to meet basic personal needs such as food, clothing and shelter.

The threshold at which **absolute poverty** is defined is considered to be more or less same, independent of the person›s permanent location or era. On the other hand, **relative poverty** occurs when a person who lives in a given country does not enjoy a certain minimum level of «living standards» as compared to the rest of the population of that very same country. Therefore, the threshold at which **relative poverty** is defined varies from one country to country or society to society.

The various constraints restrict the government's ability to provide basic needs and deliver services, such as corruption, tax avoidance, debt and loan conditionality and by the brain drain of health care and educational professionals. Strategies of increasing income to make basic needs more affordable typically include welfare, economic freedoms and transparent governance.

Causes of Poverty

The International Food Study Institute had a brief on a collection of extensive studies for examining the causes of poverty by analysing household data and reviewing empirical research in 20 countries. They found that some of the major causes of poverty were the inability of poor households to invest in property and education, limited access to credit; in some cases these instances produce more poverty via inherited debt or bondage. The systematic exclusion of ethnic minorities, scheduled castes, tribes, women and people with disabilities and health issues. Persistence of poverty is partially attributed to these classes not having access to institutions, markets, proper education, lack of information and so on.

War and violence undoubtedly has been a primary cause of poverty. Political violence and organized crime have affected 39 countries since 2000, in those countries the poverty level is twice higher than that of non-violent countries. These two items, poverty and violence may also complement each other. When the researchers asked the common people of half a dozen countries that why young people joined gangs and rebel groups, two thirds of the respondents said that unemployment was their main reason, only one tenth cited a belief in the cause.

Colonialism has been stated to be a cause of poverty; this idea was rejected by Walter E. Williams citing that not all post colonial countries have high poverty rates.

All the third World countries have low GDP per capita per cent taxed, middle income economies have average levels of GDP per capita taxed whereas the developed countries have higher per cent GDP per capita taxed.

Main causes of poverty

Primary causes

1. Lack of control in local food

It highlights the uneven and discriminated supply of food throughout the world. Though there is enough food in the world that we can feed the overall population, but the food is clubbed or scanted over a few areas.

2. Corruption:

Corruption is most likely the root cause as well as the result of poverty around the world.

3. Lack of access to education

Lack of education is one of the causes of over population. We the country which is marred in illiteracy, ignorance and over tolerance. It is very easy for the population to be absolutely unaware about their rights and from this loophole there comes the evil called 'poverty'.

4. Colonial slavery

The major cause of poverty in most counties are long years of colonial slavery along with subsequent exploitation. In the 200 years of British colonial rule, India had faced a drastic financial downsizing; which it is facing still now.

The above mentioned causes of poverty are in broader perspective, except of these there are a lot more direct and indirect reasons behind poverty like- epidemic, war, natural disaster etc.

Extreme poverty

According to World Bank's revised estimates for extreme poverty in addition with regional economic development, extreme poverty rates have subsequently fallen in China and India. In other countries, extreme poverty has increased per 2011 benchmarks compared to 2005 benchmarks.

Extreme poverty, absolute poverty, destitution, or **penury**, was originally defined by the United Nations in 1995 as "a condition characterized by severe deprivation of basic human needs, including food, safe drinking water, sanitation facilities, health, shelter, education and information. It depends not only on income but also on access to services. In 2008, "extreme poverty" widely refers to earning below the international poverty line of $1.25/day (in 2005 prices), set by the World Bank. This measure is the equivalent to earning $1.00 a day in 1996 US prices, hence the widely used expression, living on less than a dollar a day. The vast majority of those in extreme poverty i.e.96% reside in South Asia, Sub-Saharan Africa, The West Indies, East Asia and the Pacific; nearly half live in India and China alone.

Reduction of extreme poverty and hunger was the first Millennium Development Goal (MDG1), as set by 189 United Nations Member States in 2000. Specifically, MDG1 set a target of reducing the extreme poverty rate by half within 2015, a goal that was met 5 years ahead of schedule. This goal was created to end poverty in all its forms from everywhere, and the international community, including the UN, the World Bank and the United States, has set a target of ending extreme poverty by 2030.

Common criticism/alternatives

Despite of being widely accepted by most international organizations, the $1.25/day extreme poverty line has come under scrutiny from a variety of factors. For example, when used to measure headcount ratio (i.e. the percentage of people living below the line), the $1.25/day line is unable to capture other important measures such as depth of poverty, relative poverty and how people view their own financial situation (known as the "socially subjective poverty line"). Moreover, the calculation of the poverty line relies on several debatable assumptions as far as the purchasing power , homogeneity of household size , and consumer prices are concerned. Another mentionable fact is, there may be missing data from the poorest and most fragile countries which may muddle the picture even further.

Several alternative instruments for measuring extreme poverty have been suggested which incorporate other factors such as malnutrition and lack of access to a basic education. Thus, the 2010 Human Development Report introduced the Multidimensional Poverty Index (MPI), which measures not only income, but also basic needs. Using this tool, the United Nations Development Programme (UNDP) estimated that roughly 1.5 billion people remained in extreme poverty discarding the conventional figure of 1.2 billion. As this figure is considered more "holistic," it may add a new dimension in deprivation within a country. For example, in Ethiopia, 39% of the population is considered extremely poor under conventional measures, but 90% are in multidimensional poverty.

The Alkire-Foster Method is another version of the MPI which is created by Sabina Alkire and James Foster of the Oxford Poverty & Human Development Initiative (OPHI), can be broken down to reflect both the incidence and the intensity of poverty. This tool is useful as development officials, using the "M0 measure" of the method (which is calculated by multiplying "the proportion of people who are poor by the percentage of dimensions in which they are deprived"), can determine the most likely causes of poverty within a region. For example, in the Gaza Strip of Palestine, using the M0 measure of the Alkire-Foster method reveals that poverty in the region is primarily caused by the lack of access to electricity and drinking water as well as widespread overcrowding. In contrast, data from the Chhukha District of Bhutan reveals that income is a much larger contributor to poverty than the other dimensions within the region.

Current trends

Getting to zero

The worldwide poverty rate and the total number of extremely poor have shown a steady decline since the mid-1990s. In 1990, the percentage of the global population living in extreme poverty was 43%, but in 2011, that percentage had dropped down to 21%.This halving of the extreme poverty rate falls in line with the first millennium development goal (MDG1) proposed by former UN Secretary-General Kofi Annan. This reduction in extreme poverty took place in China, Indonesia, India, Pakistan and Vietnam in a notable way. These five countries accounted for the alleviation of 715 million people out of extreme poverty between 1990 and 2010 – more than the global net total of roughly 700 million. This statistical odd can be explained by the fact that the number of people living in extreme poverty in Sub-Saharan Africa rose from 290 million to 414 million over the same period. However, there have been many positive signs for extensive, global poverty reduction as well. Since 1999, the total number of extremely poor has declined by 50 million per year, on an average. Moreover, in 2005, for the first time in recorded history, poverty rates began to fall in every region of the world, including Africa. Although this is largely due to a change in the 2000 UN Millennium Declaration, extending the plan period backward to 1990, which was previously 1996. Changing the date took advantage of rapid population growth and a huge poverty reduction in China during the 1990s. The number of people living in extreme poverty has reduced from 1.9 billion to 1.2 billion over the span of the last 20–25 years. If we maintain our current trajectory, many economists predict we could reach global "zero" by 2030–2035, thus "ending" extreme poverty. Global zero entitles a world in which fewer than 3% of the global population lives in extreme poverty (projected under most optimistic scenarios to be fewer than 200 million people). This "zero" figure is set at 3% in recognition of the fact that some amount of "frictional" poverty will continue to exist, whether it is caused by political conflict or unexpected economic fluctuations, at least for the foreseeable future. However, the Brookings Institution notes that any projection about poverty more than a few years into the future runs the risk of being highly uncertain. The reason behind it is the changes in consumption and distribution throughout the developing world over the next two decades could result in monumental shifts in global poverty, for better or worse.

Scholars of other schools are more skeptical about this possibility, with many predicting a range of 193 million to 660 million people living in extreme poverty by 2035. Additionally, some believe that the rate of poverty reduction will slow down in the developing world, especially in Africa, and as such it will take closer to five decades to reach global "zero." In spite of these reservations, several prominent international and national organizations, including the UN, the World Bank and the United States Federal Government (via USAID), have set a target of reaching global zero by the end of 2030.

Exacerbating factors

A variety of factors combine to reinforce or instigate the existence of extreme poverty, such as weak institutions, cycles of violence and a low level of growth. Very recently World Bank research shows that some countries can get caught in a "fragility trap," in which the above factors prevent the poorest nations from emerging from low-level equilibrium in the long run. Moreover, most of the reduction in extreme poverty over the past twenty years has taken place in countries that have not experienced a civil conflict or have had governing institutions with a strong capacity to govern. Therefore to end extreme poverty, it is also important to focus on the interrelated problems of fragility and conflict.

The United State Agency for International Development (USAID) defines fragility as a government's lack of both legitimacy (the perception the government is adequate at doing its job) and effectiveness (how good the government is at maintaining law and order, in an equitable manner). The fragile nations are unable to equitably and effectively perform the functions of a state, hence these countries are much more prone to violent unrest and mass inequality. In addition, the countries with high levels of inequality (a common problem in countries with inadequate governing institutions), much higher growth rates are needed to reduce the rate of poverty when compared with other nations. Not to mention, after removing

China and India from the equation, up to 70% of the world's poor live in fragile states by some definitions of fragility. Looking further, some analysts project that extreme poverty will be increasingly concentrated in fragile, low-income states like Haiti, Yemen and the Central African Republic over the coming years. Whereas academicians like Andy Sumner, assert that extreme poverty will be increasingly found concentrated in Middle Income Countries, creating a "poverty paradox" – as the World's poor don't actually live in the poorest countries.

Addressing the problem of fragility remains an issue, despite of these several school of thoughts. To help low-income, fragile states make the transition towards peace and prosperity, the New Deal for Engagement in Fragile States, endorsed by roughly forty countries and multilateral institutions, was created in 2011. The idea of "New Deal," represents an important step towards redressing the problem of fragility as it was originally articulated by self-identified fragile states who called on the international community to not only "do things differently," but to also "do different things."

Civil conflict tends to be a prime cause for the perpetuation of poverty throughout the developing world. Agitations, armed movements, war can have severe effects on economic growth for a plethora of reasons – it destroys assets, creates unwanted mass migration, destroys livelihoods and diverts public resources towards war fighting. Significantly, a country that experienced major violence during 1981–2005 had extreme poverty rates 21 percentage points higher than a country with no violence. On average, a civil conflict will also cost a country roughly 30 years of GDP growth. Hence, a stong commitment from the international community is needed to address the deteriorating situation in highly fragile states is necessary for not only preventing the mass loss of lives, but to also to prevent the vicious cycle of extreme poverty.

As per the World Bank report (2013), extreme poverty is observed prevalently in low income countries. The World Bank found that in these countries progress in poverty reduction is slowest, the poor live under terrible conditions and the most affected persons are children under the age of 12.

International conferences

Millennium Summit

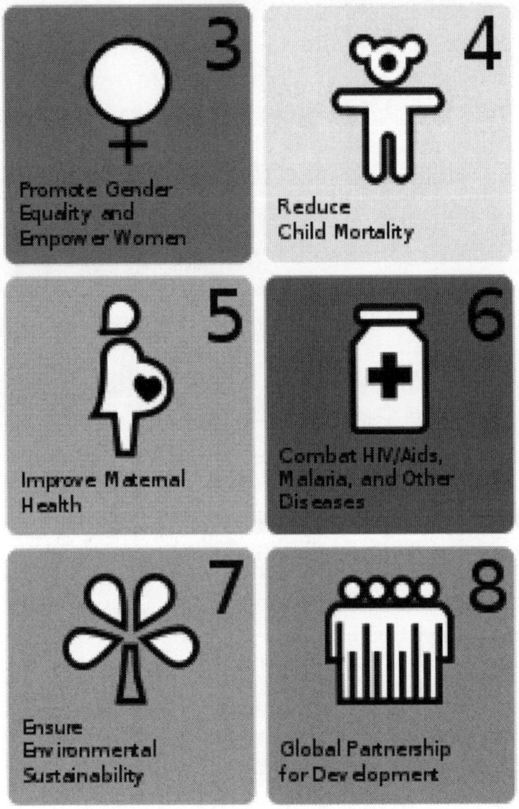

Official logos for each of the Millennium Development Goals (source internet)

World leaders gathered at the Millennium Summit held in New York in September 2000, launching the United Nations Millennium Project suggested by then UN Secretary-General Kofi Annan. Prior to the launch of the conference, the office of Secretary-General Annan released a report entitled" We The Peoples: The Role of the United Nations in the 21st Century". The document above sited (Widely known as the Millennium Report), Kofi Annan called on the international community "to adopt the target of halving the proportion of people living in extreme poverty, and so lifting more than 1 billion people out of it, by 2015." Citing studies that show "an almost perfect correlation between growth and poverty reduction in poor countries," Annan urged international leaders to indiscriminately target the problem of extreme poverty across every region. The project In charge was Jeffrey Sachs, a noted development economist, who in 2005 released a plan for action called "Investing in Development: A Practical Plan to Achieve the Millennium Development Goals."

2005 World Summit

The World Summit of 2005, held in September, was organized to measure international progress towards fulfilment of the Millennium Development Goals

(MDGs). Notably, the conference brought together more than 170 Heads of States. Although the world leaders at the summit were encouraged by the reduction of poverty in some nations, they were concerned by the uneven decline of poverty within and among different regions of the globe. However, at the end of the summit, the conference attendees reaffirmed the UN's commitment to achieve the MDGs by 2015 and urged all supranational, national and non-governmental organizations to follow through.

Post-2015 Development Agenda

As the Millennium Development Goals is approaching its target 2015, the international community is focused on accelerating their efforts to achieve the goals laid out in the original MDGs. Although, there has been notable progress towards reducing extreme poverty, with the MDG 1 target of reducing extreme poverty rates by half, met "five years ahead of the 2015 deadline...700 million fewer people lived in conditions of extreme poverty in 2010 than in 1990. However, 1.2 billion people were still living in extreme poverty in all over the globe. Sub-Saharan Africa,is the only notable region where the number of people living in extreme poverty rose from 290 million in 1990 to 414 million in 2010, comprising more than a third of those living in extreme poverty worldwide.

Keeping the importance and time-based factor into consideration, the UN convened a High Level Panel (HLP) of Eminent Persons, to advice on a Post-2015 Development Agenda. The HLP report, entitled "A New Global Partnership: Eradicate Poverty and Transform Economies Through Sustainable Development", was published in May 2013. In the report, the HLP wrote :"Ending extreme poverty is just the beginning, not the end. It is vital, but our vision must be broader: to start countries on the path of sustainable development – building on the foundations established by the 2012 UN Conference on Sustainable Development in Rio de Janeiro12, and meeting a challenge that no country, developed or developing, has met so far. We recommend to the Secretary-General that deliberations on a new development agenda must be guided by the vision of eradicating extreme poverty once and for all, in the context of sustainable development".

So, it is discernible that a central goal of the Post-Millennium Development agenda is to "eradicate extreme poverty by 2030." However, the report also emphasized that the MDGs were not enough, as they did not "focus on the devastating effects of conflict and violence on development, the importance to development of good governance and institution...nor the need for inclusive growth..." The existing synergy between the policy position papers put forward by the United States (through USAID), the World Bank and the UN itself in terms of viewing fragility and a lack of good governance as exacerbating extreme poverty. However, as the views of other organizations, the commission also proposed that the UN focus not only on extreme poverty (a line drawn at $1.25), but also on a higher target, such as $2. The report notes this change could be made to reflect the fact that escaping extreme poverty is "only a start."

The UN along with a host of other supranational and national actors such as the European Union and the African Union have published their own positions or recommendations on what should be incorporated in the Post-2015 agenda. In The European Commission's publication "A decent Life for all: from vision to collective action", the UN's commitment to "eradicate extreme poverty in our lifetime and put the world on a sustainable path to ensure a decent life for all by 2030" was affirmed. A unique vision of the report was the Commission's environmental focus (in addition to a plethora of other goals such as combating hunger and gender inequality). Specifically, the Commission argued, "long-term poverty reduction... requires inclusive and sustainable growth. Growth should create decent jobs, take place with resource efficiency and within planetary boundaries, and should support efforts to mitigate climate change." The report of the African Union, entitled "Common African Position (CAP) on the Post-2015 Development Agenda", likewise encouraged the international community to focus on eradicating the twin problems of "poverty and exclusion" in our lifetime. Moreover, the CAP pledged that it would "commit to ensure that no person – regardless of ethnicity, gender, geography, disability, race or other status – is denied universal human rights and basic economic opportunities."

UN LDC conferences

The UN organised a series of summit over the past few decades, collectively called The UN Least Developed Country (LDC) conferences, which sought to promote the development of so-called "third-world" countries.

1st UN LDC Conference

The first UN LDC Conference was held between September 1 and September 14, 1981, in Paris, this Conference was organized to finalize the UN's "Substantial New Programme of Action" for the 1980s in Least Developed Countries. This programme, which was unanimously adopted by the conference attendees, argued for internal reforms in LDCs (meant to encourage economic growth) to be complemented by strong international measures. In spite of the major economic and policy reforms initiated many of these LDCs, in addition to strong international aid, the economic situation of these countries worsened as a whole in the 1980s. This prompted UN towards the 2nd UN LDC conference almost a decade later.

2nd UN LDC Conference

The second UN LDC conference was held between September 3 and September 14, 1990, once again in Paris, this Conference was convened to measure the progress made by the LDCs towards fulfilling their development goals during the 1980s. The problems that plagued the LDCs over the past decade were recognized. The conference formulated a new set of national and international policies to accelerate the growth rates of the poorest nations. These new principles were incorporated in the "Paris Declaration and Programme of Action for the Least Developed Countries for the 1990s."

3rd UN LDC Conference

This conference was held in Brussels on 14th to 20th May, 2001. Here world's 49 least developed countries were gathered. This conference aimed to rescue cash-strapped economies of many international communities and thus end their isolation from global trade and eradication of poverty.

4th UN LDC Conference

The 4th UN LDC Conference is the most recent conference, held in May 2011 in Istanbul, it recognized the change in the nature of development since the 1st conference that took place almost 30 years earlier. The capital flow into emerging economies has increasingly become dominated by foreign direct investment and remittances in the 21st century, which is opposite to bilateral and multilateral assistance. Moreover, since the 80s, significant structural changes have taken place on the international stage. The formerly "undeveloped" nations are now able to have a much larger say in international relations with the creation of the G-20 conference of the largest economic powers, including many nations in the Global South Furthermore. This was recognized by the conference that in the midst of a deep global recession, coupled with multiple crises (energy, climate, food, etc.), the international community would have fewer resources to aid the LDCs. Thus, the UN considered the participation of a wide range of stakeholders (not least the LDCs themselves), crucial to the formulation of the conference.

ORGANIZATIONS WORKING TO END EXTREME POVERTY

International organizations

World Bank

The Board of Governors of the World Bank Group (WBG) set two overriding goals for the WBG to commit itself to in the future in 2013. First, to end extreme poverty by 2030, an objective that echoes the sentiments of the UN as well as the Obama administration. Additionally, the WBG set an interim target of reducing extreme poverty to below 9 per cent by 2020. Second, to focus on growth among the bottom 40 per cent of people, as opposed to standard GDP growth. This commitment ensures that the growth of the developing world lifts people out of poverty, rather than exacerbating inequality.

The development programs run by the World Bank are primarily commercial-based in nature because, World Bank's primary focus is on delivering economic growth in order to enable equitable prosperity, as opposed to the UN. Since the World Bank recognizes better jobs will result in higher income and thus, less poverty, the WBG seeks to support employment training initiatives, small business development programs and strong labour protection laws. However, since much of the growth in the developing world has been inequitable, the World Bank has also begun teaming with client states to map out trends in inequality and to propose public policy changes that can level the playing field.

The World Bank deals with myriad nutritional, transfer payments and transport-based initiatives. Children who experience under-nutrition from conception to two years of age have a much higher risk of physical and mental disability. Thus, they are often trapped in poverty and are unable to make a full contribution to the social and economic development of their communities as adults. The WBG estimates that as much as 3% of GDP can be lost as a result of under-nutrition among the poorest nations. To combat under nutrition, the WBG has partnered with UNICEF and the WHO to ensure all small children are fully fed. The WBG also offers conditional cash transfers to poor households who meet certain requirements such as maintaining children's healthcare or ensuring school attendance. Finally, the WBG understands investment in public transportation and better roads is key to breaking rural isolation, improving access to healthcare and providing better job opportunities for the World's poor.

1. OCHA (Office for the Coordination of Humanitarian Affairs)

The Office for the Coordination of Humanitarian Affairs (OCHA) of the United Nations works to co-ordinate the international, national and non-governmental efforts to contest poverty. The OCHA seeks to prevent "confusion" in relief operations and to ensure that the humanitarian response to disaster situations has greater accountability and predictability. To do so, OCHA has begun deploying Humanitarian Coordinators and Country Teams to provide a solid architecture for the international community to work through.

2. UNICEF (United Nations Children's Fund)

The United Nations Children's Fund (UNICEF) was created by the UN to provide food, clothing and health care to European children facing famine and disease as the immediate after effect of World War II. After the UN General Assembly extended UNICEF's mandate indefinitely in 1953, it actively worked to help children in extreme poverty in more than 190 countries and territories to overcome the obstacles that poverty, violence, disease and discrimination place in a child's path. Its current focus areas are 1) Child survival & development 2) Basic education & gender equality 3) Children and HIV/AIDS and 4) Child protection.

3. UNHCR (The UN Refugee Agency)

The UN Refugee Agency (UNHCR) was established mandatorily to lead and coordinate international action to protect refugees worldwide. Its primary purpose is to safeguard the rights of refugees by ensuring anyone can exercise the right to seek asylum in another state, with the option to return home voluntarily, integrate locally or resettle in a third country. The UNHCR operates in over 125 countries, helping approximately 33.9 million persons.

4. WFP (World Food Program)

The World Food Program (WFP) is the largest agency dedicated to fight against the hunger worldwide. On average, WFP brings food assistance to more than 90 million people in 75 countries. The WFP not only strives to prevent hunger in the present, but also in the future by developing stronger communities which will make food even more secure on their own. The WFP has a range of expertise from Food Security Analysis, Nutrition, Food Procurement and Logistics.

5. WHO (World Health Organization)

The World Health Organization (WHO) is established to provide leadership on global health matters, shaping the health research agenda, articulating evidence-based policy decisions and combating diseases that are induced from poverty, such as HIV/AIDS, malaria and tuberculosis. Moreover, the WHO deals with pressing issues ranging from managing water safety, to dealing with maternal and new born health.

BILATERAL ORGANIZATIONS

USAID

The U.S. Agency for International Development (USAID) is the pivotal government agency dedicated to ending extreme poverty. The United States channels the majority of its "development" assistance through USAID and the U.S. Department of State is currently the largest bilateral donor in the world,. In President Obama's 2013 State of the Union address, he declared "So the United States will join with our allies to eradicate such extreme poverty in the next two decades...which is within our reach." In response to Obama's call to action, USAID has made ending extreme poverty central to its mission statement. Under its New Model of Development, USAID seeks to eradicate extreme poverty through the use of innovation in science and technology, by putting a greater emphasis on evidence based decision-making, and through leveraging the ingenuity of the private sector and global citizens.

A major initiative of the Obama Administration called Power Africa, aims to bring energy to 20 million people in Sub-Saharan Africa. By reaching out to its international partners, whether commercial or public, the US has leveraged over $14 billion in outside commitments after investing only $7 billion USD of its own. To ensure that Power Africa reaches the region's poorest, the initiative engages in a transaction based approach to create systematic change. This includes expanding access to electricity to more than 20,000 additional households which already live without power.

USAID specifically works on myriad fields starting from preventing hunger, reducing HIV/AIDS, providing general health assistance and democracy assistance, to dealing with gender issues. To deal with food security, which affects roughly 842 million people (who go to bed hungry each night),USAID coordinates the Feed the Future Initiative (FtF). FtF aims to reduce poverty and under nutrition each by 20 per cent over five years. Thanks to the President's Emergency Plan for AIDS Relief (PEPFAR) and a variety of congruent actors, the incidence of AIDS and HIV, which used to ravage Africa, has reduced in scope and intensity. Through PEPFAR, the

United States has ensured over five million people have received life-saving antiviral drugs, a significant proportion of the eight million people receiving treatment in relatively poor nations.

USAID has accomplished a host of goals from working to reduce maternal mortality by 30 per cent and under-five child mortality by 35 per cent. USAID also supports the gamut of democratic initiatives, from promoting human rights and accountable, fair governance, to supporting free and fair elections and the rule of law. In pursuit of these goals, USAID has increased global political participation by training more than 9,800 domestic election observers and providing civic education to more than 6.5 million people. Since 2012, the Agency has begun integrating critical gender perspectives across all aspects of its programming to ensure all USAID initiatives work to eliminate gender disparities. To do so, USAID seeks to increase the capability of women and girls to realize their rights and determine their own life outcomes. Moreover, USAID supports additional programs to improve women's access to capital and markets, builds theirs skills in agriculture, and supports women's desire to own businesses.

DFID

The Department for International Development (DFID) is the UK's leading agency for eradicating extreme poverty. To do so, DFID focuses on the creation of jobs, empowering women and rapidly responding to humanitarian emergencies.

Governance assistance, educational initiatives, and funding cutting-edge research are some highlighted DFID projects. In 2014 alone, DFID will support "freer and fairer" elections in 13 countries. DFID will also help provide 10 million women with access to justice through strengthened judicial systems and will help 40 million people make their authorities more accountable. By 2015, DFID will have helped 9 million children attend primary school, at least half of which will be girls.[Furthermore, through the Research for Development project, DFID has funded over 35,000 projects in the name of creating new technologies to help the world's poorest. These technologies include: vaccines for diseases of African cattle, better diagnostic methods for tuberculosis, new drugs for combating malaria, and developing flood-resistant rice. In addition to technological research, the Research for Development is also used to fund projects that seek to understand what, specifically, about governance structures can be changed to help the world's poorest.

NON-GOVERNMENTAL MOVEMENTS

NGOs

Innumerable non-governmental organizations operate in the field of extreme poverty, actively working to alleviate the poorest of the poor. To name but a few notable organizations: Save the Children, The Overseas Development Institute, Concern Worldwide and Oxfam have all done a considerable amount of work in extreme poverty. Save the Children 'is the leading international organization dedicated to helping the World's indigent children. In 2013 alone, Save the Children reached over 143 million children through their work, including over 52 million children directly.[50] Save the Children also recently released their own report titled "Getting to Zero",[51] in which they argued the international community could feasibly do more than lift the world's poor above $1.25/day. The Overseas Development Institute (ODI) is the premier UK based think tank on international development and humanitarian issues. ODI is dedicated to alleviating the suffering of the world's poor by providing high-quality research and practical policy advice to the World's development officials. ODI also recently released a paper entitled, "The Chronic Poverty Report 2014–2015: The road to zero extreme poverty", in which its authors assert that though the international communities' goal of ending extreme poverty by 2030 is laudable, much more targeted resources will be necessary to reach said target. The report states that "To eradicate extreme poverty, massive global investment is required in social assistance, education and pro-poorest economic growth".

"Concern Worldwide" is an international humanitarian organization that aims to eradicate extreme poverty by persuading the policy makers at all levels of government. Concern has also produced a report on extreme poverty in which they explain their own conception of extreme poverty from a NGO's standpoint. In this paper, named "How Concern Understands Extreme Poverty", the report's creators write that extreme poverty entails more than just living under $1.25/day, it also includes having a small number of assets and being vulnerable to severe negative shocks (whether natural or man made).

The organization cofounded by Bono, named "One" is a non-profit organization funded almost entirely by foundations, individual philanthropists and corporations. ONE's goals include raising public awareness and working with political leaders to fight preventable diseases, increase government accountability and increase investment in nutrition· Finally, trickle up is a microenterprise development program targeted at those living on under $1.25/day, which provides the indigent with resources to build a sustainable livelihood through both direct financing and considerable training efforts.]

Another NGO working tirelessly to end extreme poverty is Oxfam. This non-governmental organization works prominently in Africa; their mission is to improve local community organizations and it works to reduce impediments to the development of the country. Oxfam helps families suffering from poverty receive food and healthcare to survive. There are many children in Africa experiencing growth stunting, and this is one example of an issue that Oxfam targets and aims to resolve.

Poverty and India

Poverty is a notable issue in India. In spite of having one of the fastest-growing economies in the world, clocked at a growth rate of 7.6% in 2015, and a sizable consumer economy. The World Bank reviewed and proposed revisions in May 2014, to its poverty calculation methodology and purchasing power parity basis for measuring poverty worldwide, including India. According to this revised methodology, the world had 872.3 million people below the new poverty line, of which 179.6 million people lived in India. In other words, India with 17.5% of total world's population, had 20.6% share of world's poorest in 2011. As of 2014, 58% of the total population were living on less than $3.10 per day. According to the *Modified Mixed Reference Period (MMRP)* concept proposed by World Bank in 2015, India›s poverty rate for period 2011-12 stood at 12.4% of the total population, or about 172 million people; taking the revised poverty line as $1.90.

Since the 1990's, the World Bank has been reconsidering its definition and benchmarks to measure up poverty with a $2 per day income on purchasing power parity basis as the definition in use from 2005 to 2013. Some semi-economic and non-economic indices have also been proposed to measure poverty in India; for example, the Multi-dimensional Poverty Index placed 33% weight on number of years spent in school and education and 6.25% weight on financial condition of a person, in order to determine if that a person is poor.The different definitions and different underlying small sample surveys used to determine poverty in India, have resulted in widely different estimates of poverty from 1950s to 2010s. In 2012, the Indian government stated 22% of its population is below its official poverty limit. The World Bank, in 2011 based on 2005's PPPs Program, estimated 23.6% of Indian population, or about 276 million people, lived below $1.25 per day on purchasing power parity.] According to United Nation›s Millennium Development Goals (MDG) programme 270 millions or 21.9% people out of 1.2 billion of Indians lived below poverty line of $1.25 in 2011-2012.From late 19th century through early 20th century, under British colonial rule, poverty in India intensified, peaking in 1920s. Famines and diseases killed millions each time. After India gained its independence in 1947, mass deaths from famines were prevented. Rapid economic growth since 1991, has led to sharp reductions in extreme poverties in India. However, those above poverty line live a fragile economic life.As per the methodology of the Suresh Tendulkar Committee report, that the population below the poverty line in India in 2009-2010 was 354 million (29.6% of the population) and that in 2011-2012 was 269 million (21.9% of the population). The Rangarajan Committee said in 2014 that the population below the poverty line in 2009-2010 was 454 million (38.2% of the population) and that in 2011-2012 was 363 million (29.5% of the population). Deutsche Bank Research estimated that there are nearly 300 million people who are middle class· If former trends continue, India›s share of world GDP will significantly increase from 7.3% in 2016 to 8.5% by 2020.[23] In 2015, around 170 million people, or 12.4%, lived in poverty (defined as $1.90 (Rs 123.5)), a reduction from 29.8% in 2009· The Asian Development Bank estimates India›s population to be at 1.28 billion with an average growth rate, from 2010-2015, at 1.3%. In 2014, 49.9% of the population aged 15 years and above were employed. However, there are still 21.9% of the population who live below the national poverty line· The World Poverty Clock shows real-time poverty trends in India, which are based on the latest data, of the World Bank, among others.

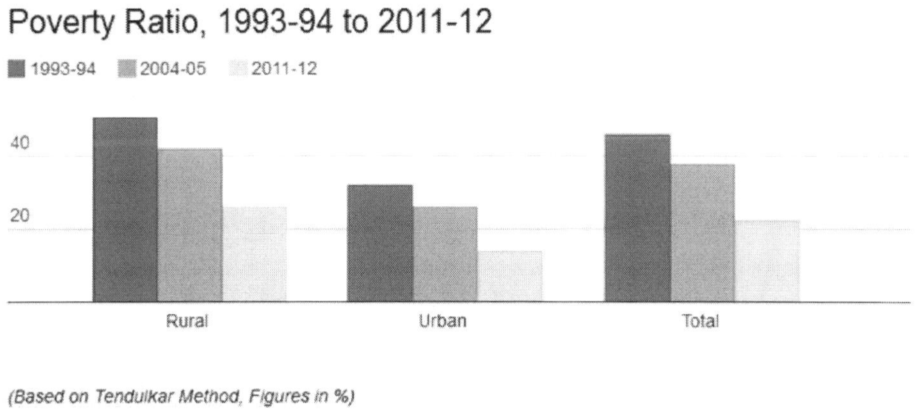

Picture source-https://www.google.co.in

Silence: the cause of hunger and poverty

In the modern civilization we basically observe two sides of the world: haves and have nots. One side is happy and prosperous with all the modern amenities and comforts of life, at the same time the other side is living in the dark and dearth of extreme poverty and hunger. But why is the discrimination? Is it because of the uneven distribution of resources? Inaccessibility of information? Lack of awareness about their rights? Or because simply of their silence?

We can site many examples from the bitter experiences where the deprived section of the society broke their silence and burst like a volcano turning the society into ashes. For example we can say about the Cuba Movement Fidel Castro, Vietnam movement, Naxal movement of India etc.

So, we can say that if hunger is to be eliminated we have to reduce poverty and if we have to eradicate both hunger and poverty, silence needs to be broken and empowerment needs to be put in place.

5 Points For A Better-fed World From Abhijit Banerjee

1. Hidden hunger is mostly invisible

Poor nutrition becomes a norm when everyone in a village is undernourished resulting to ignorance amongst the masses. This may also explain the reluctance of poor, ill-fed people to spend extra money on food, preferring instead to buy such things as televisions or a fancy wedding.

2. India can eradicate poverty and hunger

As these are not matters of resources, so the initiative to design policies that work on the ground level is not something the government experts on their offices give much importance. Policies fail when governments try to get people to do things whose value they don't appreciate. This is particularly true when, state agents are underpaid, overworked, and not well monitored.

3. Need small-scale interventions

There are chances that the poor are also misinformed about the nutritional value, resulting in consumption of more of tastier food and less of nutritional food. Small-scale interventions through local policymaking, or small but significant technological changes, can bring about all the change.

4. Direct transfers more efficient

There is an urge to stop subsidising agriculture through water, energy and fertiliser subsidies. It is better to just give farmers money. In most subsidies, the biggest beneficiary is not the poor. Cash transfers can open up the possibility of a more fulfilling, productive life, better nutrition being a part of it

5. The biggest myth is that the poor are victims

The poor have a sense of what they want, the life they want. Every time we make policies we forget that they have ambitions, aspirations.

Food Insecurity, Hunger, and Malnutrition: Necessary Policy and Technology Options

Ever since independence in 1947, agricultural development policies in India have focused mainly at reducing hunger, food insecurity, malnourishment and poverty. Keeping this overarching goal in mind, the emphasis, which was initially (for 15 years or so) on keeping food prices low, shifted to macro food-security and subsequently to household and individual food-security. Later, the food security of vulnerable, sustainable use of natural resources, and equity between rural and urban or farm and non- farm population became the issues of dominant discourse related to agricultural development. The policies and programmes related to marketing and trade were obviously guided by the overall objective sought to be achieved from the agricultural development strategy. The changes in marketing environment and production performance of the Indian agricultural sector should, therefore, be viewed in the context of weightage attached to these objectives at different points of time.

India's agricultural development strategy and approach to food security has yet again proved its resilience in the wake of recent global food crisis, which has created political and social unrest in several countries of developing world. It earlier helped India tide over the severe food crisis of mid-

1960s within a period of one and half decades, and also proved its appropriateness in the wake of economic liberalization and globalization since the early-1990s. Though, India's performance in terms of reducing hunger and malnutrition has not been as remarkable as that of China and some East Asian countries, given the political and initial socio-cultural milieu, the achievements have certainly been commendable. India's experience has provided several lessons for the countries that are struggling to come out of the poverty-malnutrition-hunger trap. In this paper, a brief review of global scenario of hunger and food prices has been presented in section

2. India's approach and status of food security have been outlined in section 3. Some details of policy instruments and programmes impacting on marketing and production environment for farmers have been discussed in section 4. Current agricultural scenario, emerging concerns relating to production performance and new policy initiatives and programmes launched to revive the agricultural sector have been highlighted section 5. Some specific suggestions emerging from the analysis of the current scenario are given in the last concluding section.

Global Scenario of Hunger and Food Prices

India, with a population of 1.11 million, accounts for 17 per cent of world's population. Its size, in terms of consumers, is many-times larger than the average size of rest of the countries, except China. The state of food insecurity and hunger in India is of considerable significance for the global situation and, therefore, figures prominently in such discussions at the world fora.

For the world community, hunger and food insecurity has been given highest priority ever since the World Food Summit (WFS) of 1996, that had set a target of reducing the number of hungry and food insecure to half (842 million in the base period of 1990-92) by 2015. Later, the UN Millennium Declaration of 2000 set eight Millennium Development Goals (MDGs), the first of which was to reduce the proportion of hungry (from a base level of around 20 per cent) to half by 2015. Other seven MDGs were also closely linked to the first goal of reducing hunger (reducing malnutrition, IMR and MMR; increasing school enrolment; improving gender equality; increasing access to safe drinking water and sanitation; and combating HIV/AIDS, malaria and other diseases) (FAO, 2005). Obviously, the MDG was milder than WFS target.

FAO's assessment for the period 2003-05 shows that the number of people suffering from chronic hunger worldwide has increased to 848 million, representing an increase of six million over the base period estimates. However, the proportion of hungry people (MDG) during 2003-05 came down to around

16 per cent. During this period, in Sub-Saharan Africa (SSA), the number of hungry people went up from 169 million to 212 million, but the proportion dropped from 34 per cent to 30 per cent.

The rise in food prices since the end of 2006 has further aggravated the situation of hunger worldwide. The number of hungry in 2007 went up to 923 million (75 million plus over 2003-05 figure) and is likely to have increased further currently as the food price rise was much steeper in the early-2008. The proportion of hungry people is reported to have also increased to 17 per cent in 2007. Thus, meeting the internationally agreed hunger-reduction goals in the few years remaining to 2015 is becoming an enormous challenge.

At the regional level, the largest increases in the number of hungry people, as a result of rising food prices have taken place in Asia and Sub-Saharan Africa (SSA). The two regions combined already accounted for 750 million, or 89 per cent, of the hungry people in the world in 2003-05. The rising food prices have driven an additional 41 million in Asia and 24 million in SSA below the hunger threshold. In other regions also, although the numbers are smaller, the hunger has increased due to rising food prices. Even in Latin America, which has shown a decade of steady progress towards the WFS target, there has been a sharp reversal of trend since 2006 (FAO, 2008).

Driving forces behind soaring food prices are innumerable and complex, with both supply-side and demand-side factors playing crucial a part. Long-term structural trends and underlying growth in demand for food have coincided with short-term cyclical or temporary factors affecting food supply. On the supply side, while cereals stocks were at the lowest levels in three decades, as a result of unfavourable weather, cereals production fell by 3.6 per cent in 2005 and 6.9 per cent in 2006, before recovering slightly in 2007. Further, the rapid rise in petroleum prices exerted an upward pressure on food prices as fertilizer prices nearly tripled and transport costs doubled during 2006-08. On the demand side, increasing use of sugar, maize, cassava, oilseeds and palm oil by the emerging biofuels industry has been the leading factor behind the increase in their prices. About 100 Mt of cereals (4.7 per cent of global production) were used in the production of biofuels in 2007-08. These apart, adoption of export restrictions by some countries, speculative re- stocking or pre-stocking by large importers, entry of new types of investors in derivatives markets based on agricultural commodities, and increased speculative activities have been the factors that exacerbated the rise in food prices globally. It is being predicted that while global rise in food prices may abate slightly, these may not come down to the levels of 2004. It is in this context that India's situation of food security and policy response to the global challenges assumes added importance.

In some international quarters, one of the major factors that led to rise in global food prices was increase in demand by the middle class in India (and also China), leading to food shortages. India's dietary pattern atleast cannot substantiate this argument. During 2004-06, the average per capita consumption of cereals in India was 175 kg per year as against 953 kg in USA, 288 kg in China, and 316 kg in the world as a whole. Simplistic calculation would show that if the world adopts India's dietary pattern, only 56 per cent of the existing global cereals output is enough to feed the world.

INDIA'S FOOD SECURITY: APPROACH AND STATUS

Approach to Food Security

Food security, at both the national and household levels, has been a major focus of agricultural development strategy in India ever since mid-1960s when import dependence for cereals had gone up to 16 per cent and the country faced severe droughts persistently for two years. The new strategy launched at that point of time was aimed at 'maximizing the production of cereals', and involved building a solid foundation of food security on three key elements, viz. (a) provision of an improved technology package to the farmers; (b) delivery of modern farm inputs, technical know-how and institutional credit to the farmers; and (c) assurance of a remunerative marketing and pricing environment for farmers. For achieving these objectives, several policy instruments were used that influenced production potential, and marketing system of agricultural commodities in general and food products in particular. Some of the main policy instruments that reflect the policy package adopted by India are as follows:

(i) Creation, strengthening and expansion of the national agricultural research system (NARS) for developing and perfecting new production technologies for food grains and other agricultural commodities.

(ii) Establishment, strengthening and expansion of agricultural education and training system for agricultural extension workers and farmers for transfer of new technologies at the farm level.

(iii) An arrangement for the production and/or import and distribution of high-yielding farm inputs like improved seeds, fertilizers, agro-chemicals and improved implements/farm machines to the farmers.

(iv) Planning and execution of major, medium, and micro irrigation schemes for increasing the area under irrigation.

(v) Nationalization of commercial banks, creation of cooperative credit institutions, and reorientation of monetary policy to increase institutional credit flow to the farmers.

(vi) Creation and expansion of physical and institutional infrastructure (primary market yards, roads, storage facilities, farmers' cooperatives and public sector organizations) for improvement of the marketing system to handle and distribute the emerging marketed surplus.

(vii) Regulation of traders' exploitative marketing practices through a series of legal and regulatory measures such as licensing, levies, stocking limits and movement restrictions.

(viii) Fixation of minimum support prices (MSPs) and arrangements for price support purchases and procurement by public or cooperative agencies.

(ix) Building-up and maintenance of buffer stocks of cereals and distribution of cereals through public distribution system (PDS).

(x) Provision of food and input subsidies, explicit or implicit, for reconciling the conflicting objectives of cereal producers and consumers in terms of prices.

(xi) Canalization, licensing, minimum export prices (MEPs), and custom duties for maintaining price stability in the domestic market.

Dimensions of Food Security

Food security can be achieved when 'all people at all times have physical and economic access to food that is sufficient to meet dietary needs for a healthy and productive life'. In this sense, achievement of food security implies producing (or importing) sufficient food and making it accessible to all individuals throughout the year and on a sustainable basis from year to year. Further, fulfilling dietary needs for a productive and healthy life implies physical and economic access of all people to nutritive food, according to each individual's requirement. Food security thus connotes freedom from hunger and malnutrition.

The status of food security of a country needs to be assessed at three levels. First is the availability of food at national level on sustainable basis, which depends on level and growth of food production, or adequate capacity to import food (if availability elsewhere is assured). Second is the physical and economic access of all households to food. Physical access requires efficient marketing, transport, and storage system to carry the food within an easy reach or at a reasonable distance from human settlements (villages). Economic access of every household to food depends on its purchasing power and prices of food at which it is available. And third is the utilization of available food by individuals, which depends on intra-family allocation of food, and maintenance of reasonable level of health of all individuals to consume and absorb required level of food. Social factors like education, primary healthcare, gender bias, and role of women in household decisions affect food security at the individual level.

It is in this context that India tackled the hunger and food-insecurity situation through both long-term and short-term measures. As a part of long-term strategy, it adopted a development strategy encompassing maintenance of adequate growth of national food production, and employment and incomes of masses, improvement in marketing infrastructure and access to education and primary healthcare. These apart, the short-term strategy involved selective market intervention and targeted distribution of subsidized food to reduce hunger and food insecurity. Further, as the nutritional status is also influenced by non-food factors such as clean water and sanitation, it was recognised that attention to these aspects will help translate food security into good nutrition.

Macro Food Security

(i) The most remarkable change has been in the increase in national production of staple food, i.e. cereals. The production of cereals went up from 44 Mt in triennium ending (TE) 1951-52 to 203 Mt during TE 2007-08. The average incremental production was around 4 Mt per year continuously for two decades from TE 1974-75 to TE 1994-95 (Table 1). The rate of increase came down during the later decade but has picked up again during the recent three years.

(ii) The growth rate of cereal production has kept pace with the growth of population and cereals demand. The production of cereals as a group increased at the annual compound rate of more than 3 per cent per annum up to 1991 and around 2.4 per cent up to the mid-1990s.

Household and Individual Food-Security

(i) Several factors have contributed in considerable improvement in physical access of households to food in different parts of the country. First, the share of rice, which is more geographically dispersed, in total staple food, continues to be quite high at around 45 per cent. Second, the expansion of network of public distribution system helped in reaching cereals to deficit and geographically difficult regions (hilly and desert areas). And third, expansion of road networks, creation of primary market yards, and building-up of storage facilities in the rural areas increased physical access of rural households to food even in otherwise deficit areas.

(ii) Yet, another important development has been the persistent improvement in the economic access of consumers to food. The increase in retail prices of two staple food items (rice and wheat) has been lower than the increase in per capita income, and thus the proportion of consumer income required to buy a unit quantity of rice or wheat has continued to decline. For example, the price of 100 kg wheat as a proportion of annual per capita income in rural areas declined from 15.4% in 1973-74 to 8.7 per cent in 1983-84, 5.9 per cent in 1990-01,5.0 per cent in 1994-95, 4.4 per cent in 1999-2000 and 4.0 per cent in January 2008. A similar declining trend has been noticed for urban communities, as well as in the case of rice for both rural and urban areas.

(iii) Improvement in the availability of staple food at declining real prices has contributed to improved nutritional-security. Farmers have shifted from the low-yielding coarse cereals to non-cereal food products since the middle of 1980s, a fact which has *inter alia* helped to increase production and availability of sugar, fruits, vegetables, spices, milk, eggs, meat and fish/fish products. As the production growth of all these food items was considerably higher than the population growth, per capita production of nutritive foods went up substantially. India now produces 58 Mt fruits, 112 Mt vegetables, 102 Mt milk, 46 billion eggs and 7 Mt fish.

(iv) Despite expansion in the availability of cereals and decline in their relative prices vis-à-vis incomes, the per capita consumption of cereals has also tended to drop in recent years (Dev,2003). The decline in consumption has been sharper in coarse cereals, and has occurred even among the lowest 30 per cent of consumers, reflecting a shift towards more nutritive foods like fruits, vegetables and livestock products. Long-term data from National Sample Survey Organization (NSSO) has also indicated a declining trend in the per capita consumption of cereals in both rural and urban areas, accompanied by a decrease in the proportion of expenditure on cereals and an increase in that on milk, meat, eggs, fruits and vegetables. This shows improvement in the nutritional levels.

(v) There has been drastic reduction in hunger or non-availability of food at the household level. As per large sample surveys of NSSO, the percentage of households not getting enough food daily even in some months declined from 16.2 per cent in 1983 to 4.2 per cent in 1993-94, to 2.6 per cent in 1999-00 and to 1.9 per cent only in 2004-05. During 2004-05, only 0.3per cent households reported inadequate food in all the months of a year.

(vi) Economic poverty is a defining factor affecting food security at the household level. Over the years, the incidence of poverty, in both rural and urban areas, has declined considerably. The percentage of population below poverty line declined from 51 per cent in 1977-78 to 39 per cent in 1987-88, to 26 per cent in 1999-00 and further to 22 per cent in 2004-05.

Food for Work, Employment Guarantee and Right to Food

Besides the direct food assistance programmes, an increase in the purchasing power of rural households was attempted through food for work and rural wage employment programmes. The Food for Work Programme was started in 1977-78. Since then, there have been several modifications of rural wage employment programmes, with foodgrain component built into these as a part of wages. In September 2001, different programmes were merged into an umbrella programme called SGRY (*Sampoorna Gramin Rozgar Yojana*– Comprehensive Rural Employment Scheme). Under SGRY, the works taken up were labour-intensive, wages were equal to statutorily prescribed minimum wages by provincial (state) In February 2006, a National Rural Employment Guarantee (NREG) scheme was launched in 200 most backward rural districts under NREG Act passed by the Parliament. Under the NREG Scheme, at least one member of a rural family is guaranteed employment for 100 days in a year. In case, the local agency fails to provide employment within 15 days of application, he/she becomes entitled for cash compensation. The scheme was extended to 300 districts in 2007 and since April 2008, it has been extended to the entire country (588 rural districts). The SGRY has been subsumed in NREGS from April 2008.

A development, which focuses towards the better implementation of food assistance and employment schemes is related to 'right to food' and 'right to information' campaigns in the country. These campaigns were started in 2001 and got momentum subsequently. Grassroots Civil Society Organizations became active in the implementation of public distribution system and wage employment programmes. Even the Supreme Court of India intervened in the matter of effective implementation of PDS and wage-employment schemes. As a consequence, the demand or off-take of foodgrains from PDS outlet was increased and became almost equal to the entitled quota. However, due to shortfall in production, relatively low support prices, and big/ corporate traders being active in the market, the government could not procure sufficient quantities of rice and wheat to meet the PDS commitment. Consequently, it had to import 5.5 Mt of wheat at a price higher by 100 per cent or more than the support price in 2005-06. There was lot of resentment among farmers and criticism of government policy on this count.

Current Marketing and Trade Environment

Agricultural marketing system and trade policy have a massive effect on the prices received by the farmers and, in turn, influence the profitability of agriculture. Profitability of farming is the sum total of profitability of individual farm enterprises that include crops, horticultural enterprises, livestock and fisheries. For each enterprise, the profitability is the net outcome of physical input-output ratio (or total factor governments, and wages were paid in the form of 5 kg of foodgrains (at subsidized prices) plus cash. productivity) and relative prices of inputs and outputs. Physical input-output ratio depends on the state of technology available and adopted by the farmers (varieties, quality of seeds or planting material or feeds, dosages of inputs, and agronomic and protection practices) and weather conditions. The relative prices of inputs and output are determined by the market structure and efficiency of the marketing system for essential inputs and output.

Price Policy and Support for Farmers

Price support for farmers has been an important inclusion in the agricultural development and food policy since the mid-1960s. The main objectives of price policy are: (a) to provide incentives to farmers for adopting new technology and maximizing production, (b) to safeguard the interests of consumers or users of farm products by maintaining market prices at reasonable levels, and (c) to keep the fluctuations in prices within certain limits. The main instruments of price policy, *inter alia* are minimum support prices, buffer stocking, and operation of a public distribution system of cereals. The main challenge of the policy has always been to reconcile the conflicting price interests of farmers and consumers. It is partly achieved through the provision of food subsidy and supply of essential farm inputs (fertilizers, electricity and canal water) to farmers at reasonable prices or user charges. Currently, minimum support prices (MSPs) are announced for 25 farm products, that include cereals, pulses, oilseeds, raw cotton, raw jute, sugarcane and copra (dried coconut). Buffer stocking and public distribution system are operated for rice, wheat and to some extent for sugar. Commission for Agricultural Costs and Prices (CACP) is the advisory body of Government of India in all matters relating to agricultural price policy. The quantities that the government agencies need to purchase at support prices depend on the behaviour of market prices and private trade, and fluctuate from year to year. For example, price support purchases of rice and wheat accounted for 15.8 per cent of the production during TE 1992-93, 24.6 per cent during TE 2002-03 and 22.7 per cent during TE 2006-07. In terms of absolute quantities, these varied between 20 Mt and 39 Mt at these points of time. About 25 per cent is retained by the producer farmers for self-consumption and rest, i.e. more than half of the production is handled by private trade.

Several committees in recent years have reviewed the current price policy regime. These include Long-Term Grain Policy Committee (Abhijit Sen); Repositioning of CACP Committee (Y.K. Alagh); Planning Commission's Working Group for XI Five-Year Plan (S.S. Acharya); Foodgrain Policy Review Committee

(Ramesh Chand); and National Commission for Farmers (M.S. Swaminathan). Some of the important suggestions of these committees are:

(i) Continuation of the policy of maintaining self- sufficiency in cereals.

(ii) Continuation of the system of minimum support prices and making it effective in all the states/areas of the country.

(iii) Continuation of the policy of buffer stocking of cereals and their subsidised distribution for maintaining price stability.

(iv) Retaining Food Corporation of India as the national foodgrain handling agency on behalf of the Government of India.

(v) Fixing the minimum support prices at a level 50 per cent higher than the cost of production.

(vi) Dovetailing of domestic price policy with trade policy by renaming the CACP and redefining its terms of reference.

(vii) Assigning statutory status to CACP.

(viii) Phasing out of levy on rice millers and sugar factories.

(ix) Switching over to universal PDS from the present system of targeted public distribution of cereals.

Considering these recommendations, the government is continuing the policy of minimum support prices, buffer stocking of cereals, and distribution of subsidised food grains. In addition, the FCI is continuing to perform its critical role of food management on behalf of the government. Implicitly, the need for maintaining a high degree of self- sufficiency in cereals is also recognised. The suggestion to fix MSPs at levels 50 per cent higher than the cost of production has rightly been not accepted by the government because there are several issues involved in this suggestion. As regards other suggestions, there is perhaps no firm decision on either side.

Public Distribution System (PDS) of Cereals

Cereals being essentially a staple food in India, supply of these at affordable prices has been an essential component of food-security policy. Chronic food- insecurity is being addressed through subsidized food distribution, food for work, and employment generation and guarantee programmes. Transitory food-insecurity is addressed through short-term relief programmes. And nutritional insecurity of women and children is addressed through supplementary nutrition and mid-day meals programme in schools. As per the assessment of World Food Programme (UNWFP, 2002), food assistance programmes in India have moved from 'food for the nation' to 'food for the people' and recently to 'food security for the vulnerable'. For a clear understanding of India's PDS of foodgrains, one needs to look at buffer stocking, provisions of targeted PDS, supplementary nutrition and mid-day meals programme, size of PDS, and food subsidy involved.

Buffer Stocking of Rice and Wheat

Government of India maintains stocks of rice and wheat to meet the requirements of public distribution system and also for open market sales to reduce the fluctuations in prices. The stocks are built up mainly through price support operations. Occasionally, import route is also used to build-up the stocks. The size of minimum normal stocks is determined every five years by an expert group and is guided by the degree of inter-year fluctuations in production and government's commitment for PDS. The actual stocks at a point of time differ from the norms due to a variety of factors.

Distribution of Subsidized Cereals

The distribution of subsidized rice/wheat is supplemental in nature and does not intend on meeting the entire requirement of a family. The system is operated under the control of state governments. The food grains are distributed to target groups at different prices through a network of 4,62,000 shops spread throughout the country. The target groups have been issued ration cards for buying subsidized grains. There are four categories of entitled citizens (non- income tax payee) under PDS: (a) 20 million poorest of the poor families are supplied 35 kg of rice/wheat per month at a price of Rs 3/Rs 2 per kg. (b) Remaining 61.6 million poor families (BPL) are supplied 35 kg for rice/wheat per month at half the economic cost of rice and wheat (purchase price plus handling costs). Since July 2002, central issue prices are Rs 5.65/ kg for rice and Rs 4.15/ kg for wheat. (c) The families above poverty line are eligible to receive grains under PDS at a price close to the economic cost. For this group of families, the central issue prices are Rs 8.30/ kg for rice and Rs 6.10/ kg for wheat. (d) In addition, indigent senior citizens without any means of income or family support are provided 10 kg of rice/wheat per month free of cost. About 65000 persons are covered under this scheme.

Supplementary Nutrition Programme (SNP)

The main objective of SNP is to alleviate or prevent malnutrition among the vulnerable children below 6 years and expectant or nursing mothers. The programme was launched in 1975 and now covers 4.8 million mothers and 22.9 million children through a network of 4200 projects covering 75 per cent of development blocks in the rural areas and 273 slum pockets in the urban areas. According to nutritional needs, hot meals or snacks, along with other items (vitamins and iron tablets, etc.) are provided through childcare (*Anganwadi*) centres established in the locality with a local lady as the in-charge.

Mid-Day Meals (MDM) for School Children

The MDM programme started as a national programme with a vision to provide nutritional support to children undergoing primary education in 1995. It has twin objectives of improving the nutritional status of primary school children and of increasing enrolment, regular attendance and retention in schools. The foodgrains are supplied by the central government free of cost to the state governments, while the transport and cooking costs are borne by the state governments. Under this programme, three options are available to the states, viz. (a) providing a hot, cooked meal consisting of 100 gram of rice/wheat per day per child for 200 school days, or (b) distributing pre-cooked ready-to-eat meals, or (c) dispensing 3 kg of rice or wheat per child per month for 10 months. Presently, 120 million children in almost one million schools are covered under this programme (GoI, 2006-07).

Size of PDS

The quantities of subsidized cereals distributed under PDS have increased drastically during the current decade (Table 3). During 1960s and 1970s, the distribution was around 10 Mt of foodgrains per year. It was around 15 Mt per year during 1980s and 1990s. Since 2002, the commitment under PDS has been increasing with annual average distribution going up sharply to around 39 Mt during 2002-05. Out of the total distribution of subsidized foodgrains during the past four years, nearly 82 per cent went to below poverty line families.

Food Subsidy

Food subsidy is the difference between MSP plus handling/distribution expenses incurred by Food Corporation of India (FCI) and the issue prices of food grains under PDS. This is the amount disbursed by the government to FCI for its procurement, handling and distribution activities. In India, the food subsidy has served the multiple objectives of minimum guaranteed prices to the farmers, maintenance of buffer stocks, supply of subsidized foodgrains under identified schemes .

Farm Inputs Delivery

There is a well-defined system that works towards supplying quality certified seeds to the farmers. While the breeder seed is produced mainly by the ICAR Institutes and SAUs, the production of foundation and certified seeds is done by the private, cooperative as well as government agencies. Over the years, the share of private sector in total seed business has been increasing. The private seed companies accounted for nearly 58 per cent of the total seed produced and supplied to the farmers in 2006-07. During this year, 1.55 Mt of quality seed was supplied to the farmers, contrary to 0.58 Mt in 1991-92. The fertilizer-use in terms of nutrients has gone up to 22Mt, which is 113 kg/ha (2006-07), as against 12.7 Mt(70 kg/ha) in 1991-92. While the use of quality seed and fertilizers is increasing, the use of plant protection chemicals has been going down due to several developments, including use of disease-resistant varieties, bio-control practices and integrated pest management techniques. The use of chemical pesticides decreased from 72000 tonnes in 1991-92 to 38000 tonnes in 2006-07. The outlets for supply or sale of these inputs are mainly with farmer's cooperatives and private sector.

Apart from these mechanization of farm operations has expanded manifold. For example, the use of irrigation water lifting pumps (both diesel engines and electric motors) went up from 0.1 million in 1951 to3.2 million in 1972 and 15.7 million in 2003. The number of farm tractors increased from 0.15 million in 1972 to 2.4 million in 2003 and of threshers increased from 0.2 million to 9.1 million during this 30-year period. The electricity use in agriculture increased from 96 billion kWh in 1982-83 to 386 billion kWh in 2004-05, accounting for around 23 per cent of total electricity used in the country. However, the actual use was much lower than the demand. The demand for electricity in agriculture as well as other sectors of the economy is increasing at a very rapid rate.

In order to facilitate the use of high-yielding inputs, credit delivery system was reorganised and geared towards farmers and rural areas through several monetary policy measures. The credit flow to farmers during 2007-08 was of Rs 1.4 trillion (GOI, 2007-08). In addition, for reducing the production risks of farmers, the provisions of crop and livestock insurance were rigorously reviewed and made more farmer- friendly. In 2006-07, 18 million farmers were covered under the crop insurance programmes. Now, the scheme of weather-based crop insurance has also been introduced and both public and private sector companies are trying to increase the coverage, to help farmers and cover their weather-induced risks (Chand and Raju, 2018).

Regulation of Marketing System

As mentioned earlier, regulation of food marketing system has been an important part of food policy instruments ever since Independence. Till the middle of 1990s, several regulations were in place at different points of time. Some of these are legal restrictions on activities of traders and processors, including licensing, stocking limits, movement restrictions on foodgrains, levy obligation and size restrictions on grain milling; restrictions on bank credit for traders; ban on futures trading, canalization of imports and exports; and restrictions on setting up of private market yards. During late-1990s and first half of the current decade, several official committees reviewed the marketing situation and came out with recommendations for deregulation and liberalization of domestic markets for food commodities. Consequently, many steps were taken between 2000 and 2003 towards liberalization of foodgrain markets. The situation of domestic market deregulation in 2004 has been as follows:

- Movement restrictions – Lifted
- Storage controls – Lifted
- Small scale reservation – Lifted
- Credit control – Lifted
- Ban on futures trading – Lifted
- Bulk handling and storage (BHS) by private trade – Allowed
- Ban on foreign investment in BHS – Lifted

- Licensing system – Lifted
- Export and import – Liberalized
- Ban on set up of private wholesale markets – Lifted
- Contract farming – Allowed
- Direct purchase from farmers outside market yards – Allowed
- Minimum support prices – Continue
- Levy on rice mills and sugar factories – Continues
- Entry of organized retail trade – Allowed

Trade Policy Instruments

In order to maintain domestic supply and price stability at reasonable levels the import and export policy of food grains has played an important role. The policy instruments till mid-1980s included canalization through public agencies, quota restrictions, licensing, minimum export prices (MEP), and devaluation of currency for maintaining balance of payments. Liberalization of general trade policy began in mid-1980s. Since then, trade policy is usually announced on a five-yearly basis, but import duties and other specific instruments are announced yearly or whenever the need arises. Since 1997, MEP has been abolished, stocking limits for exporters have been relaxed, levy on non-basmati rice meant for export was withdrawn and QRs (Quantitative Restrictions) were withdrawn. The export of rice was allowed freely, but recently ban on exports of non-basmati rice and MEP on basmati rice was reimposed, which was in response to sharp rise in their domestic prices.

Current Agricultural Scenario

Growth Performance and Instability

The share of agriculture in GDP of India has registered a steady decline from 36.4 per cent in 1982-83 to 17.8 per cent in 2007-08. Yet, this sector continues to support more than half a billion people providing employment to 52 per cent of the workforce (GoI, 2007-08). The gap between the growth of agricultural and non-agricultural sectors began to widen since 1981-82 and more particularly since mid-1990s, because of acceleration in the growth of industry and service sectors. During 1950s, 1960s and 1970s, the India's economy was growing at the rate of around 3.6 per cent. The economic growth rate accelerated during the 1980s and 1990s to around 5.6 per cent per annum. It further accelerated during the current decade, recording a growth rate of 7.6 per cent during 2002-07 and around 9.2 per cent during the past two years. The growth rate of agricultural sector, which was around 2.5 per cent during the first three decades after Independence, accelerated to 3.6 per cent during 1980s and up to the middle of 1990s. However, since then it has decelerated to less than 2.5 per cent per annum. It has picked up again in the past two years but whether it is being sustained is yet to be seen (Table 1). In 2007-08, it has been 4.5 per cent.

Table 1.2- Growth rates of Indian economy and agricultural sector (% per annum)

Period	Total economy (GDP)	Agriculture and allied sectors
1951-52 to 1967-68	3.69	2.54
1968-69 to 1980-81	3.52	2.44
1981-82 t0 1990-91	5.40	3.52
1991-92 to 1996-97	5.69	3.66
1997-98 to 2001-02	5.52	2.50
2002-03 to 2006-07	7.64	2.29
2005-06 to 2006-07	9.17	4.35
2007-08 to 2008-2009	7.86	14.6
2009-10 to 2010-11	8.50	16.0
2011-12 to 2012-13	5.46	13.7
2013-2014 to 2014-15	7.41	17.1
2015-16 to 2016-17	7.11	17.0
2017-18 to 2018-19	7.23	17.32

Source: Planning Commission (2018)

Concerns Relating to Production Performance

The concerns related to production performance can be looked at from different angles.

The first is the reasons for sharp deceleration in the growth of agricultural sector since the early- 1990s. This happened on account of (a) growth rates of net sown area, gross cropped area, and electricity use in agricultural turning negative; (b) deceleration occurring in growth of fertilizer-use in agriculture; (c) growth of terms of trade for agriculture turning negative; and (d) decline in the growth rate of total factor productivity in agriculture. Thus, there was nothing favourable to the agricultural sector during mid-1990s to 2004-05.

The second aspect of the current agricultural crisis is the weakening of macro food (cereals) security during TE 1994-95 to TE 2004-05. It was mainly owing to a sense of complacency in the matter of cereal production that prevailed since mid-1990s.

Table 1.3: Year wise Growth rate of prices of crops in India

	(% per annum at constant prices)			
Period	**Cereals**	**Pulses & oilseeds**	**Fruits & vegetables**	**Other crops**
All crops				
1951-68	4.19	2.98	2.70	2.41

1968-81	3.43	0.97	4.80	2.98
3.00				
1981-91	3.52	5.41	2.84	1.73
2.97				
1991-97	2.36	2.92	6.07	2.25
3.09				
1997-02	1.49	(-)1.43	3.68	4.14
2.25				
2002-07	0.66	3.69	1.19	3.76
1.88				
2005-07	3.52	0.47	3.12	6.83
4.12				

Source: **Planning Commission (2007)**

The atmosphere against cereals had its toll in several forms. The growth rate of area under cereals decelerated and even turned negative for rice. The research investment in cereals slowed down. The growth rates of yield and total factor productivity of rice and wheat slowed down. As a consequence, the decline in real cost of production of rice and wheat either slowed down or turned the other way. There was also a set back on the front of cereals trade balance. On quinquennial basis, India could reduce its net import of cereals continuously till the second half of 1980s and later increased its net exports till the middle of current decade, but during 2005-07, our import dependence has again resurfaced.

As regards the wheat imports during 2005-07, two important factors have led to the situation. One, the large traders are now competing with the public agencies for buying wheat and rice during the peak marketing period and there is nothing wrong in it as farmers face a more competitive market situation. And two, the government has now larger commitment of distribution of subsidised foodgrains under various food security and food safetynet programmes and, therefore, needs larger public stocks of grains. But, the government's strategy to build up these stocks continues to rely only on price support purchases. The price support operations, by very definition, need not and are not expected to serve the objective of meeting enlarged PDS needs. There is a need for more prudent management of the cereals sector to avoid the criticism, which the government has to face in the wake of imports at very high prices.

The third aspect of the current agricultural crisis is largely due to declining profitability of agriculture and consequently the loss of farmers' faith in farming activities. Aslo-down in the growth of TFP is bound to slow down the rate of decline in cost of production. In the case of rice and wheat, the growth rate of TFP in Punjab and Haryana has turned negative. As a consequence, the cost of production has tended to increase. The terms of trade for agriculture, which is the ratio of prices received by the farmers (for products sold by them) to the prices paid by them (for inputs and consumption goods purchased by them) has turned against the farmers. The average income of Indian farmers vis-à-vis those engaged in other occupations has continued to be substantially lower and has deteriorated further during the past decade. This is evident from several facts. The growth of agricultural sector has been substantially lower than that of non-farm sector, whereas agricultural sector sustains more than 60 per cent of the population. The incidence of rural poverty is substantially higher in land-owning households than landless households. The production of staple food is turning out to be less profitable relative to other competing enterprises.

For a better understanding of the income levels of farmers vis-a-vis income in other occupations, take the case of Punjab state, which has assured irrigation for at least two crops a year. The per hectare yields of wheat and rice, which are the major crops of Punjab, are the highest in the country. Both these crops are commercial crops for the Punjab farmers and an effective price support system is in place. The average farm size in Punjab is 3.8 ha, which is 2.5- times the average size of Indian farms. In spite of these favourable factors, an average Punjab farmer (with 3.8 ha land and growing two crops of rice and wheat) is able to earn an income for his labour and management, that is less than the starting salary of a Class IV employee in the government sector. The situation of farmers in the other states can be easily visualized from this fact. The questions that arise are: (a) How can Indian farmers have faith in cereal production?; (b) Why to question the supply of fertilizers at lower prices and electricity and irrigation to farmers at lower user charges?; (c) Why not to substantially enhance the level of minimum support prices for various crops?; and (d) At what level to stabilize the declining trend in real prices of staple foods like rice and wheat?

The perception about the level of staple food prices for consumers has continued to be outdated. Whether the price is higher or not should be seen in relation to the consumers' income. In the early-1970s, the average retail price of one quintal of rice/wheat was equivalent to 16 per cent of the India's average per capita income and one-day's wages of a rural labourer fetched one kg of rice or wheat. The real prices of rice and wheat have continued to decline since then. And this is a welcome development. However, we should not forget the plight of staple food producers. Undoubtedly, there is a need to pay a higher price to Indian farmers producing staple food for maintaining a high degree of macro food-security of the country. The perception about what is a high price for rice and wheat must change. Apart from the prices of farm products, farmers' profits also depend on availability of new technology, trade policy of the government, freedom to lease-in or lease-out land without fear of losing ownership rights and supplementing income from non-farm occupations, either within the rural areas or in nearby larger settlements where investment is growing fast. Several steps are necessary pertaining to these aspects for improving farm household incomes.

The fourth aspect of current agricultural crisis persists at the unfinished agenda of the marketing system improvement. The main objective of the marketing system improvement is to enhance its efficiency by saving Rs 50000 crore to Rs 80000 crore that are annually lost and is an avoidable national wastage. These are not intended, as commonly understood, to eliminate market functionaries but to link farm-gate to the retailer or consumers through efficient supply chain management. Saving of huge avoidable losses would provide higher price to farmers, lower price to consumers and create new employment opportunities along the supply chain. Agricultural marketing system improvement, through encouraging technologically superior supply chains, is a win-win situation for all stakeholders. Recent steps taken by the government in this regard is a welcome development, but considerably more steps are necessary to speed up the process of agricultural market reforms. An amendment in State APMR Acts is yet to be adopted by all the states. Wherever Act has been passed, rules for its implementation have not been adopted. The notion of one national market for farm products is yet to be grounded. The infrastructure in the existing market yards continues to be inadequate. Adhocism in marketing policies and reimposition of marketing regulations continue to keep the investors away from the marketing investments. The role and attitudes ofAPMCs and State Marketing

Board have not changed substantially. Market fees are not fully ploughed back. Contract farming and direct purchases from the farmers continue to be on a limited scale. The progress towards organising farmers into marketing groups continues to be inadequate. The import policy for agricultural products (edible oils, pulses and recently wheat) is mainly oriented towards consumers, without caring for the farmers who produce these products.

Fifth several areas in the country, particularly intensive farming areas are now suffering from soil and land degradation, depletion of groundwater, and micronutrient deficiency.

Sixth is the considerable increase in the imports of edible oils and pulses since mid-1990s. The import of edible oils, which accounted for around 13 per cent (1.2 Mt) of total consumption in 1995-96, has sharply increased to 37.5 per cent (4.7 Mt) during 2006-07. The imports help in keeping consumer prices low, but adversely affect the producer-farmers, who are mostly resource-poor and operate in drylands or rainfed areas. While the net social gain from imports is insignificant, there is considerable redistribution of income from dryland resource-poor farmers to consumers.

New Policy Initiatives and Programmes

Several new measures have been taken during the past few years to tackle the situation and bring back farmers' confidence in farming in general and cereals production in particular.

(i) A National Food Security Mission (NFSM) was launched with the specific objective of increasing the production of rice, wheat and pulses in targeted 305 districts, with an outlay of Rs 48.8 billion during the five-year period. The focus is on providing quality seeds of high-yielding varieties and all possible efforts to transfer improved technology to farmers, with enough flexibility to chose the interventions at district level (GoI, 2007-08).

(ii) *Rashtriya Krishi Vikas Yojana* (RKVY) (National Agricultural Development Scheme) is another new scheme, which aims at incentivising the state governments to increase the share of investment in agriculture in their state plans. The states have been given complete flexibility to plan on the basis of agro-climatic conditions of each region. The allocation under RKVY is of Rs 250 billion for a period of five years (GoI, 2007-08).

(iii) The Government of India has approved and adopted a National Policy for Farmers in 2007. It covers several areas but focuses on the economic well-being of farmers. It includes asset reforms, use of biotechnology and ICT, bio-security system, seed and soil health, credit, insurance, higher support prices for farmers, and enlargement of food-security basket.

(iv) Under the farm credit package, the flow of institutional credit to the farmers was doubled within two years (2005-06 to 2006-07). More than 72 million farmers have been issued credit cards by commercial and cooperative banks.

(v) The crop, weather and livestock insurance schemes for farmers have been redesigned and coverage has been expanded.

(vi) All the rural districts have been equipped with a 'Farm Science Centre', for increasing the reach of farmers to new technologies and agricultural scientists. These centres are a part of either national agricultural research institutes or state agricultural universities.

(vii) The investment in agricultural research has been increased considerably. Similarly, the allocation for Horticultural Mission and several other agricultural development schemes has been enhanced.

(viii) Recently, outstanding loans of farmers (from commercial and cooperative banks) amounting to Rs 710 billion have been waived by the government to provide relief to 40 million farmers and to make these farmers eligible for fresh loans.

(ix) In addition, the support prices for rice and wheat have been substantially hiked during the current season.

These apart, following new initiatives and expansion programmes are at different stages of implementation (GoI, 2007):

(i) Bharat Nirman (India connectivity and infrastructure programme)

(ii) Watershed Development and Micro Irrigation Programme

(iii) Establishment of National Rainfed Area Authority

(iv) Establishment of National Fisheries Development Board

(v) Establishment of National Bee Board

(vi) Revitalization of Cooperative Sector

(vii) Enactment of Integrated Food Law and setting up of Food Safety and Standards Authority of India (FSSA)

(viii) Agri-business Development through Venture Capital Participation Schemes

(ix) Legislative Framework for Warehousing Development and Regulation

(x) Protection of Plant Varieties Regulation and Farmers' Rights Act

(xi) Establishment of Bamboo Mission

(xii) Increasing Knowledge Connectivity through Common Service Centres and IT Initiatives

(xiii) National Rural Health Mission

The outcome of these medium- and short-term measures has been positive. With favourable behaviour of monsoon, the production of cereals increased from 185.2 Mt during 2004-05 to 212.2

Mt during 2007-08. The procurement of wheat and rice increased considerably during the current year. Wheat procurement has been more than 23 Mt and of rice around 27 Mt, which is more than sufficient to meet the requirements of PDS.

CONCLUSION AND SUGGESTION

Production Environment and Policy

Based on a brief review of the current policy regime relating to pricing, marketing and trade and current agricultural scenario, has paved a path for making agriculture profitable and efficient and for restoring the farmers' faith in farming as an occupation. In predominantly agricultural and rural economies, accelerated agricultural growth and an efficient agricultural sector is the key for reducing hunger, food insecurity and malnutrition at a rapid rate. Acceleration of agricultural growth and improving profitability and efficiency of agriculture in general, requires the following:

(i) Investment in agricultural research for expanding the productivity potential.

(ii) Investment in productivity-raising infrastructure like irrigation, soil and water conservation, and integrated natural resource management.

(iii) Persuading farmers to increasingly shift to land-saving enterprises (livestock and fisheries) and high-value crops (horticulture).

(iv) Adequate and efficient mechanism for transfer of new technology to farmers and supply of key farm-inputs at affordable prices or user charges.

(v) Efficient system of post-harvest and on-farm handling of farm-products (training of farmers and provision of appropriate technology).

(vi) Efficient system of entire chain of marketing activities from farm-gate to consumers.

(vii) Adequate investment in rural roads and marketing infrastructure.

(viii) Farmer-friendly marketing, pricing and trade policies.

(ix) Continuously pursuing the policy of self- sufficiency in staple cereals, which is a necessary condition for household and individual food-security.

Marketing Policy and System Improvement

We should initiate on moving towards one national market for agricultural products, which *inter alia* would require shifting the subject of agricultural marketing from the 'state' list to the 'Concurrent' list in the Constitution.

The amendments in State APMR Acts should be speed up and model rules/ regulations should be adopted in all the states to promote contract farming and direct marketing/purchase arrangements from farmers.

The role of APMCs and State Agricultural Marketing Boards should be redefined to *inter- alia* to encourage value addition in primary markets, rather than just collection of fees and undertaking of construction activities.

The investment on research and training in post- harvest handling of farm products should be given priority because the internal rate of return (IRR) for such investments is very high.

Training of farmers, farm women and rural youth in post-harvest handling and value addition should be stepped-up and promoted by increasing the scientific staff strength of KrishiVigyanKendras (KVK), which now exist in all the rural districts of the country. Every KVK should be provided with an agricultural marketing or agri-business specialist.

Rural roads being a critical link in marketing of farm products should be emphasised more so that adequate investment in all the areas, especially in hilly, tribal and desert regions. The returns to investment in rural roads, in terms of accelerating rural growth and reducing rural poverty, have been estimated to be very high in India.

The efforts and emphasis towards organizing farmers' marketing groups should be scaled-up in order to increase the farm-gate and retail linkages, through contract farming or otherwise. Atleast 100,000 such groups should be organized and financially supported with the technical support from KVKs, ATMAs or NGOs. Each such group should be financially supported to engage an agri-business manager.

Evolution and development of supply chains and organized retail outlets should be promoted to save the huge national wastage of farm products that occur frequently under the present system and in this process, involvement of the existing players should not be endangered.

The aphorism in marketing and trade policies, has been persistently observed during the past two year should be stopped because fears of reposition of marketing restrictions keep the potential investors away and as a consequence, farmers suffer.

Pricing and Trade Policy

The policy of minimum support prices should be pursued rather rigorously and its effective implementation should be ensured and emphasised for all the crops covered under the scheme and in all the areas, including the eastern states.

The recent hike in minimum support prices of cereals and other crops is an appreciable development, but the hike is inadequate. The MSPs should be further raised in the subsequent seasons. This is justified because the interests of vulnerable sections in terms of cheap staple food are safeguarded under various schemes of PDS, which now cover 50 per cent of India's population. However, effective implementation of all PDS programmes should be ensured.

The trade policy for agricultural commodities should be dovetailed with domestic price policies by renaming the Commission for Agricultural Costs and Prices as the 'Commission for Agricultural Prices and Trade', and suitably redefining its terms of reference, as suggested by the Alagh Committee.

Oilseeds and pulses are grown by resource-poor farmers, mostly in the rainfed and dry areas. The policy of liberal imports at low duties has come in the way of growth of agriculture and livelihood of farmers in these areas. The reduced import duties on edible oils imply transfer of around Rs 10000 crore from dryland farmers to edible oil consumers. The import duties on edible oils and pulses should be raised within the bound rates committed by India in WTO. The fundamental question involved is the balancing of interests between farmers and consumers.

Other Policies

We should continue pursuing and not disrupt the policy of very high degree of self-sufficiency in cereals. The objective of self-sufficiency in cereals should be explicitly announced by the government and be explicitly mentioned in all official documents.

As 95 per cent of implicit input subsidies (fertilizers, canal water and electricity for irrigation) go to food crops and the share of marginal and small farmers in these subsidies is proportionally more than their share in operated land, we should not intervene with these subsidies at the current stage of Indian agriculture. Their withdrawal will further squeeze farm incomes and reduce profitability in agriculture. A categorical announcement in this regard will boost the morale of Indian farmers.

Marginal farmers should be treated as living under safety net as part time farmers and should be provided with cheap inputs and improved technology to improve their food security and livelihoods. Further, they should be encouraged to engage themselves in non-farm activities by improving their skills, both hard and soft.

The land-lease market should be legalised to help in diversification of rural household incomes, bring existing unused land under productive use, and encourage tiny landholders to engage in other occupations, without fear of losing ownership rights of land.

Rural-urban migration should be reinforced by migrant-friendly policies and should be treated as a approach for improving the incomes of rural households, alleviation of rural poverty and reducing pressure on rural resources. Rural- urban migration has been globally recognised as an important pathway out of rural poverty.

Famine: The worst manifestation of hunger and poverty: Its definition, proliferation and prosecution

From top-left to bottom-right or from top to bottom (mobile): child victims of famines in the Netherlands (1944-45), India (1943-44), and Nigeria (1967-70)

A famine is an extensive shortage of food as a result of several factors including war, inflation, crop failure, population imbalance, or government policies. This phenomenon is usually accompanied or followed by regional malnutrition, starvation, epidemic, and increased mortality. Every inhabited continent in the world has experienced a period of famine throughout history. In the 19th and 20th century, it was generally Southeast and South Asia, as well as Eastern and Central Europe that suffered the most deaths from famine. The numbers dying from famine began to fall sharply from the 1970s.

Some countries, specifically in sub-Saharan Africa, continue to have extreme cases of famine. Since 2010, Africa has been the most affected continent in the world. As of 2017, the United Nations has warned some 20 million are at risk in South Sudan, Somalia, Nigeria and Yemen. Agricultural conditions have been fluctuating more and more due to variations in weather, and the distribution of food has been affected by conflict. Most programmes now direct their aid towards Africa.

Definitions

According to the United Nations humanitarian criteria, even if there are food shortages with large numbers of people lacking nutrition, a famine is declared only when certain measures of mortality, malnutrition and hunger are met. The criteria are:

- At least 20% of households in an area face extreme food shortages with a limited ability to cope
- The prevalence of acute malnutrition in children exceeds 30%
- The death rate exceeds two persons per 10,000 persons per day

The declaration of a famine carries no binding obligations on the UN or member states, but serves to focus global attention on the problem.

Weeping of Voiceless

The cyclical occurrence of famine has been a deeply engraved in the societies engaged in subsistence agriculture since the dawn of agriculture itself. The frequency and intensity of famine has fluctuated throughout history, depending on changes in food demand, such as population growth, and supply-side shifts caused by changing climatic conditions. Famine was first eliminated in Holland and England during the 17th century, due to the commercialization of agriculture and the implementation of improved techniques to increase crop yields.

Decline of famine

In the 16th and 17th century, the feudal system saw a drastic change, it was observed that the more prosperous farmers began to enclose their own land and improve their yields to sell the surplus crops for a profit. These capitalist landowners paid their labourers with money, thereby increasing the commercialization of rural society. In the emerging competitive labour market, better techniques for the improvement of labour productivity were increasingly valued and rewarded. It was in the farmer's interest to produce as much as possible on their land in order to sell it to areas that demanded that product. They produced guaranteed surpluses of their crop every year if they could.

Subsistence peasants were also vehemently forced to commercialize their activities because of the increasing burden of the taxes. Taxes that had to be paid to central governments in money forced the peasants to produce crops to sell. Sometimes they produced industrial crops, but they would find ways to increase their production in order to meet both their subsistence requirements as well as their tax obligations. Peasants also used the new money to purchase manufactured goods. The agricultural and social developments encouraging increased food production were gradually taking place throughout the 16th century, but took off in the early 17th century.

By the 1590s, these trends were sufficiently developed in the rich and commercialized province of Holland thereby allowing its population to withstand a general outbreak of famine in Western Europe at that time. By that time, the Netherlands had one of the most commercialized agricultural systems in Europe. They grew many industrial crops such as flax, hemp and hops. Agriculture became increasingly specialized and efficient. The efficiency of Dutch agriculture allowed for much more rapid urbanization in the late sixteenth and early seventeenth centuries than anywhere else in Europe. As a result, productivity and wealth increased, allowing the Netherlands to maintain a steady food supply.

By 1650, English agriculture had also become commercialized on a wider scale. The last peace-time famine in England was in 1623–24. There were still periods of hunger, as in the Netherlands, but no more famines ever occurred. Common areas for pasture were enclosed for private use and large scale, efficient farms were consolidated. Other technical developments included the draining of marshes, more efficient field use patterns, and the wider introduction of industrial crops. These agricultural developments led to wider prosperity in England and increasing urbanization. By the end of the 17th century, English agriculture was the most productive in Europe.In both England and the Netherlands, the population stabilized between 1650 and 1750, the same time period in which the sweeping changes to agriculture occurred. Famine still occurred in other parts of Europe, however. In East Europe, famines occurred as late as the twentieth century.

Attempts at famine alleviation

Skibbereen, Ireland, during the Great Famine, 1847 illustration by James Mahony for the *Illustrated London News*

As per the severity of famine, it was a chief concern for governments and other authorities. In pre-industrial Europe, preventing famine, and ensuring timely food supplies, was one of the chief concerns of many governments, although they were severely limited in their options due to limited levels of external trade and an infrastructure and bureaucracy generally too rudimentary to effect real relief.

Most governments were concerned by famine because it could lead to revolt and other forms of social disruption.

By the mid-19th century and the onset of the Industrial Revolution, it became possible for governments to alleviate the effects of famine through price controls, large scale importation of food products from foreign markets, stockpiling, rationing, regulation of production and charity. The Great Famine of 1845 in Ireland was one of the first famines to feature such intervention, although the government response was often lacklustre. The initial response of the British government to the early phase of the famine was «prompt and relatively successful,» according to F. S. L. Lyons.Confronted by widespread crop failure in the autumn of 1845, Prime Minister Sir Robert Peel purchased £100,000 worth of maize and cornmeal secretly from America. Baring Brothers & Co initially acted as purchasing agents for the Prime Minister. The government hoped that they would not «stifle private enterprise» and that their actions would not act as a disincentive to local relief efforts. Due to weather conditions, the first shipment did not arrive in Ireland until the beginning of February 1846.The maize corn was then re-sold for a penny a pound.[8]

In 1846, Peel moved to repeal the Corn Laws, tariffs on grain which kept the price of bread artificially high. The famine situation worsened during 1846 and the repeal of the Corn Laws in that year did little to help the starving Irish; the measure split the Conservative Party, leading to the fall of Peel›s ministry.In March, Peel set up a programme of public works in Ireland.

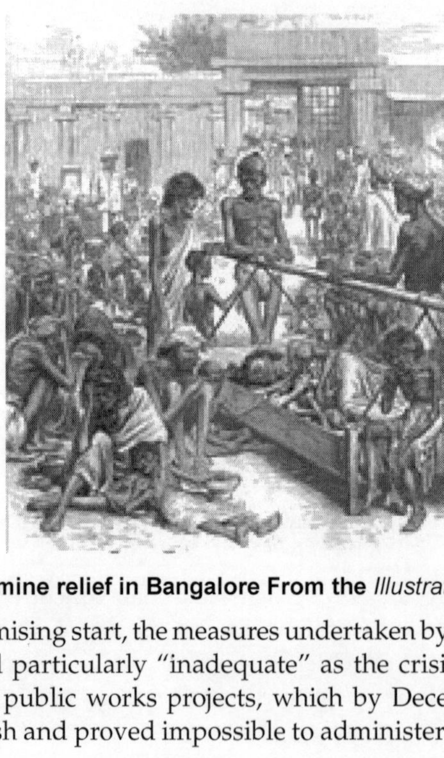

People waiting for famine relief in Bangalore From the *Illustrated London News*, 1877

Despite this promising start, the measures undertaken by Peel's successor, Lord John Russell, proved particularly "inadequate" as the crisis deepened. Russell's ministry introduced public works projects, which by December 1846 employed some half million Irish and proved impossible to administer. The government was influenced by a laissez-faire belief that the market would provide the food needed. It halted government food and relief works, and turned to a mixture of «indoor» and «outdoor» direct relief; the former administered in work houses through the Poor Law, the latter through soup kitchenss.

A systematic approach to creating a necessary regulatory framework for dealing with famine was developed by the British Raj in the 1880s. In order to comprehensively address the issue of famine, the British created an Indian Famine commission to recommend steps that the government would be required to take in the event of a famine.The Famine Commission issued a series of government guidelines and regulations on how to respond to famines and food shortages called the Famine Code. The famine code was also one of the first attempts to scientifically predict famine in order to mitigate its effects. These were finally passed into law in 1883 under Lord Ripon.

The Code introduced the first famine scale: three levels of food insecurity were defined: near-scarcity, scarcity, and famine. «Scarcity» was defined as three successive years of crop failure, crop yields of one-third or one-half normal, and large populations in distress. «Famine» further included a rise in food prices above 140% of «normal», the movement of people in search of food, and widespread mortality. The Commission identified that the loss of wages from lack of employment of agricultural labourers and artisans were the cause of famines. The Famine Code applied a strategy of generating employment for these sections of the population and relied on open-ended public works to do so.

20th Century

During the 20th century, an estimated 70 million people died from famines across the world, of whom an estimated 30 million died during the famine of 1958–61 in China.The other most notable famines of the century included the Bengal famine of 1943, famines in China in 1928 and 1942, and a sequence of famines in Russia and elsewhere in the Soviet Union, including the Soviet famine of 1932–1933, caused by the policies of Stalin.

Feed The World logo designed for Band Aid.

The great famines of the late 20th century were: the Biafran famine in the 1960s, the Khmer Rouge-caused famine in Cambodia in the 1970s, the North Korean famine of the 1990s and the Ethiopian famine of 1983–85.

The latter event was reported on television reports around the world, carrying footage of starving Ethiopians whose plight were centred on a feeding station near the town of Korem. This stimulated the first mass movements to end famine across the world.

BBC newsreader Michael Buerk gave moving commentary of the tragedy on 23 October 1984, which he described as a «biblical famine». This prompted the Band Aid single, which was organized by Bob Geldof and featured more than 20 pop stars. The Live Aid concerts in London and Philadelphia raised even more funds for the cause. An estimated 900,000 people died within one year as a result of the famine, but the tens of millions of pounds raised by Band Aid and Live Aid are believed to have saved the lives of some Ethiopians who were in danger of dying.

21st Century

Until 2017, the worldwide death due to famine has drastically fallen. The World Peace Foundation reported that from the 1870s to the 1970s, great famines killed an average of 928,000 people a year.Since 1980, annual deaths had dropped to an average of 75,000, less than 10% of what they had been until the 1970s. That reduction was achieved despite the approximately 150,000 lives lost in the 2011 Somalia famine. Yet in 2017, the UN officially declared famine had returned to Africa, with about 20 million people at risk of death from starvation in Nigeria, in South Sudan, in Yemen, and in Somalia.

REGIONAL HISTORY

Africa

Early history

In the mid-22nd century BC, an abrupt and short-spanned climatic change resulted in reduced rainfall leading to several decades of drought in Upper Egypt. The resulting famine and civil strife is believed to have been a major cause of the collapse of the Old Kingdom. An account from the First Intermediate Period states, «All of Upper Egypt was dying of hunger and people were eating their children.» In 1680s, famine extended across the entire Sahel, and in 1738 half the population of Timbuktu died of famine.In Egypt, between 1687 and 1731, there were six famines. The famine that afflicted Egypt in 1784 cost it roughly one-sixth of its population. The Maghreb experienced famine and plague in the late 18th century and early 19th century. There was famine in Tripoli in 1784, and in Tunis in 1785.

According to John Iliffe, "Portuguese records of Angola from the 16th century show that a great famine occurred on average every seventy years; accompanied by epidemic disease, it might kill one-third or one-half of the population, destroying the demographic growth of a generation and forcing colonists back into the river valleys.

The first documentation of weather in West-Central Africa occurred around the mid-16th to 17th centuries in areas such as Luanda Kongo, however, not much data had been recorded on the issues of weather and disease except for a few notable documents. The only records obtained are of violence between Portuguese and Africans during the Battle of Mbilwa in 1665. In these documents the Portuguese wrote of African raids on Portuguese merchants solely for food, giving clear signs of famine. Additionally, instances of cannibalism by the African Jaga were also more prevalent during this time frame, indicating an extreme deprivation of a primary food source.

Colonial period

A 1906 *Punch* cartoon depicting King Leopold II as a rubber vine entangling a Congolese man.

A notable period of famine occurred during the 20th century in the Congo Free State. In forming this state, Leopold used mass labour camps to finance his empire. This period resulted in the death of up to 10 million Congolese from brutality, disease and famine. Some colonial «pacification» efforts often caused severe famine, notably with the repression of the MajiMaji revolt in Tanganyika in 1906. The introduction of cash crops such as cotton, and forcible measures to impel farmers to grow these crops, sometimes impoverished the peasantry in many areas, such as northern Nigeria, contributing to greater vulnerability to famine when severe drought struck in 1913.

A large-scale famine outbreak occurred in Ethiopia in 1888 and succeeding years, as the rinder pest epizootic, introduced into Eritrea by the infected cattle, spread southwards reaching ultimately as far as South Africa. In Ethiopia it was estimated that as much as 90 percent of the national herd died, rendering rich farmers and herders destitute overnight. This coincided with drought associated with an el Nino oscillation, human epidemics of smallpox, and in several countries, intense war. The Ethiopian Great famine that afflicted Ethiopia from 1888 to 1892 cost it roughly one-third of its population. In Sudan the year 1888 is remembered as the worst famine in history, on account of these factors and also the exactions imposed by the Mahdist state.

Records compiled for the Himba recall two droughts from 1910 to 1917. They were recorded by the Himba through a method of oral tradition. From 1910 to 1911 the Himba described the drought as "drought of the omutati seed" also called *omangowi*, which means the fruit of an unidentified vine that people ate during the time period. From 1914 to 1916 droughts brought *katur' ombanda* or *kari' ombanda* which means «the time of eating clothing».

20th century

Malnourished children in Niger, during the 2005 famine

For the middle part of the 20th century, agriculturalists, economists and geographers did not specifically consider Africa to be famine prone. From 1870 to 2010, 87 per cent of deaths from famine occurred in Asia and Eastern Europe, with only 9.2 per cent in Africa. There were notable counter-examples, such as the famine in Rwanda during World War II and the Malawi famine of 1949, but most famines were localized and brief food shortages. Although the drought was brief the main cause of death in Rwanda was due to Belgian prerogatives to acquisition grain from their colony (Rwanda). The increased grain acquisition was related to WW2. This and the drought caused 300,000 Rwandans to perish.

From 1967 to 1969 large scale famine occurred in Biafra and Nigeria due to a government blockade of the Breakaway territory. It is estimated that 1.5 million people died of starvation due to this famine. Additionally, drought and other government interference with the food supply caused 500 thousand Africans to perish in Central and West Africa.

Famine recurred in the early 1970s, when Ethiopia and west African Sahel suffered drought and famine. The Ethiopian famine of that time was closely linked to the crisis of feudalism in that country, and in due course helped to bring about the downfall of the Emperor Haile Selassie. The Sahelian famine was associated with the slowly growing crisis of pastoralism in Africa, which has seen livestock herding decline as a viable way of life over the last two generations.

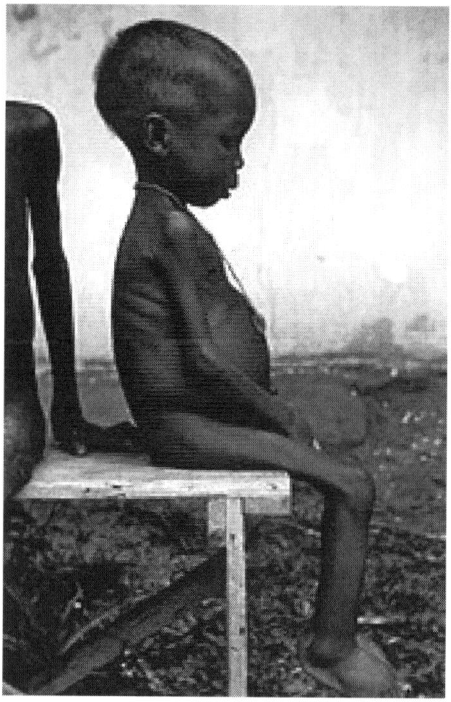

A girl during the Nigerian Civil War of the late 1960s. Pictures of the famine caused by Nigerian blockade garnered sympathy for the Biafrans worldwide.

Famines occurred in Sudan in the late-1970s and again in 1990 and 1998. The 1980 famine in Karamoja, Ugandawas, in terms of mortality rates, one of the worst in history. 21% of the population died, including 60% of the infants. In the 1980s, large scale multilayer drought occurred in the Sudan and Sahelian regions of Africa. This caused famine because even though the Sudanese Government believed there was a surplus of grain, there were local deficits across the region.

In October 1984, television reports describing the Ethiopian famine as "biblical", prompted the Live Aid concerts in London and Philadelphia, which aided in collecting large sums to alleviate the suffering. A primary cause of the famine (one of the largest seen in the country) is that Ethiopia (and the surrounding Horn) was still recovering from the droughts which occurred in the mid-late 1970s. Compounding this problem was the intermittent fighting due to civil war, the government's lack of organization in providing relief, and hoarding of supplies to control the population. Ultimately, over 1 million Ethiopians died and over 22 million people suffered due to the prolonged drought, which lasted roughly 2 years.

In 1992 Somalia became a war zone with no effective government, police, or basic services after the collapse of the dictatorship led by Siad Barreand the split of power between warlords. This coincided with a massive drought, causing over 300,000 Somalis to perish.

Recent years

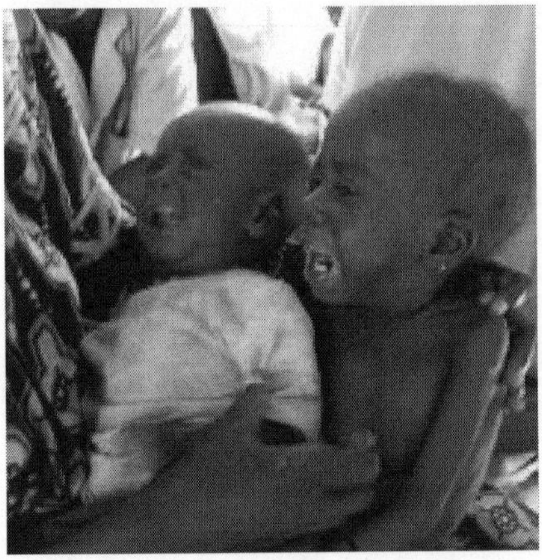

Laure Souley holds her three-year-old daughter and an infant son at a MSF aide center during the 2005 famine, Maradi Niger

Since the start of the 21st century, more effective early warning and humanitarian response actions have considerably reduced the number of deaths due to famine. That said, many African countries are not self-sufficient in food production, relying on income from cash crops to import food. Agriculture in Africa is susceptible to climatic fluctuations, especially droughts which can reduce the amount of food produced locally. Other agricultural problems include soil infertility, land degradation and erosion, swarms of desert locusts, which can destroy whole crops, and livestock diseases. Desertification is increasingly problematic: the Sahara reportedly spreads up to 48 kilometres (30 mi) per year.The most serious famines have been caused by a combination of drought, misguided economic policies, and conflict. The 1983–85 famine in Ethiopia, for example, was the outcome of all these three factors, made worse by the Communist government›s censorship of the emerging crisis. In Sudan at the same date, drought and economic crisis combined with denials of any food shortage by the then-government of President GaafarNimeiry, to create a crisis that killed perhaps 250,000 people—and helped bring about a popular uprising that overthrew Nimeiry.

Numerous factors led to making the food security situation in Africa grave, including political instability, armed conflict and civil war, corruption and mismanagement in handling food supplies, and trade policies that harm African agriculture. An example of a famine created by human rights abuses is the 1998 Sudan famine. AIDS is also having long-term economic effects on agriculture by reducing the available workforce, and is creating new vulnerabilities to famine by overburdening poor households. On the other hand, in the modern history of Africa on quite a few occasions famines acted as a major source of acute political instability.[38] In Africa, if current trends of population growth and soil degradation continue, the continent might be able to feed just 25% of its population by 2025, according to United Nations University (UNU)›s Ghana-based Institute for Natural Resources in Africa.

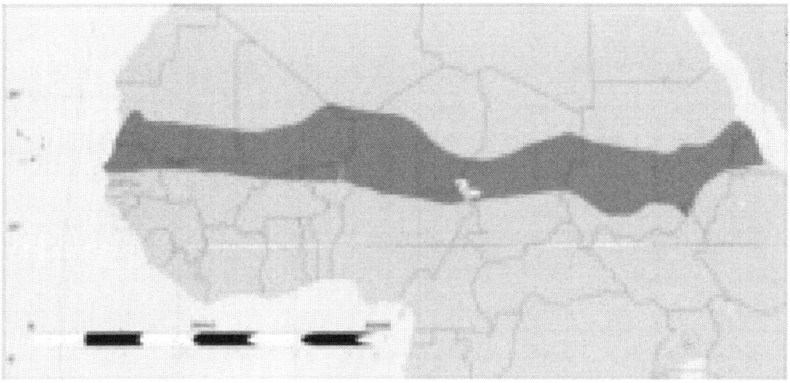

Famine-affected areas in the western Sahel belt during the 2012 drought.

Recent famines in Africa include the 2005–06 Niger food crisis, the 2010 Sahel famine and the 2011 East Africa drought, where two consecutive missed rainy seasons resulted in the worst drought in East Africa in 60 years. An estimated 50,000 to 150,000 people are reported to have died during the period. In 2012, the Sahel drought put more than 10 million people in the western Sahel at risk of famine (according to a Methodist Relief & Development Fund (MRDF) aid expert), due to a month-long heat wave.

Today, famine is most widespread in Sub-Saharan Africa, but as a consequence to exhaustion of food resources, over extraction of groundwater, wars, internal struggles, and economic failure, famine continues to be a worldwide phenomena with hundreds of millions of people suffering. These famines causes widespread malnutrition and impoverishment. The famine in Ethiopia in the 1980s had an immense death toll, although Asian famines of the 20th century have also produced extensive death tolls. Modern African famines are characterized by widespread destitution and malnutrition, with heightened mortality confined to young children.

Current initiatives

Alternative initiatives have been pioneered to address the problem of food security, as against a backdrop of conventional interventions through the state or markets One pan-African example is the Great Green Wall. Another example is the "Community Area-Based Development Approach" to agricultural development ("CABDA"), an NGO programme with the objective of providing an alternative approach to increasing food security in Africa. CABDA proceeds through specific areas of intervention such as the introduction of drought-resistant crops and new methods of food production such as agro-forestry. Piloted in Ethiopia in the 1990s it has spread to Malawi, Uganda, Eritrea and Kenya. In an analysis of the programme by the Overseas Development Institute, CABDA's focus on individual and community capacity-building is highlighted. This enables farmers to influence and drive their own development through community-run institutions, bringing food security to their household and region.

The role of African Unity organization

The organization of African unity and its role in the African crisis, has been keen with regard to the political aspects of the continent, particularly the liberation of the occupied parts of it and the elimination of racism. The organization has succeeded in this area but the economic field and development has not succeeded in these fields. African leaders have agreed to waive the role of their organization in the development to the United Nations through the Economic Commission for Africa "ECA".

Far East

Chinese officials engaged in famine relief, 19th-century engraving

Chinese scholars had kept count of 1,828 instances of famine from 108 BC to 1911 in one province or another—an average of close to one famine per year. From

1333 to 1337 a terrible famine killed 6 million Chinese. The four famines of 1810, 1811, 1846, and 1849 are said to have killed no fewer than 45 million people.

Japan experienced more than 130 famines between 1603 and 1868.

The period from 1850 to 1873 saw, as a result of the Taiping Rebellion, drought, and famine, the population of China drop by over 30 million people. China's Qing Dynasty bureaucracy, which devoted extensive attention to minimizing famines, is credited with averting a series of famines following El Niño-Southern Oscillation-linked droughts and floods. These events are comparable, though somewhat smaller in scale, to the ecological trigger events of China's vast 19th-century famines. Qing China carried out its relief efforts, which included vast shipments of food, a requirement that the rich open their storehouses to the poor, and price regulation, as part of a state guarantee of subsistence to the peasantry (known as *ming-sheng*).

When a stressed monarchy shifted from state management and direct shipments of grain to monetary charity in the mid-19th century, the system broke down. Thus the 1867–68 famine under the Tongzhi Restoration was successfully relieved but the Great North China Famine of 1877–78, caused by drought across northern China, was a catastrophe. The province of Shanxi was substantially depopulated as grains ran out, and desperately starving people stripped forests, fields, and their very houses for food. Estimated mortality is 9.5 to 13 million people.

Great Leap Forward

The largest famine of the 20th century, and most certainly of all time, was the 1958–61 Great Leap Forward famine in China. The immediate causes of this famine lay in Mao Zedong›s ill-fated attempt to transform China from an agricultural nation to an industrial power in one huge leap. Communist Party cadres across China insisted that peasants abandon their farms for collective farms, and begin to produce steel in small foundries, often melting down their farm instruments in the process. Collectivisation undermined incentives for the investment of labour and resources in agriculture; unrealistic plans for decentralized metal production sapped needed labour; unfavourable weather conditions; and communal dining halls encouraged overconsumption of available food. Such was the centralized control of information and the intense pressure on party cadres to report only good news—such as production quotas met or exceeded—that information about the escalating disaster was effectively suppressed. When the leadership did become aware of the scale of the famine, it did little to respond, and continued to ban any discussion of the cataclysm. This blanket suppression of news was so effective that very few Chinese citizens were aware of the scale of the famine, and the greatest peacetime demographic disaster of the 20th century only became widely known twenty years later, when the veil of censorship began to lift.

The exact number of famine deaths during 1958–61 is difficult to determine, and estimates range from 18to at least 42 million people, with a further 30 million cancelled or delayed births. It was only when the famine had wrought its worst that Mao reversed agricultural collectivisation policies, which were effectively dismantled in 1978. China has not experienced a famine of the proportions of the Great Leap Forward since 1961.

Khmer Rouge

Picture :Skulls of Khmer Rouge murder victims at ChoeungEk

In 1975, the Khmer Rouge took control of Cambodia. The new government was led by Pol Pot, who desired to turn Cambodia into a communist, agrarian utopia. His regime emptied the cities, abolished currency and private property, and forced Cambodia's population into slavery on communal farms. In less than four years, the Khmer Rouge had executed nearly 1.4 million people, mostly those believed to be a threat to the new ideology.

Due to the failure of the Khmer Rouge's agrarian reform policies, Cambodia experienced widespread famine. As many as one million more died from starvation, disease, and exhaustion resulting from these policies. In 1979 Vietnam invaded Cambodia and removed the Khmer Rouge from power. By that time about one quarter of Cambodia›s population had been killed.

North Korean famine in the 1990s

Famine struck North Korea in the mid-1990s, due to unprecedented floods. This autarkic urban, industrial state depended on massive inputs of subsidised goods, including fossil fuels, primarily from the Soviet Union and the People's Republic of China. When the Soviet collapse and China's marketization switched trade to a hard currency, full-price basis, North Korea's economy collapsed. The vulnerable agricultural sector experienced a massive failure in 1995–96, expanding to full-fledged famine by 1996–99.

Estimates based on the North Korean census suggest that 240,000 to 420,000 people died as a result of the famine and that there were 600,000 to 850,000 unnatural deaths in North Korea from 1993 to 2008.[62] North Korea has not yet regained food self-sufficiency and relies on external food aid from China, Japan, South Korea, Russia and the United States. While Woo-Cumings have focused on the FAD side of the famine, Moon argues that FAD shifted the incentive structure of the authoritarian regime to react in a way that forced millions of disenfranchised people to starve to death (Moon, 2009).

According to the UN's Food and Agriculture Organisation (FAO), North Korea is facing a serious cereal shortfall in 2017 after the country's crop harvest was diminished because of severe drought. The FAO estimated that early-season production fell by over 30 percent compared to agricultural output from the previous year, leading to the country's worst famine since 2001.

Vietnam

Various famines have occurred in Vietnam. Japanese occupation during World War II caused the Vietnamese Famine of 1945, which caused 2 million deaths, or 10% of the population then.Following the unification of the country after the Vietnam War, Vietnam experienced a food shortage in the 1980s, which prompted many people to flee the country.

India

India's overdependence on the vagaries of monsoon rains makes it even more vulnerable to crop failures, which occasionally lead to famine. There were 14 famines in Indiabetween the 11th and 17th centuries (Bhatia, 1985). For example, during the 1022–1033 Great famines in India entire provinces were depopulated. Famine in Deccan killed at least two million people in 1702–1704. B.M. Bhatia believes that the earlier famines were localised, and it was only after 1860, during the British rule, that famine came to signify general shortage of foodgrains in the country. There were approximately 25 major famines spread through states such as Tamil Nadu in the south, and Bihar and Bengal in the east during the latter half of the 19th century.

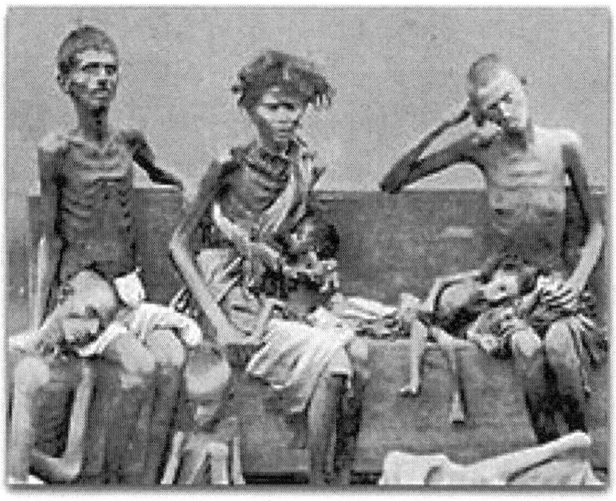

Victims of the Great Famine of 1876–78 in India during British rule, pictured in 1877.

Romesh Chunder Dutt argued as early as 1900, and present-day scholars such as Amartya Sen agree, that some historic famines were a product of both fluctuating rainfall and British economic and administrative policies, which since 1857 had led to the seizure and conversion of local farmland to foreign-owned plantations, restrictions on internal trade, heavy taxation of Indian citizens to support British expeditions in Afghanistan (see The Second Anglo-Afghan War), inflationary measures that increased the price of food, and substantial exports of staple crops from India to Britain. (Dutt, 1900 and 1902; Srivastava, 1968; Sen, 1982; Bhatia, 1985.)

Some British citizens, such as William Digby, agitated for policy reforms and famine relief, but Lord Lytton, the governing British viceroy in India, refused such changes in the belief that it would stimulate Indian workers to evade. The first, the Bengal famine of 1770, is estimated to have taken around 10 million lives—one-third of Bengal's population at the time. Other notable famines include the Great Famine of 1876–78, in which 6.1 million to 10.3 million people died[68] and the Indian famine of 1899–1900, in which 1.25 to 10 million people died. The famines were ended by the 20th century with the exception of the Bengal famine of 1943 killing an estimated 2.1 million Bengalis during World War II.

The observations of the Famine Commission of 1880 supported the notion that food distribution is more to blame for famines than food scarcity. They observed that each province in British India, including Burma, had a surplus of foodgrains, and the annual surplus was 5.16 million tons (Bhatia, 1970). At that time, annual export of rice and other grains from India was approximately one million tons.

- Population growth worsened the plight of the peasantry. As a result of peace and improved sanitation and health, the Indian population rose from perhaps 100 million in 1700 to 300 million by 1920. While encouraging agricultural productivity, the British also provided economic incentives to have more children to help in the fields. Although a similar population increase occurred in Europe at the same time, the growing numbers could be absorbed by industrialization or emigration to the Americas and Australia. India enjoyed neither an industrial revolution nor an increase in food growing. Moreover, Indian landlords had a stake in the cash crop system and discouraged innovation. As a result, population numbers far outstripped the amount of available food and land, creating dire poverty and widespread hunger.

The Maharashtra drought in which there were zero deaths and one which is known for the successful employment of famine prevention policies, unlike during British rule.

Middle East

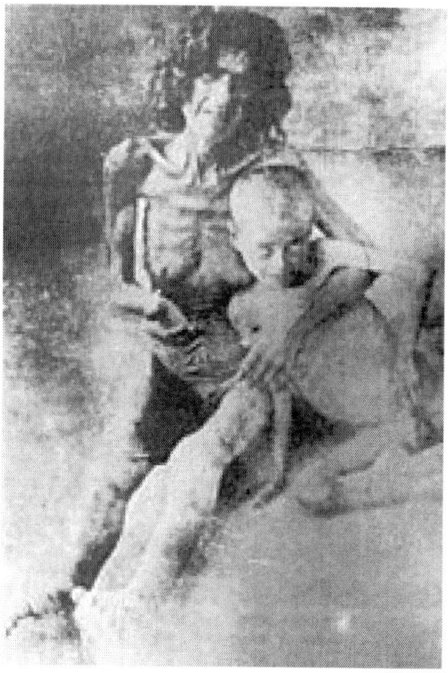

A starving woman and child during the Assyrian Genocide. Ottoman Empire, 1915

The Great Persian Famine of 1870–1871 is believed to have caused the death of 1.5 million persons (20–25 percent of the population) in Persia(present-day Iran).

In the early 20th century an Ottoman blockade of food being exported to Lebanon caused a famine which killed up to 450,000 Lebanese (about one-third of the population). The famine killed more people than the Lebanese Civil War. The blockade was caused by uprisings in the Syrian region of the Empire including one which occurred in the 1860s which lead to the massacre of thousands of Lebanese and Syrian by Ottoman Turks and local Druze

EUROPE

Middle Ages

Further information: Medieval demography and Crisis of the Late Middle Ages

The Great Famine of 1315–1317 (or to 1322) was regarded as the first major food crisis to strike Europe in the 14th century. Millions in northern Europe died over an extended number of years, marking a clear end to the earlier period of growth and prosperity during the 11th and 12th centuries. An unusually cold and wet spring of 1315 led to widespread crop failures, which lasted until at least the summer of 1317; some regions in Europe did not fully recover until 1322. Most nobles, cities, and states were slow to respond to the crisis and when they realized its severity, they had little success in securing food for their people. In 1315, in Norfolk England, the price of grain soared from 5 shillings/quarter to 20 shillings/quarter. It was a period marked by extreme levels of criminal activity, disease and mass death, infanticide, and cannibalism. It had consequences for Church, State, European society and future calamities to follow in the 14th century. There were 95 famines in medieval Britain,and 75 or more in medieval France. More than 10% of England's population, or at least 500,000 people, may have died during the famine of 1315–1316.

Famine was a very disruptive and catastrophic occurrence. The prospect of starvation led people to take desperate measures. When scarcity of food became apparent to peasants, they would sacrifice long-term prosperity for short-term survival. They would kill their draught animals, leading to lowered production in subsequent years. They would eat their seed corn, sacrificing next year›s crop in the hope that more seed could be found. Once those means had been exhausted, they would take to the road in search of food. They migrated to the cities where merchants from other areas would be more likely to sell their food, as cities had a stronger purchasing power than did rural areas. Cities also administered relief programs and bought grain for their populations so that they could keep order. With the confusion and desperation of the migrants, crime would often follow them. Many peasants resorted to banditry in order to acquire enough to eat.

One famine would often lead to difficulties in the following years because of lack of seed stock or disruption of routine, or perhaps because of less-available labour. Famines were often interpreted as signs of God's displeasure. They were seen as the removal, by God, of His gifts to the people of the Earth. Elaborate religious processions and rituals were made to prevent God's wrath in the form of famine.

16th century

An engraving from Goya's *Disasters of War*, showing starving women, doubtless inspired by the terrible famine that struck Madrid in 1811–1812.

During the 15th century to the 18th century, famines in Europe frequently occurred due to the Little Ice Age. The colder climate resulted in harvest failures and shortfalls that led to a rise in conspiracy theories concerning the causes behind these famines, such as the Pacte de Famine in France.

The worst famines in centuries across all of Europe was likely observed in the 1590. Famine had been relatively rare during the 16th century. The economy and population had grown steadily as subsistence populations tend to when there is an extended period of relative peace (most of the time). Although peasants in areas of high population density, such as northern Italy, had learned to increase the yields of their lands through techniques such as promiscuous culture, they were still quite vulnerable to famines, forcing them to work their land even more intensively.

The great famine of the 1590s began a period of famine and eventually declined in the 17th century. The price of grain, all over Europe was high, as was the population. Various types of people were vulnerable to the succession of bad harvests that occurred throughout the 1590s in different regions. The increasing number of wage labourers in the countryside were vulnerable because they had no food of their own, and their meager living was not enough to purchase the expensive grain of a bad-crop year. Town labourers were also at risk because their wages would be insufficient to cover the cost of grain, and, to make matters worse, they often received less money in bad-crop years since the disposable income of the wealthy was spent on grain. Often, unemployment would be the result of the increase in grain prices, leading to ever-increasing numbers of urban poor.

All areas of Europe were badly affected by the famine in these periods, especially rural areas. The Netherlands was able to escape most of the damaging effects of the famine, though the 1590s were still difficult years there. Amsterdam's grain trade with the Baltic, guaranteed a food supply.

17th century

The years around 1620 saw another period of famine sweep across Europe. These famines were generally less severe than the famines of twenty-five years earlier, but they were nevertheless quite serious in many areas. Perhaps the worst famine since 1600, the great famine in Finland in 1696, killed one-third of the population.

Devastating harvest failures afflicted the northern Italian economy from 1618 to 1621, and its total recovery took centuries. There were serious famines in the late-1640s and less severe ones in the 1670s throughout northern Italy.

Over two million people died in two famines in France between 1693 and 1710. Both famines were made worse by ongoing wars.

Illustration of starvation in northern Sweden, Finnish famine of 1866–1868

As late as the 1690s, Scotland experienced famine which reduced the population of parts of Scotland by at least 15%.

The Great Famine of 1695–1697 estimated to have killed a third of the Finnish population and roughly 10% of Norway's population. Death rates rose in Scandinavia between 1740 and 1800 as the result of a series of crop failures. For instance, the Finnish famine of 1866–1868 killed 15% of the population.

18th century

The period of 1740–43 saw frigid winters and summer droughts, which further led to famine across Europe and a major spike in mortality. The winter 1740–41 was unusually cold, possibly because of volcanic activity.

According to Scott and Duncan (2002), "Eastern Europe experienced more than 150 recorded famines between AD 1500 and 1700 and there were 100 hunger years and 121 famine years in Russia between AD 971 and 1974."

The Great Famine, which lasted from 1770 until 1771, killed about one tenth of Czech lands' population, or 250,000 inhabitants, and radicalised country sides leading to peasant uprisings.

There were sixteen good harvests and 111 famine years in northern Italy from 1451 to 1767.According to Stephen L. Dyson and Robert J. Rowland, "The Jesuits of Cagliari [in Sardinia] recorded years during the late 1500s «of such hunger and so sterile that the majority of the people could sustain life only with wild ferns and other weeds» ... During the terrible famine of 1680, some 80,000 persons, out of a total population of 250,000, are said to have died, and entire villages were devastated...»

According to Bryson (1974), there were thirty-seven famine years in Iceland between 1500 and 1804. In 1783 the volcano Laki in south-central Iceland erupted. The lava caused little direct damage, but ash and sulphur dioxide spewed out over most of the country, causing three-quarters of the island's livestock to perish. In the following famine, around ten thousand people died, one-fifth of the population of Iceland. [Asimov, 1984, 152–53]

19th century

Depiction of victims of the Great Famine in Ireland, 1845–1849

Other areas of Europe have recently been accustomed to famines. France saw famines as recently as the 19th century. The Great Famine in Ireland, 1846–1851, caused by the failure of the potato crop over a few years, resulted in 1,000,000 dead and another 2,000,000 refugeesfleeing to Britain, Australia and the United States.

20th century

Famine further persisted in Eastern Europe during the 20th century. Droughts and famines in Imperial Russia are known to have happened every 10 to 13 years, with average droughts happening every 5 to 7 years. Russia experienced eleven major famines between 1845 and 1922, one of the worst being the famine of 1891–92.The Russian famine of 1921–22 killed an estimated 5 million.

Victims of the Russian famine of 1921–22 during the Russian Civil War

Famines continued in the Soviet era, the most notorious being the *Holodomor* in various parts of the country, especially the Volga, and the Ukrainian and northern Kazakh SSR›s during the winter of 1932–1933. The Soviet famine of 1932–1933 is nowadays reckoned to have cost an estimated 6 million lives.The last major famine in the USSR happened in 1947 due to the severe drought and the mismanagement of grain reserves by the Soviet government.

The Hunger Plan, i.e. the Nazi plan to starve large sections of the Soviet population, succeed in causing the deaths of many. The Russian Academy of Sciences in 1995 reported civilian victims in the USSR at German hands, including Jews, totalled 13.7 million dead, 20% of the 68 million persons in the occupied USSR. This included 4.1 million famine and disease deaths in occupied territory. There were an additional estimated 3 million famine deaths in areas of the USSR not under German occupation.

The 872 days of the Siege of Leningrad (1941–1944) caused unparalleled famine in the Leningrad region through disruption of utilities, water, energy and food supplies. This resulted in the deaths of about one million people.

Famine even struck in Western Europe during the Second World War. In the Netherlands, the *Hongerwinter* of 1944 killed approximately 30,000 people. Some other areas of Europe also experienced famine at the same time.

Latin America

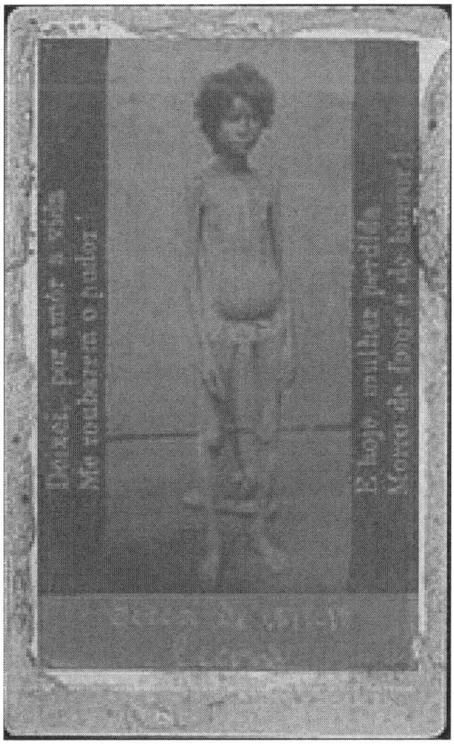

Malnourished child during Brazil's 1877–78 *Grande Seca* (Great Drought).

The pre-Columbian Americans often dealt with severe food shortages and famines.The persistent drought around 850 AD coincided with the collapse of Classic Maya civilization, and the famine of One Rabbit (AD 1454) was a major catastrophe in Mexico.

Brazil's 1877–78 *Grande Seca* (Great Drought), the worst in Brazil›s history, caused approximately half a million deaths. The one from 1915 was devastating too.

Oceania

Easter Island was hit by a destructive famine between the 15th and 18th centuries. Hunger and subsequent cannibalism was caused by overpopulation and depletion of natural resources as a result of deforestation, partly because work on megalithic monuments required a lot of wood.

There are other documented episodes of famine in various islands of Polynesia, such as occurred in Kau, Hawaii in 1868.

According to Daniel Lord Smail, "'Famine cannibalism' was until recently a regular feature of life in the islands of the Massim near New Guinea and of some other societies of Southeast Asia and the Pacific.»

Risk of future famine

The Guardian reports that in 2007 around 40% of the world's agricultural land is remarkably degraded. If current trends of soil degradation continue in Africa, the continent might be able to feed just 25% of its population by 2025, according to UNU's Ghana-based Institute for Natural Resources in Africa. As of late 2007, increased farming for use in bio fuels, along with world oil prices at nearly $100 a barrel,has pushed up the price of grain used to feed poultry and dairy cows and other cattle, causing higher prices of wheat (up 58%), soybean (up 32%), and maize (up 11%) over the year. In 2007 Food riots have taken place in many countries across the world.An epidemic of stem rust, which is destructive to wheat and is caused by race Ug99, has in 2007 spread across Africa and into Asia.

Beginning in the 20th century, nitrogen fertilizers, new pesticides, desert farming, and other agricultural technologies came to the surface and were used vehemently to increase food production in order to combat famine. Between 1950 and 1984, as the Green Revolution influenced agriculture, world grain production increased by 250%. Developed nations have shared these technologies with developing nations with a famine problem. However, as early as 1995, there were signs that these new developments may contribute to the decline of arable land (e.g. persistence of pesticides leading to soil contamination, salt accumulation due to irrigation, erosion).

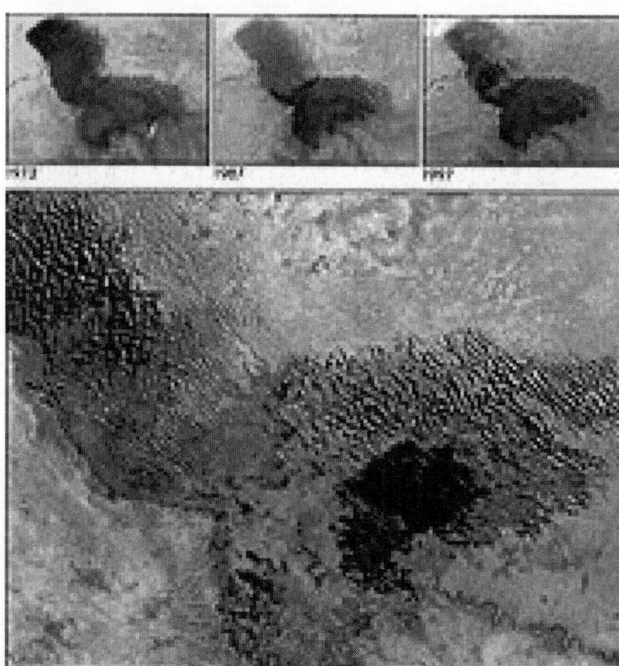

Lake Chad in a 2001 satellite image, with the actual lake in blue. The lake has shrunk by 95% since the 1960s.[

In 1994, David Pimentel, professor of ecology and agriculture at Cornell University, and Mario Giampietro, senior researcher at the National Research Institute on Food and Nutrition (INRAN), estimated the maximum U.S. population for a sustainable economy at 200 million.

According to geologist Dale Allen Pfeiffer, the coming decades could likely see a considerable rise in food prices without relief and a massive starvation on a global level. Water deficits, which are already spurring heavy grain imports in numerous smaller countries, may soon do the same in larger countries, such as China or India. The water tables are falling in many countries (including Northern China, the US, and India) due to widespread overconsumption. Other countries affected include Pakistan, Iran, and Mexico. This will eventually lead to water scarcity and cutbacks in grain harvest. Even while overexploiting its aquifers, China has developed a grain deficit, contributing to the upward pressure on grain prices. Most of the three billion people projected to be added worldwide by mid-century will be born in countries already experiencing water shortages.

After China and India, there is a second tier of smaller countries with large water deficits – Algeria, Egypt, Iran, Mexico, and Pakistan. Four of these already import a large share of their grain. Only Pakistan remains marginally self-sufficient. But with a population expanding by 4 million a year, it will also soon turn to the world market for grain. According to a UN climate report, the Himalayan glaciers that are the principal dry-season water sources of Asia's biggest rivers – Ganges, Indus, Brahmaputra, Yangtze, Mekong, Salween and Yellow – could disappear by 2350 as temperatures rise and human demand rises. Approximately 2.4 billion people live in the drainage basin of the Himalayan rivers.[India, China, Pakistan, Afghanistan, Bangladesh, Nepal and Myanmar could experience floods followed by severe droughts in coming decades.]In India alone, the Ganges provides water for drinking and farming for more than 500 million people.

Evan Fraser, a geographer at the University of Guelph in Ontario, Canada, explores the ways in which climate change may affect future famines. To do this, he draws on a range of historic cases where relatively small environmental problems triggered famines as a way of creating theoretical links between climate and famine in the future. Drawing on situations as diverse as the Great Irish Potato Famine,a series of weather induced famines in Asia during the late 19th century, and famines in Ethiopia during the 1980s, he concludes there are three «lines of defense» that protect a community›s food security from environmental change. The first line of defense is the agro-ecosystem on which food is produced: diverse ecosystems with well managed soils high in organic matter tend to be more resilient. The second line of defense is the wealth and skills of individual households: If those households affected by bad weather such as drought have savings or skills they may be able to do all right despite the bad weather. The final line of defense is created by the formal institutions present in a society. Governments, churches, or NGOs must be willing and able to mount effective relief efforts. Pulling this together, Evan Fraser argues that if an ecosystem is resilient enough, it may be able to withstand weather-related shocks. But if these shocks overwhelm the ecosystem›s line of defense, it is necessary for the household to adapt using its skills and savings. If a problem is too big for the family or household, then people must rely on the third line of defense, which is whether or not the formal institutions present in a society are able to provide help. Evan Fraser concludes that in almost every situation where an environmental problem triggered a famine you see a failure in each of these three lines of defense.[131] Hence, understanding how climate change may cause famines in the future requires combining both an assessment of local socio-economic and environmental factors along with climate models that predict where bad weather may occur in the future·

Causes

A victim of starvation in besieged Leningrad suffering from dystrophia in 1941.

Definitions of famines are based on three different categories—these include food supply-based, food consumption-based and mortality-based definitions. Some definitions of famines are:

- Blix – Widespread food shortage leading to significant rise in regional death rates.
- Brown and Eckholm – Sudden, sharp reduction in food supply resulting in widespread hunger.
- Scrimshaw – Sudden collapse in level of food consumption of large numbers of people.
- Ravallion – Unusually high mortality with unusually severe threat to food intake of some segments of a population.
- Cuny – A set of conditions that occurs when large numbers of people in a region cannot obtain sufficient food, resulting in widespread, acute malnutrition.

Food shortages in a population are majorly due to either lack of food or difficulty in food distribution; it may be worsened by natural climate fluctuations and by extreme political conditions related to oppressive government or warfare. The conventional explanation until 1981 for the cause of famines was the Food availability decline (FAD) hypothesis. The assumption was that the central cause of all famines was a decline in food availability. However, FAD could not explain why only a certain section of the population such as the agricultural labourer was affected by famines while others were insulated from famines. Based on the studies of some recent famines, the decisive role of FAD has been questioned and it has been suggested that the causal mechanism for precipitating starvation includes many variables other than just decline of food availability. According to this view, famines are a result of entitlements, the theory being proposed is called the "failure of exchange entitlements" or FEE. A person may own various commodities that can be exchanged in a market economy for the other commodities he or she needs. The exchange can happen via trading or production or through a combination of the two. These entitlements are called trade-based or production-based entitlements. Per this proposed view, famines are precipitated due to a breakdown in the ability of the person to exchange his entitlements. An example of famines due to FEE is the inability of an agricultural labourer to exchange his primary entitlement, i.e., labour for rice when his employment became erratic or was completely eliminated.

According to the Physicians for Social Responsibility (PSR), global climate change is additionally challenging the Earth›s ability to produce food, potentially leading to famine.

Some elements make a particular region more vulnerable to famine. These include poverty, population growth, an inappropriate social infrastructure, a suppressive political regime, and a weak or under-prepared government.

According to a FEWSNET report, "Famines are not natural phenomena, they are catastrophic political failures."

Climate and population pressure

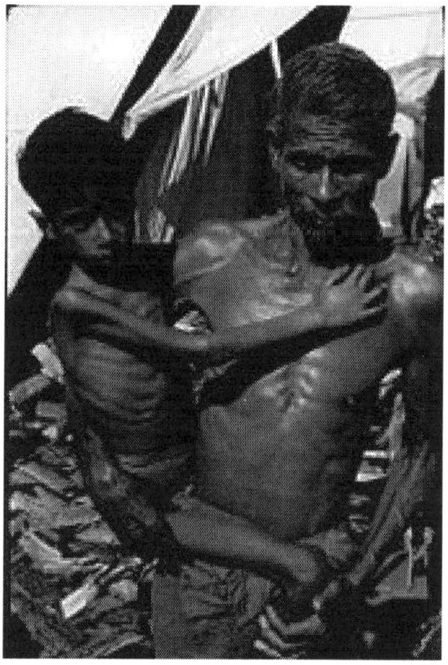

A child suffering extreme starvation in India, 1972

Many famines are a consequence of imbalance in food production as compared to the large populations of the countries whose population most likely exceeds the regional carrying capacity. Historically, famines have occurred from agricultural problems such as drought, crop failure, or pestilence. Changing weather patterns, the ineffectiveness of medieval governments in dealing with crises, wars, and epidemic diseases such as the Black Death helped to cause hundreds of famines in Europe during the Middle Ages, including 95 in Britain and 75 in France. In France, the Hundred Years' War, crop failures and epidemics reduced the population by two-thirds.

The failure of a harvest or change in conditions, such as drought, results in a situation where large numbers of people continue to live despite of the radical drop in the carrying capacity of the land. Famine is often associated with subsistence agriculture. The total absence of agriculture in an economically strong area does not cause famine; Arizona and other wealthy regions import the vast majority of their food, since such regions produce sufficient economic goods for trade.

Famines have also been caused by volcanism. The 1815 eruption of the Mount Tambora volcano in Indonesia caused crop failures and famines worldwide and caused the worst famine of the 19th century. The current consensus of the scientific community is that the aerosols and dust released into the upper atmosphere causes cooler temperatures by preventing the sun›s energy from reaching the ground. The same mechanism is theorized to be caused by very large meteorite impacts to the extent of causing mass extinctions.

State-sponsored famines

In certain cases, such as the Great Leap Forward in China (which produced the largest famine in absolute numbers), North Korea in the mid-1990s, or Zimbabwe in the early-2000s, famine can occur because of government policy.

The government's forced collectivization of agriculture was one of the main causes of the Soviet famine of 1932–1933.

In 1932, under the rule of the USSR, Ukraine experienced one of its most destructive famines between 2.4 and 7.5 million peasants died as an outcome of a state sponsored famine. It was termed the Holodomor, suggesting that it was a deliberate campaign of repression designed to eliminate resistance to collectivization. Forced grain quotas imposed upon the rural peasants and a brutal reign of terror contributed to the widespread famine. The Soviet government continued to deny the problem and it did not provide aid to the victims nor did it accept foreign aid.

In 1958 in China, Mao Zedong's Communist Government launched the Great Leap Forward campaign, that focused on rapidly industrializing the country.[149]

The government forcibly took control of agriculture. Barely enough grain was left for the peasants, and starvation occurred in many rural areas. Exportation of grain continued despite the famine and the government attempted to conceal it. While the famine is attributed to unintended consequences, it is believed that the government refused to acknowledge the problem, thereby further contributing to the deaths. In many instances, peasants were persecuted. About 20 to 45 million people perished in this famine, making it one of the deadliest famines till date.

Malawi ended its famine by subsidizing farmers despite the strictures imposed by the World Bank. During the 1973 Wollo Famine in Ethiopia, food was shipped out of Wollo to the capital city of Addis Ababa, where it could command higher prices. In the late-1970s and early-1980s, residents of the dictatorships of Ethiopia and Sudan suffered massive famines, but the democracy of Botswana avoided them, despite also suffering a severe drop in national food production. In Somalia, famine occurred because of a failed state.

The famine in Yemen was a direct result of the Saudi Arabian-led intervention in Yemen and blockade imposed by Saudi Arabia and its allies, including the United States.

According to Amartya Sen (1999) "there has never been a famine in a functioning multiparty democracy". Hasell and Roser have demonstrated that while there have been a few minor exceptions, famines rarely occur in democratic system but are strongly correlated with autocratic and colonial systems.

Famine prevention

A starving child during the 1869 famine in Algeria.

Relief technologies, including immunization, improved public health infrastructure, general food rations and supplementary feeding for vulnerable children, has succeeded in providing temporary mitigation to the mortality impact of famines, while leaving their economic consequences unchanged, and not solving the underlying issue of too large a regional population relative to food production capability. Humanitarian crises may also arise from genocide campaigns, civil wars, refugee flows and episodes of extreme violence and state collapse, creating famine conditions among the affected populations.

Despite repeated stated intentions by the world's leaders to end hunger and famine, famine remains a chronic threat in much of Africa, Eastern Europe, the Southeast, South Asia, and the Middle East. In July 2005, the Famine Early Warning Systems Network labelled Nigerwith emergency status, as well as Chad, Ethiopia, South Sudan, Somalia and Zimbabwe. In January 2006, the United Nations Food and Agriculture Organization warned that 11 million people in Somalia, Kenya, Djibouti and Ethiopia were in danger of starvation due to the combination of severe drought and military conflicts. In 2006, the most serious humanitarian crisis in Africa was in Sudan's region Darfur.

Frances Moore Lappé, later co-founder of the Institute for Food and Development Policy (Food First) argued in *Diet for a Small Planet* (1971) that vegetarian diets can provide food for larger populations, with the same resources, compared to omnivorous diets.

Modern famines are aggravated by misguided economic policies to some extent, political design to impoverish or marginalize certain populations, or acts of war, political economists have investigated the political conditions under which famine is prevented. Economist Amartya Sen states that the liberal institutions that exist in India, including competitive elections and a free press, have played a major role in preventing famine in that country since independence. Alex de Waal has developed this theory to focus on the «political contract» between rulers and people that ensures famine prevention, noting the rarity of such political contracts in Africa, and the danger that international relief agencies will undermine such contracts through removing the locus of accountability for famines from national governments.

A woman, a man and a child, all three dead from starvation. Russia, 1921.

The demographic impacts of famine are sharp. Mortality is concentrated among children and the elderly. A consistent demographic fact is that in all recorded famines, male mortality exceeds female, even in those populations (such as northern India and Pakistan) where there is a male longevity advantage during normal times. Reasons for this may include greater female resilience under the pressure of malnutrition, and possibly female's naturally higher percentage of body fat. Famine is also accompanied by lower fertility. Famines therefore leave the reproductive core of a population—adult women—lesser affected compared to other population categories, and post-famine periods are often characterized a "rebound" with increased births.

Even though the theories of Thomas Malthus would predict that famines reduce the size of the population commensurate with available food resources, in fact even the most severe famines have rarely dented population growth for more than a few years. The mortality in China in 1958–61, Bengal in 1943, and Ethiopia in 1983–85 was all made up by a growing population over just a few years. Of greater long-term demographic impact is emigration: Ireland was chiefly depopulated after the 1840s famines by waves of emigration.

Food security

Long term measures to improve food security, include investment in modern agriculture techniques, such as fertilizers and irrigation, but also to focus in the strategic national food storage.

World Bank strictures restrict government subsidies for farmers, and increasing use of fertilizers is opposed by some environmental groups because of its unintended consequences: adverse effects on water supplies and habitat.

Norman Borlaug, father of the Green Revolution, is often credited with saving over a billion people worldwide from starvation.

The effort to bring modern agricultural techniques found in the Western world, such as nitrogen fertilizers and pesticides, to the Indian Sub-continent, called the Green Revolution, that lead to a decreases in malnutrition similar to those seen earlier in Western nations. This was possible because of existing infrastructure and institutions that are in short supply in Africa, such as a system of roads or public seed companies that made seeds available. Supporting farmers in areas of food insecurity, through such measures as free or subsidized fertilizers and seeds, increases food harvest and reduces food prices.

The World Bank and some rich nations press nations that depend on them for aid to cut back or eliminate subsidized agricultural inputs such as fertilizer, in the name of privatization even as the United States and Europe extensively subsidized their own farmers.

Relief

There is a growing awareness amongst aid groups that giving cash or cash vouchers instead of food is a cheaper, faster, and more efficient way to deliver efficient help to the hunger struck, particularly in areas where food is available but unaffordable. The United Nations' World Food Program (WFP), the biggest non-governmental distributor of food, announced that it will begin distributing cash and vouchers instead of food in some areas, which Josette Sheeran, the WFP›s executive director, described as a «revolution» in food aid.The aid agency Concern Worldwide is piloting a method through a mobile phoneoperator, Safaricom, which runs a money transfer program that allows cash to be sent from one part of the country to another.

However, in case of drought inflicted area due to its limited access to markets, delivering food may be considered a more appropriate way to help. Fred Cuny stated that «the chances of saving lives at the outset of a relief operation are greatly reduced when food is imported. By the time it arrives in the country and gets to people, many will have died.»US Lawwhich requires buying food at home rather than where the hungry live, is inefficient because approximately half of what is spent goes for transport.Fred Cuny further pointed out «studies of every recent famine have shown that food was available in-country—though not always in the immediate food deficit area» and «even though by local standards the prices are too high for the poor to purchase it, it would usually be cheaper for a donor to buy the hoarded food at the inflated price than to import it from abroad.»

Deficient micronutrients can be provided through fortifying foods. Fortifying foods such as peanut butter sachets have revolutionized emergency feeding in humanitarian emergencies because they can be eaten directly from the packet, do not require refrigeration or mixing with scarce clean water, can be stored for years and, vitally, can be absorbed by extremely ill children.

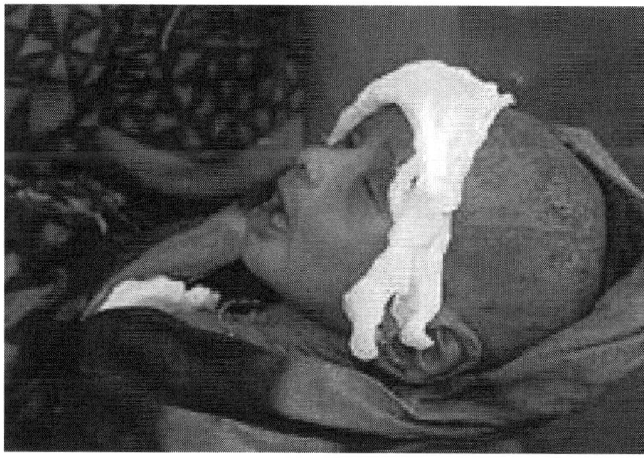

A Somali boy receiving treatment for malnutrition at a health facility in Hilaweyn during the drought of 2011.

WHO and other sources recommend that the malnourished children—and adults who also have diarrhoea—drink rehydration solution, and continue to eat, in addition to antibiotics, and zinc supplements. There is a special oral rehydration solution called ReSoMal which has less sodium and more potassium than standard solution. However, if the diarrhea is severe, the standard solution is preferable as the person needs the extra sodium. Obviously, this is a judgment call best made by a physician, and using either solution is better than doing nothing. Zinc supplements often can help reduce the duration and severity of diarrhea, and Vitamin A can also be helpful.The World Health Organization underlines the importance of a person with diarrhea continuing to eat, with a 2005 publication for physicians stating: «Food should *never* be withheld and the child›s usual foods should not be diluted. Breastfeeding should *always* be continued.»

Ethiopia has been pioneering a program that has now become a part of the World Bank's prescribed recipe for coping with a food crisis and had been observed by aid organizations as a model on how to best help hungry nations. Through the country's main food assistance program, the Productive Safety Net Program, Ethiopia has been giving rural residents who are chronically short of food, a chance to work for food or cash. Foreign aid organizations like the World Food Program were then able to buy food locally from surplus areas to distribute in areas with a shortage of food.

The Green Revolution was widely regarded as a solution to famine in the 1970s and 1980s. Between 1950 and 1984, hybrid strains of high-yielding crops transformed agriculture around the globe and world grain production increased by 250%.Somecriticize the process, stating that these new high-yielding crops require more chemical fertilizers and pesticides, which can harm the environment Although these high-yielding crops make it technically possible to feed more people, there are indications that regional food production has peaked in many world sectors, due to certain strategies associated with intensive agriculture such as groundwater overdrafting and overuse of pesticides and other agricultural chemicals.

Levels of food insecurity

In modern times, it is observed that the local and political governments and non-governmental organizations that deliver famine relief have limited resources not sufficient enough to address the multiple situations of food insecurity that are occurring simultaneously. Various methods of categorizing the gradations of food security have thus been used in order to most efficiently allocate food relief. One of the earliest were the Indian Famine Codes devised by the British in the 1880s. The Codes listed three stages of food insecurity: near-scarcity, scarcity and famine, and were highly influential in the creation of subsequent famine warning or measurement systems. The early warning system developed to monitor the region inhabited by the Turkana peoplein northern Kenya also has three levels, but links each stage to a pre-planned response to mitigate the crisis and prevent its deterioration

The experiences of famine relief organizations throughout the world over the 1980s and 1990s resulted in two major developments: the "livelihoods approach" and the increased use of nutrition indicators in order to determine the severity of a crisis. Individuals and groups in food stressful situations will attempt to cope by rationing consumption, finding alternative means to supplement income, etc., before taking desperate measures, such as selling off plots of agricultural land. When all means of self-support are exhausted, the affected population begins to migrate in search of food or fall victim to outright mass starvation. Famine may thus be viewed partially as a social phenomenon, involving markets, the price of food, and social support structures. A second lesson drawn was the increased use of rapid nutrition assessments, in particular of children, to give a quantitative measure of the famine's severity.

Since 2003, many of the most important organizations in famine relief, such as the World Food Programme and the U.S. Agency for International Development, have adopted a five-level scale to measure intensity and magnitude. The intensity scale uses both livelihoods' measures and measurements of mortality and child malnutrition to categorize a situation as food secure, food insecure, food crisis, famine, severe famine, and extreme famine. The number of deaths determines the magnitude designation, with under 1000 fatalities defining a "minor famine" and a "catastrophic famine" resulting in over 1,000,000 deaths.

The Bengal Famine: The Agony and Destiny

The **Bengal famine of 1943** was a devastating famine in the Bengal province of British India during World War II. An estimated 2.1–3 million, out of a population of 60.3 million, died of starvation, malaria, or other diseases aggravated by malnutrition, population displacement, unsanitary conditions and lack of health care. Millions were impoverished as the crisis overwhelmed large segments of the economy and catastrophically disrupted the social fabric. Eventually, families disintegrated; men sold their small farms and left home to look for work or to join the army, and women and children became homeless migrants, often travelling to Calcutta or another large city in search of organised relief. Historians have frequently characterised the famine as "man-made", asserting that wartime colonial policies created and then exacerbated the crisis. A minority view holds that the famine arose instead from natural causes.

Bengal's economy had been predominantly agrarian, with between half and three-quarters of the rural poor surviving in a "semi-starved condition". Stagnant agricultural productivity and a stable land base were unable to cope with a rapidly increasing population, resulting in both long-term decline in per capita availability of rice and growing numbers of the land-poor or landless labourers. A high proportion laboured beneath a chronic and spiralling cycle of debt that ended in debt bondage and the loss of their landholdings due to land grabbing.

The financing of military escalation led to war-time inflation, as land was appropriated from thousands of peasants. Many workers received monetary wages rather than payment in kind with a portion of the harvest. When prices rose sharply, their wages failed to follow suit; this drop in real wages left them less able to purchase food. During the Japanese occupation of Burma, many rice imports were lost as the region's market supplies and transport systems were disrupted by British "denial policies" for rice and boats (a "scorched earth" response to the occupation). The Bengal Chamber of Commerce (composed mainly of British-owned firms), with the approval of the Government of Bengal, devised a Foodstuffs Scheme to provide preferential distribution of goods and services to workers in high-priority war industries, to prevent them from leaving their positions. gave preferential treatment in the distribution of supplies to the armed forces, civil servants and other "priority classes". These factors were compounded by restricted access to grain: domestic sources were constrained by emergency inter-provincial trade barriers, while access to international imports was largely denied by Churchill's War Cabinet, arguably due to a wartime shortage of shipping. More proximate causes included large-scale natural disasters in south-western Bengal (a cyclone, tidal waves and flooding, and rice crop disease). The relative importance of each of these factors on the death toll is a matter of controversy.

The provincial government denied the existence of famine, and the humanitarian aid was ineffective and inefficient throughout the worst months of the crisis. The government first attempted to influence the price of rice paddy, but instead created a black market which encouraged sellers to withhold stocks, leading to hyperinflation from speculation and hoarding after controls were abandoned. Aid increased significantly when the Indian Army took control of funding in October 1943, but effective relief arrived after a record rice harvest that December. Deaths from starvation declined, yet over half the famine-related deaths occurred in 1944, as a result of disease, after the food security crisis had abated.

Background

Main articles: Economy of India under the British Raj, Agriculture in India, and Timeline of major famines in India during British rule

From the late nineteenth century through the Great Depression, social and economic forces exerted a disruptive effect on the structure of Bengal's income distribution and the capability of its agricultural sector to sustain its natives. These included a rapidly growing population, increasing household debt, stagnant agricultural productivity, increased social stratification, and alienation of the peasant class from their landholdings. These processes left social and economic groups mired in poverty and indebtedness, unable to cope with the economic shocks they faced in 1942 and 1943, in the context of World War II.

Rice farmers ploughing a rice field with water buffaloes near Gushkara, Bengal, 1944

The Government of India's Famine Inquiry Commission report (1945) described Bengal as a "land of rice growers and rice eaters". Rice dominated the agricultural output of the province, accounting for nearly 88% of its arable land use and 75% of its crops.Overall, Bengal produced one third of India's rice – more than any other single province. Rice accounted for 75–85% of daily food consumption,with fish the second major food source,supplemented by small amounts of wheat.

There are three seasonal rice crops in Bengal. By far the most important is the winter crop of *aman* rice. Sown in May and June and harvested in November and December, it produces about 70% of the total annual crop. Crucially, the (debated) shortfall in rice production in 1942 occurred during the all-important *aman* harvest.

Rice yield per acre had been stagnant since the beginning of the twentieth century; further coupled along with a rising population, creating a pressure leading to the occurrence of the famine. Bengal had a population of about 60 million in an area of 77,442 square miles, according to a 1941 census. Declining mortality rates, induced in part by the pre-1943 success of the British Raj in famine reduction caused its population to increase by 43% between 1901 and 1941 – from 42.1 million to 60.3 million. Over the same period India's population as a whole increased by 37%.The economy was almost solely agrarian, but agricultural productivity was among the lowest in the world. Agricultural technology was undeveloped, access to credit was limited and expensive, and any potential for government aid was hampered by political and financial constraints. Land quality and fertility had been deteriorating in Bengal and other regions of India, but the loss was especially severe here. Agricultural expansion required deforestation and land reclamation. These activities damaged the natural drainage courses, silting up rivers and the channels that fed them, leaving them and their fertile deltas moribund. The combination of these factors caused stubbornly low agricultural productivity.

Prior to about 1920, the food demands of Bengal's growing population could be met partly by cultivating unused scrub lands. No later than the first quarter of the twentieth century, Bengal began to experience an acute shortage of such land, leading to a chronic and growing shortage of rice. Its inability to keep pace with rapid population growth changed it from a net exporter of foodgrains to a net importer. Imports were a small portion of the total available food crops, however, and did little to alleviate problems of food supply. Bengali doctor and chemist Chunilal Bose, a professor in Calcutta's medical college, estimated in 1930 that both the ingredients and the small total amount of food in the Bengali diet made it among the least nutritious in India and the world, and greatly harmful to the physical health of the populace.Economic historian Cormac Ó Gráda writes, «Bengal›s rice output in normal years was barely enough for bare-bones subsistence ... the province›s margin over subsistence on the eve of the famine was slender.» These conditions left a large proportion of the population continually on the brink of malnutrition or even starvation.

Land-grabbing

Structural changes in the credit market and land transfer rights further led Bengal into a recurring danger of famine, thus resulting in the economic groups to suffer greater hardship. The Indian system of land tenure, particularly in Bengal, was very complex, with rights unequally divided among three diverse economic and social groups: traditional absentee large landowners or *zamindars*; the upper-tier "wealthy peasant" *jotedars*; and, at the lower socioeconomic level, the *ryot* (peasant) smallholders and dwarfholders, *bargadars* (sharecroppers), and agricultural labourers. *Zamindar* and *jotedar* landowners were protected by law and custom,but those who cultivated the soil, with small or no landholdings, suffered persistent and increasing losses of land rights and welfare. During the late nineteenth and early twentieth centuries, the power and influence of the landowners fell and that of the *jotedars* rose. Particularly in less developed regions, *jotedars* gained power as grain or jute traders and, more importantly, by making loans to sharecroppers, agricultural labourers and ryots. They gained power over their tenants using a combination of debt bondage through the transfer of debts and mortgages, and parcel-by-parcel land-grabbing.

Land-grabbing predominantly took place via informal credit markets. Many financial entities had disappeared during the Great Depression; peasants with small landholdings generally had to resort to informal local lenders to purchase basic necessities during lean months between harvests.As influential Bengali businessman M. A. Ispahani testified, «...the Bengal cultivator, before the war, had three months of feasting, five months of subsistence diet and four months of starvation». Moreover, if a labourer did not possess goods recoverable as cash, such as seed or cattle for ploughing, he would go into debt. Particularly during poor crops, smallholders fell into cycles of debt, often eventually forfeiting land to creditors.

Small landholders and sharecroppers acquired debts swollen by usurious rates of interest. Any poor harvest exacted a heavy toll; the accumulation of consumer debt, seasonal loans and crisis loans began a cycle of spiralling, perpetual indebtedness. It was then relatively easy for the *jotedars* to use litigation to force debtors to sell all or part of their landholdings at a low price or forfeit them at auction. Debtors then became landless or land-poor sharecroppers and labourers, usually working the same fields they had once owned. The accumulation of household debt to a single, local, informal creditor bound the debtor almost inescapably to the creditor/ landlord; it became nearly impossible to settle the debt after a good harvest and simply walk away. In this way, the *jotedars* effectively dominated and impoverished the lowest tier of economic classes in several districts of Bengal.

Further, the Muslim inheritance practices that divided land among multiple siblings, widened inequalities in land ownership. At the time, millions of Bengali agriculturalists held little or no land. In absolute terms, the social group which suffered by far the most of every form of impoverishment and death during the Bengal famine of 1943 were the landless agricultural labourers.

Transport

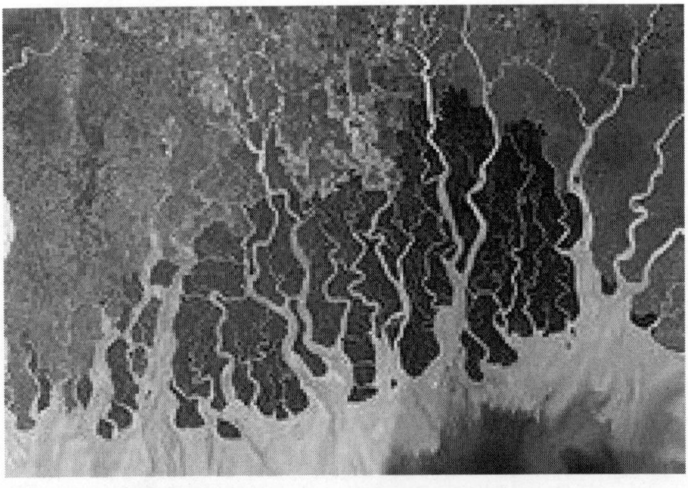

Satellite view of the Sundarbans

Water was the primary source of transport during rainy seasons, and throughout the year in areas such as the vast delta of the coastal south eastern Sundarbans. River transport was integral to Bengal's economy, an irreplaceable factor in the production and distribution of rice. Roads were generally scarce and in poor condition,and Bengal's extensive railway system was employed largely for military purposes until the very late stages of the crisis.

The development of railways in Bengal in the 1890s disrupted natural drainage and divided the region into innumerable poorly drained «compartments». Rail indirectly brought about excessive silting, which increased flooding and created stagnant water areas, damaging crop production and sometimes contributing to a partial shift away from the productive *aman* rice cultivar towards less productive cultivars, and also created a more hospitable environment for water-borne diseases such as cholera and malaria.

Soil and water supply

The soil profile in Bengal differs between east and west. The sandy soil of the east, and the lighter sedimentary earth of the Sundarbans, tended to drain more rapidly after the monsoon season than the laterite or heavy clay regions of western Bengal. Soil exhaustion necessitated that large tracts in western and central Bengal be left fallow; eastern Bengal had far fewer uncultivated fields. The annual flooding of these fallow fields created a breeding place for malaria-carrying mosquitoes;malaria epidemics lasted a month longer in the central and western areas with slower drainage.

Rural areas lacked access to safe water supplies. Water came primarily from large earthen tanks, rivers and tube wells. In the dry season, partially drained tanks became a further breeding area for malaria-vector mosquitoes.Tank and river water was susceptible to contamination by cholera; with tube wells being much safer.However, as many as one-third of the existing wells in wartime Bengal were in disrepair.

Pre-famine shocks and distress

Throughout 1942 and early 1943, military and political events combined with natural disasters and plant disease to place widespread stress on Bengal's economy. While Bengal's food needs rose from increased military presence and an influx of refugees from Burma, its ability to obtain rice and other grains was restricted by inter-provincial trade barriers.

Japanese invasion of Burma

Main articles: Japanese conquest of Burma, Indian Ocean raid, and Arakan Campaign 1942–43

Indian refugees flee Burma along the Prome Road from Rangoon to Mandalay and eventually on to India, January 1942

The Japanese campaign for Burma set off an exodus of more than half of the one million Indians from Burma for India. The flow began after the bombing of Rangoon (1941–1942), and for months thereafter desperate people poured across the borders, escaping into India through Bengal and Assam. On 26 April 1942, all Allied forces were ordered to retreat from Burma into India. Military transport and other supplies were dedicated to military use, and unavailable for use by the refugees. By mid May 1942, the monsoon rains became heavy in the Manipur hills, further inhibiting civilian movement.

The number of refugees who successfully reached India totalled at least 500,000; tens of thousands died along the way. In later months, 70 to 80% of these refugees were afflicted with diseases such as dysentery, smallpox, malaria, or cholera, with 30% "desperately so".The influx of refugees created several conditions that may have contributed to the famine. Their arrival created an increased demand for food,clothing and medical aid, further straining the resources of the province. The poor hygienic conditions of their forced journey sparked official fears of a public health risk due to epidemics caused by social disruption. Finally, their distraught state after their struggles bred foreboding, uncertainty, and panic amongst the populace of Bengal; this aggravated panic buying and hoarding that may have contributed to the onset of the famine.

By April 1942, Japanese warships and aircraft had sunk approximately 100,000 tons of merchant shipping in the Bay of Bengal. According to General Archibald Wavell, Commander-in-Chief of the army in India, both the War Office in London and the commander of the British Eastern Fleet acknowledged that the fleet was powerless to mount serious opposition to Japanese naval attacks on Ceylon, southern or eastern India, or on shipping in the Bay of Bengal. For decades, rail transport had been integral to successful efforts by the Raj to forestall famine in India.However, Japanese raids put additional strain on railways, which also endured flooding in the Brahmaputra, a malaria epidemic, and the Quit India movement targeting road and rail communication.Throughout, transportation of civil supplies were compromised by the railways› increased military obligations, and the dismantling of tracks carried out in areas of eastern Bengal in 1942 to hamper a potential Japanese invasion.

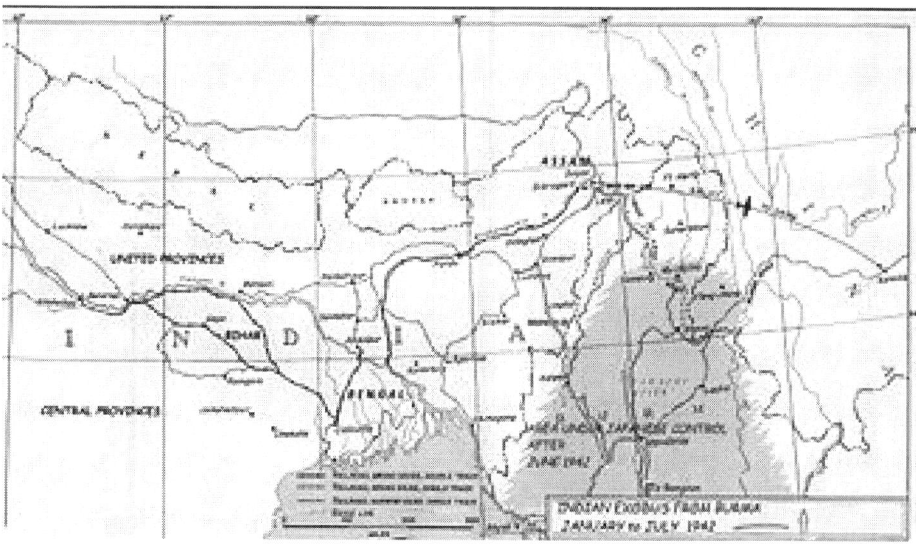

Map of Indian civilian exodus from Burma into Manipur, Bengal, and Assam, between January and July 1942

The fall of Rangoon in March 1942 further cut off the import of the Burmese rice into India and Ceylon. Due in part to rises in local populations, prices for rice were already 69% higher in September 1941 than in August 1939.The loss of Burmese imports led to further increased demand on the rice producing regions. This, according to the Famine Commission, was in a market in which the "progress of the war made sellers who could afford to wait reluctant to sell". The loss of imports from Burma provoked an aggressive scramble for rice across India, which sparked a dramatic and unprecedented surge in demand-pull price inflation in Bengal and other rice producing regions of India. Across India and particularly in Bengal, this caused a «derangement» of the rice markets.Particularly in Bengal, the price effect of the loss of Burmese rice was vastly disproportionate to the relatively modest size of the loss in terms of total consumption.Despite this, Bengal continued to export rice to Ceylon for months afterwards, even as the beginning of a food crisis began to become apparent. All this, together with transport problems created by the government›s "boat denial" policy, were the direct causes of inter-provincial trade barriers on the movement of foodgrains,and contributed to a series of failed government policies that further exacerbated the food crisis.

1942–45: Military build-up, inflation, and displacement

American soldiers at the Calcutta Jain Temple, July 1943. Calcutta became a hub for hundreds of thousands of Allied troops.

The fall of Burma brought Bengal even closer to the war front; the thrust fell more strongly on Bengal than elsewhere in India. Major urban areas, especially Calcutta, drew increasing numbers of workers into military industries and troops from many nations. Unskilled labourers from Bengal and nearby provinces were employed by military contractors, particularly for the construction of American and British airfields. Hundreds of thousands of American, British, Indian, and Chinese troops arrived in the province, straining domestic supplies and leading to scarcities across wide ranges of daily necessities. The general inflationary pressures of a war-time economy caused prices to rise rapidly across the entire spectrum of goods and services. The rise in prices was "not disturbing" until 1941, when it became more alarming. Then in early 1943, the rate of inflation for foodgrains in particular took an unprecedented upward turn.

Nearly the full output of India's cloth, wool, leather and silk industries were sold to the military. In the system that the British Government used to procure goods through the Government of India, industries were left in private ownership rather than facing outright requisitioning of their productive capacity. Firms were required to sell goods to the military on credit and at fixed, low prices. However, firms were left free to charge any price they desired in their domestic market for whatever they had left over. In the case of the textiles industries that supplied cloth for the uniforms of the British military, for example, they charged a very high price in domestic markets. By the end of 1942, cloth prices had more than tripled from their pre-war levels; they had more than quadrupled by mid-1943.Much of the goods left over for civilian use were purchased by speculators. As a result, "civilian consumption of cotton goods fell by more than 23% from the peace time level by 1943/44".The hardships that were felt by the rural population through a severe "cloth famine" were alleviated when military forces began distributing relief supplies between October 1942 and April 1943.

Hawker Hurricane Mark IIBs and IICs of No. 67 Squadron RAF lined up at Chittagong. Construction of airfields displaced the civilian population and increased inflation.

The method of credit financing was tailored according to the UK wartime needs. Britain agreed to pay for defence expenditures above the amount that India had paid in peacetime (adjusted for inflation). However, their purchases were made entirely on credit accumulated in the Bank of England and not redeemable until after the war. At the same time, the Bank of India was permitted to treat those credits as assets against which it could print currency up to two and a half times more than the total debt incurred. India's money printing presses then began running overtime, printing the currency that paid for all these massive expenditures. The tremendous rise in nominal money supply coupled with a scarcity of consumption goods spurred monetary inflation, reaching its peak in 1944–45.The accompanying rise in incomes and purchasing power fell disproportionately into the hands of industries in Calcutta (in particular, munitions industries).

Military build-up led to the massive displacement of Bengalis from their homes. Farmland purchased for airstrip and camp construction is "estimated to have driven between 30,000 and 36,000 families (about 150,000 to 180,000 persons) off their land", according to the historian Paul Greenough. They were paid for the land, but they had lost their employment.The urgent need for housing for the immense influx of workers and soldiers from 1942 onward created further problems. Military barracks were scattered around Calcutta. The Famine Commission report of 1945 stated that the owners had been paid for these homes, but "there is little doubt that the members of many of these families became famine victims in 1943".

March 1942: Denial policies

To curtail the Japanese invasion of British India via the eastern border of Bengal, the British military strategically launched a pre-emptive, two-pronged scorched-earth initiative in eastern and coastal Bengal. Its goal was to deny the expected invaders access to food supplies, transport and other resources.[

First, a "denial of rice" policy was executed in three southern districts along the coast of the Bay of Bengal – Bakarganj (or Barisal), Midnapore and Khulna – that were expected to have surpluses of rice. John Herbert, the governor of Bengal, issued an urgent directive in late March 1942 immediately requiring stocks of paddy (unmilled rice) deemed surplus, and other food items, to be removed or destroyed in these districts.Official figures for the amounts impounded were relatively small and would have contributed only modestly to local scarcities. However, evidence that fraudulent, corrupt and coercive practices by the purchasing agents removed far more rice than officially recorded, not only from designated districts, but also in unauthorised areas, suggests a greater impact.Far more damaging were the policy›s disturbing impact on regional market relationships and contribution to a sense of public alarm.

The second prong, a "boat denial" policy, was strategized to deny Bengali transport to any invading Japanese army. It applied to districts readily accessible via the Bay of Bengal and the larger rivers that flow into it. Implemented on 1 May after an initial registration period, the policy authorised the Army to confiscate, relocate or destroy any boats large enough to carry more than ten people, and allowed them to requisition other means of transport such as bicycles, bullock carts, and elephants. Under this policy, the Army confiscated approximately 46,000 rural boats, severely disrupting river-borne movement of labour, supplies and food, and compromising the livelihoods of boatmen and fishermen. Leonard G. Pinnell, a British civil servant who headed the Bengal government›s Department of Civil Supplies, told the Famine Commission that the policy «completely broke the economy of the fishing class». Transport was generally unavailable to carry seed and equipment to distant fields or rice to the market hubs. Artisans and other groups who relied on boat transport to carry goods to market were offered no recompense; neither were rice growers nor the network of migratory labourers. The large-scale removal or destruction of rural boats caused a near-complete breakdown of the existing transport and administration infrastructure and market system for movement of rice paddy. No steps were taken to provide for the maintenance or repair of the confiscated boats,and many fishermen were unable to return to their trade.The Army took no steps to distribute food rations to make up for the interruption of supplies.

These policies had important political repercussions. The Indian National Congress, among other groups, staged protests denouncing the denial policies for placing draconian burdens on Bengali peasants; these were part of a nationalist sentiment and outpouring that later peaked in the "Quit India" movement. The policies' wider impact – the extent to which they compounded or even caused the famine to occur one year later – has been the subject of much discussion.

Provincial trade barriers

Many Indian provinces and princely states imposed inter-provincial trade barriers from mid-1942, in order to prevent the trade in domestic rice. Anxiety and soaring rice prices, triggered by the fall of Burma, were one underlying reason for the trade barriers. Trade imbalances brought on by price controls were another. The power to restrict inter-provincial trade was given provincial governments in November 1941 under the Defence of India Act, 1939.Provincial governments began erecting trade barriers that prevented the flow of foodgrains (especially rice) and other goods between provinces. These barriers reflected a desire to see that local populations were well fed, thus forestalling local emergencies.

In January 1942, Punjab banned exports of wheat; this increased the food insecurity and led the enclave of wheat-eaters in Greater Calcutta to increase their demand for rice prarticularly when an impending rice shortage was feared. The Central Provinces prohibited the export of foodgrains outside the province two months later. Madras banned rice exports in June, followed by export bans in Bengal and its neighbouring provinces of Bihar and Orissa that July.

The Famine Inquiry Commission of 1945 characterised this «critical and potentially most dangerous stage» as a key policy failure. As one deponent to the Commission put it: «Every province, every district, and every [administrative division] in the east of India had become a food republic unto itself. The trade machinery for the distribution of food [between provinces] throughout the east of India was slowly strangled, and by the spring of 1943 was dead." Bengal was unable to import domestic rice; this policy helped transform market failures and food shortage into famine and widespread death.

Mid-1942: Prioritised distribution

The loss of Burma reinforced the strategic importance of Calcutta as the hub of heavy industry and the main supplier of armaments and textiles for the entire Asian theatre. To support its wartime mobilisation, the Indian Government categorised the population into socioeconomic groups of "priority" and "non-priority" classes, according to their relative importance to the war effort. Members of the "priority" classes were largely composed of *bhadraloks*, who were upper-class or bourgeois middle-class, socially mobile, educated, urban, and sympathetic to Western values and modernisation. Protecting their interests was a major concern of both private and public relief efforts. This placed the rural poor in direct competition for scarce basic supplies with workers in public agencies, war-related industries, and in some cases even politically well-connected middle-class agriculturalists.

As food prices spiked and the signs of famine became more apparent from July 1942, the Bengal Chamber of Commerce (composed mainly of British-owned firms) designed a Foodstuffs Scheme to provide preferential distribution of goods and services to workers in high-priority war industries, in order to prevent them from leaving their positions. The scheme was approved by Government of Bengal. Rice was directed away from the starving rural districts to workers in industries considered vital to the military effort – particularly in the area around Greater Calcutta.Workers in prioritised sectors – private and government wartime industries, military and civilian construction, paper and textile mills, engineering firms, the Indian Railways, coal mining, and government workers of various levels– were given significant advantages and benefits. Essential workers received subsidised food, and were frequently paid in part in weekly allotments of rice sufficient to feed their immediate families, further protecting them from inflation. Essential workers also benefited from ration cards, a network of "cheap shops" which provided essential supplies at discounted rates, and direct, preferential allocation of supplies such as water, medical care, and antimalarial supplies. They also received subsidised food, free transportation, access to superior housing, regular wages and even "mobile cinema units catering to recreational needs".By December of that year, the total number of individuals covered (workers and their families) was approximately a million. Medical care was directed to the priority groups – particularly the military. Public and private medical staff at all levels were transferred to military duty, while medical supplies were monopolised.

Rural labourers and civilians not members of these groups received severely reduced access to food and medical care, generally available only to those who migrated to selected population centres. Otherwise, according to medical historian Sanjoy Bhattacharya, "vast areas of rural eastern India were denied any lasting state-sponsored distributive schemes".For this reason, the policy of prioritised distribution is sometimes discussed as one cause of the famine.

Civil unrest

Main articles: Quit India Movement and Indian independence movement

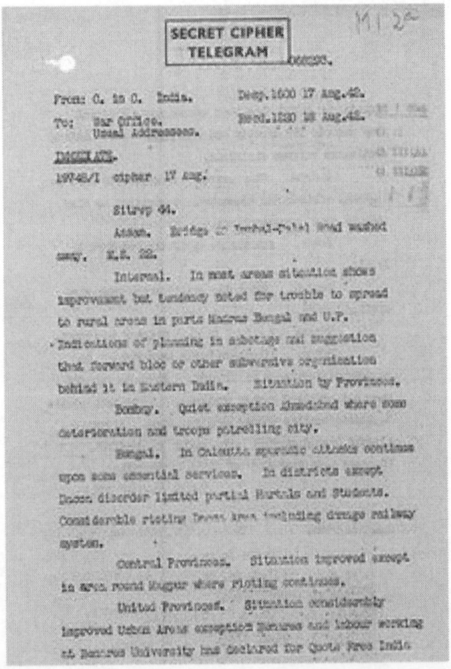

SECRET CIPHER TELEGRAM

From: C. in C. India. Desp. 1600 17 Aug. 42.

To: War Office. Recd. 1230 18 Aug. 42.
Usual Addressees.

[illegible]

19748/I cipher 17 Aug.

Sitrep 44.

Assam. Bridge [illegible] Road washed away. [illegible] 22.

Internal. In most areas situation shows improvement but tendency noted for trouble to spread to rural areas in parts Madras Bengal and U.P. Indications of planning in sabotage and suggestion that forward bloc or other subversive organisation behind it in Eastern India. Situation by Provinces.

Bombay. Quiet exception Ahmedabad where some deterioration and troops patrolling city.

Bengal. In Calcutta sporadic attacks continue upon some essential services. In districts except Dacca disorder limited partial Hartals and Students. Considerable rioting [illegible] area including damage railway system.

Central Provinces. Situation improved except in area round Nagpur where rioting continues.

United Provinces. Situation considerably improved Urban Areas exception Benares and labour working at Benares University has declared for Quote Free India

Secret Cipher Telegram from C. in C. India to the UK War Office, dated 17 August 1942, describing the civil unrest in wake of the Quit India Resolution, 9 August 1942

The war escalated resentment and fear of the Raj among rural agriculturalists and business and industrial leaders in Greater Calcutta. The unfavourable military situation of the Allies after the fall of Burma led the US and China to urge the UK to enlist India's full cooperation in the war by negotiating a peaceful transfer of political power to an elected Indian body; this goal was also supported by the Labour Party in Britain. Winston Churchill, the British prime minister, responded to the new pressure through the Cripps' mission, broaching the post-war possibility of an autonomous political status for India in exchange for its full military support, but negotiations collapsed in early April 1942.

On 8 August 1942, the Indian National Congress launched the Quit India movement as a nationwide display of nonviolent resistance. The British authorities reacted by imprisoning the Congress leaders. Without its leadership, the movement changed its character and took to sabotaging factories, bridges, telegraph and railway lines, and other government property,thereby threatening the British Raj's war enterprise. The British acted forcefully to suppress the movement, taking around 66,000 in custody (of whom just over 19,000 were still convicted under civil law or detained under the Defence of India Act in early 1944). More than 2,500 Indians were shot when police fired upon protesters, many of whom were killed.In Bengal, the movement was strongest in the Tamluk and Contai subdivisions of Midnapore district, where rural discontent was well-established and deep. In Tamluk, by April 1942 the government had destroyed some 18,000 boats in pursuit of its denial policy, while war-related inflation further alienated the rural population, who became eager volunteers when local Congress recruiters proposed open rebellion.

The violence during the "Quit India" movement was internationally condemned, and hardened some sectors of British opinion against India; The historians Christopher Bayly and Tim Harper believe it reduced the British War Cabinet's willingness to provide famine aid at a time when supplies were also needed for the war effort. In several ways the political and social disorder and distrust that were the effects and after-effects of rebellion and civil unrest placed political, logistical, and infrastructural constraints on the Government of India that contributed to later famine-driven woes.

1942–43: Price chaos

Throughout April 1942, British and Indian refugees fled Burma, many through Bengal, as the cessation of Burmese imports continued to drive up rice prices. In June, the Bengal government established price controls for rice, and on 1 July fixed prices at a level considerably lower than the prevailing market price. The principal result of the fixed low price was to make sellers reluctant to sell; stocks disappeared, either on to the black market or into storage. The government then let it be known that the price control law would not be enforced except in the most egregious cases of war profiteering. This easing of restrictions plus the ban on exports created about four months of relative price stability. In mid-October, though, south-west Bengal was struck by a series of natural disasters that destabilised prices again,causing another rushed scramble for rice, greatly to the benefit of the Calcutta black market. Between December 1942 and March 1943 the government made several attempts to «break the Calcutta market» by bringing in rice supplies from various districts around the province; however, these attempts to drive down prices by increasing supply were unsuccessful.

On 11 March 1943, the provincial government revoked its price controls, resulting in a drastic rises in the price of rice, due in part to increasing levels of speculation. The period of inflation between March and May 1943 was especially intense; May was the month of the first reports of death by starvation in Bengal. The government attempted to re-establish public confidence by insisting that the crisis was being caused almost solely by speculation and hoarding,but their propaganda failed to dispel the widespread belief that there was a shortage of rice. The provincial government never formally declared a state of famine, even though its Famine Code would have mandated a sizable increase in aid. In the early stages of the famine, the rationale for this was that the provincial government was expecting aid from the Government of India. It felt then its duty lay in maintaining confidence through propaganda that asserted that there was no shortage. After it became clear that aid from central government was not forthcoming, the provincial government felt they simply did not have the amount of food supplies that a declaration of famine would require them to distribute, while distributing more money might make inflation worse.

When inter-provincial trade barriers were abolished on 18 May, prices temporarily fell in Calcutta, but soared in the neighbouring provinces of Bihar and Odisha when traders rushed to purchase stocks. The provincial government's attempts to locate and seize any hoarded stocks failed to find significant hoarding. In Bengal, prices were soon five to six times higher than they had been before April 1942.Free trade was abandoned in July 1943,and price controls were reinstated in August. Despite this, there were unofficial reports of rice being sold in late 1943 at roughly eight to ten times the prices of late 1942. Purchasing agents were sent out by the government to obtain rice, but their attempts largely failed. Prices remained high, and the black market was not brought under control.

October 1942: Natural disasters

See also: North Indian Ocean tropical cyclone

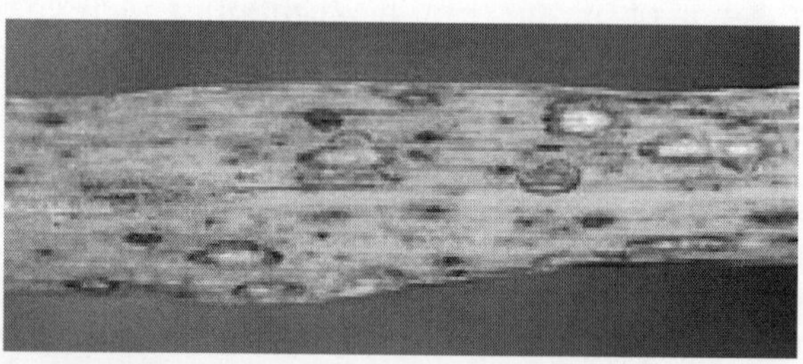

Brown spot disease: symptoms of *Cochliobolusmiyabeanus* on rice

Bengal was affected by a series of natural disasters late in 1942. The winter rice crop was afflicted by a severe outbreak of fungal brown spot disease, while, on 16–17 October a cyclone and three storm surges ravaged croplands, destroyed houses and killing thousands, at the same time dispersing high levels of fungal spores across the region and increasing the spread of the crop disease.The fungus reduced the crop yield even more than the cyclone. After describing the horrific conditions he had witnessed, the mycologist S.Y. Padmanabhan wrote that the outbreak compared to the potato blight of the Irish Great Famine: "Though administrative failures were immediately responsible for this human suffering, the principal cause of the short crop production of 1942 was the [plant] epidemic ... nothing as devastating ... has been recorded in plant pathological literature".

The Bengal cyclone came through the Bay of Bengal, landing on the coastal areas of Midnapore and 24 Parganas. It killed 14,500 people and 190,000 cattle, whilst rice paddy stocks in the hands of cultivators, consumers, and dealers were destroyed. It also created local atmospheric conditions that contributed to an increased incidence of malaria.The three storm surges which followed the cyclone destroyed the seawalls of Midnapore and flooded large areas of Contai and Tamluk. Waves swept an area of 450 square miles (1,200 km^2), floods affected 400 square miles (1,000 km^2), and wind and torrential rain damaged 3,200 square miles (8,300 km^2). For nearly 2.5 million Bengalis, the accumulative damage of the cyclone and storm surges to homes, crops and livelihoods was catastrophic:

Corpses lay scattered over several thousand square miles of devastated land. 7,400 villages were partly or wholly destroyed by the storm, and standing flood waters remained for weeks in at least 1,600 villages. Cholera, dysentery and other water-borne diseases flourished. 527,000 houses and 1,900 schools were lost. Over 1,000 square miles of the most fertile paddy land in the province was entirely destroyed, and the standing crop over an additional 3,000 square miles was damaged.[188]

The cyclone, floods, plant disease, and warm, humid weather reinforced each other and combined to have a substantial impact on the *aman* rice crop of 1942. Their impact was felt in other aspects as well, as in some districts the cyclone was responsible for an increased incidence of malaria, with deadly effect.

October 1942: Unreliable crop forecasts

At about the same time, official forecasts of crop yields predicted a significant shortfall. However, crop statistics of the time were scant and unreliable. Administrators and statisticians had known for decades that India's agricultural production statistics were completely inadequate and "not merely guesses, but frequently demonstrably absurd guesses". There was little or no internal bureaucracy for creating and maintaining such reports, and the low-ranking police officers or village officials charged with gathering local statistics were often poorly supplied with maps and other necessary information, poorly educated, and poorly motivated to be accurate. The Bengal Government thus did not act on these predictions,doubting their accuracy and observing that forecasts had predicted a shortfall several times in previous years, while no significant problems had occurred.

Air raids on Calcutta

The Famine Inquiry Commission's 1945 report singled out the first Japanese air raids on Calcutta in December 1942 as causation. The attacks, largely unchallenged by Allied defences, continued throughout the week, triggering an exodus of thousands from the city. As evacuees travelled to the countryside, food-grain dealers closed their shops. To ensure that workers in the prioritised industries in Calcutta would be fed, the authorities seized rice stocks from wholesale dealers, breaking any trust the rice traders had in the government. "From that moment", the 1945 report stated, "the ordinary trade machinery could not be relied upon to feed Calcutta. The [food security] crisis had begun".

1942–43: Shortfall and carryover

Whether the famine resulted from crop shortfall or failure of land distribution has been much debated .According to Amartya Sen: "The ... [rice paddy] supply for 1943 was only about 5% lower than the average of the preceding five years. It was, in fact, 13% higher than in 1941, and there was, of course, no famine in 1941."The Famine Inquiry Commission report concluded that the overall deficit in rice in Bengal in 1943, taking into account an estimate of the amount of carryover of rice from the previous harvest, was about three weeks' supply. In any circumstances, this was a significant shortfall requiring a considerable amount of food relief, but not a deficit large enough to create widespread deaths by starvation. According to this view, the famine "was not a crisis of food availability, but of the [unequal] distribution of food and income". There has been very considerable debate about the amount of carryover available for use at the onset of the famine.

Several contemporary experts provided evidences of a much larger shortfall. Commission member Wallace Aykroyd argued in 1975 that there had been a 25% shortfall in the harvest of the winter of 1942,while L. G. Pinnell, responsible to the Government of Bengal from August 1942 to April 1943 for managing food supplies, estimated the crop loss at 20%, with disease accounting for more of the loss than the cyclone; other government sources privately admitted the shortfall was 2 million tons.The economist George Blyn argues that with the cyclone and floods of October and the loss of imports from Burma, the 1942 Bengal rice harvest had been reduced by one-third.

1942–44: Refusal of imports

Beginning as early as December 1942, high-ranking government officials and military officers (including John Herbert, the Governor of Bengal; Viceroy Linlithgow; Leo Amery the Secretary of State for India; General Claude Auchinleck, Commander-in-Chief of British forces in India,and Admiral Louis Mountbatten, Supreme Commander of South-East Asiabegan requesting food imports for India through government and military channels, but for months these requests were either rejected or reduced to a fraction of the original amount by Churchill's War Cabinet. The colony was also not permitted to spend its own sterling reserves, or even use its own ships, to import food. Although Viceroy Linlithgow appealed for imports from mid-December 1942, he did so on the understanding that the military would be given preference over civilians. The Secretary of State for India, Leo Amery, was on one side of a cycle of requests for food aid and subsequent refusals from the British War Cabinet that continued through 1943 and into 1944.Amery did not mention worsening conditions in the countryside, stressing that Calcutta's industries must be fed or its workers would return to the countryside. Rather than meeting this request, the UK promised a relatively small amount of wheat that was specifically intended for western India (that is, not for Bengal) in exchange for an increase in rice exports from Bengal to Ceylon.

The tone of Linlithgow's warnings to Amery grew increasingly grave over the first half of 1943, as did Amery's requests to the War Cabinet; on 4 August 1943 Amery noted the spread of famine, and specifically stressed the effect upon Calcutta and the potential effect on the morale of European troops. The cabinet again offered only a relatively small amount, explicitly referring to it as a token shipment. The explanation generally offered for the refusals included insufficient shipping, particularly in light of Allied plans to invade Normandy. The Cabinet also refused offers of food shipments from several different nations. When such shipments did begin to increase modestly in late 1943, the transport and storage facilities were understaffed and inadequate. When Viscount Archibald Wavell replaced Linlithgow as Viceroy in the latter half of 1943, he too began a series of exasperated demands to the War Cabinet for very large quantities of grain. His requests were again repeatedly denied, causing him to decry the current crisis as "one of the greatest disasters that has befallen any people under British rule, and the damage to our reputation both among Indians and foreigners in India is incalculable".

Experts' disagreement over political issues can be found in differing explanations of the War Cabinet's refusal to allocate funds to import grain. Lizzie Collingham holds the massive global dislocations of supplies caused by World War II virtually guaranteed that hunger would occur somewhere in the world, yet Churchill's animosity and perhaps racism toward Indians decided the exact location where famine would fall. Similarly, Madhusree Mukerjee makes a stark accusation: "The War Cabinet's shipping assignments made in August 1943, shortly after Amery had pleaded for famine relief, show Australian wheat flour travelling to Ceylon, the Middle East, and Southern Africa – everywhere in the Indian Ocean but to India. Those assignments show a will to punish."In contrast, Mark Tauger strikes a more supportive stance: "In the Indian Ocean alone from January 1942 to May 1943, the Axis powers sank 230 British and Allied merchant ships totalling 873,000 tons, in other words, a substantial boat every other day. British hesitation to allocate shipping concerned not only potential diversion of shipping from other war-related needs but also the prospect of losing the shipping to attacks without actually bringing help to India at all."

Famine, disease, and the death toll

An estimated 2.1–3 million Bengalis died, out of a population of 60.3 million. However, contemporary mortality statistics were to some degree under-recorded, particularly for the rural areas, where data collecting and reporting was rudimentary even in normal times. Thus, many of those who died or migrated were unreported. The principal causes of death also changed as the famine progressed in two waves.

Early on, conditions drifted towards famine at different rates in different Bengal districts. The Government of India dated the beginning of the Bengal food crisis from the air raids on Calcutta in December 1942, blaming the acceleration to full-scale famine by May 1943 on the effects of price decontrol. However, in some districts the food crisis had begun as early as mid-1942.The earliest indications were somewhat obscured, since rural poor were able to draw upon various survival strategies for a few months. After December 1942 reports from various commissioners and district officers began to cite a "sudden and alarming" inflation, nearly doubling the price of rice; this was followed in January by reports of distress caused by serious food supply problems.In May 1943, six districts – Rangpur, Mymensingh, Bakarganj, Chittagong, Noakhali and Tipperah – were the first to report deaths by starvation. Chittagong and Noakhali, both "boat denial" districts in the Ganges Delta (or Sundarbans Delta) area, were the hardest hit. In this first wave – from May to October 1943 – starvation was the principal cause of excess mortality (that is, those attributable to the famine, over and above the normal death rates), filling the emergency hospitals in Calcutta and accounting for the majority of deaths in some districts.According to the Famine Inquiry Commission report, many victims on the streets and in the hospitals were so emaciated that they resembled «living skeletons».While some districts of Bengal were relatively less affected throughout the crisis,no demographic or geographic group was completely immune to increased mortality rates caused by disease – but deaths from starvation were confined to the rural poor.

Deaths due to extreme starvation had reached the summit by November 1943. Disease began its sharp upward turn around October 1943 and overtook starvation as the most common cause of death around December. Disease-related mortality then continued to take its toll through early-to-mid 1944.Among diseases, malaria was the biggest killer. From July 1943 to June 1944, the monthly death toll from malaria averaged 125% above rates from the previous five years, reaching 203% above average in December 1943.Malaria parasites were found in nearly 52% of blood samples examined at Calcutta hospitals during the peak period, November–December 1944.Statistics for malaria deaths are almost certainly inaccurate, since the symptoms often resemble those of other fatal fevers, but there is little doubt that it was the main killer. Other famine-related deaths resulted from dysentery and diarrhoea, typically through consumption of poor-quality food or deterioration of the digestive system caused by malnutrition. Cholera is a waterborne disease associated with social disruption, poor sanitation, contaminated water, crowded living conditions (as in refugee camps), and a wandering population – problems brought on after the October cyclone and flooding and then continuing through the crisis. The epidemic of smallpox largely resulted from a result of lack of vaccinations and the inability to quarantine patients, caused by general social disruption. According to social demographer Arup Maharatna, statistics for smallpox and cholera are probably more reliable than those for malaria, since their symptoms are more easily recognisable.

Map of Bengal districts 1943

The mortality statistics present a shady picture of the distribution of deaths among age and gender groups. Although very young children and the elderly are usually more susceptible to the effects of starvation and disease, overall in Bengal it was adults and older children who suffered the highest proportional mortality rises. However, this picture was inverted in some urban areas, perhaps because the cities attracted large numbers of very young and very old migrants. In general, males suffered generally higher death rates than females,although the rate of female infant death was higher than for males, perhaps reflecting a discriminatory bias. A relatively lower death rate for females of child-bearing age may have reflected a reduction in fertility, brought on by malnutrition, which in turn reduced maternal deaths.

Regional differences in mortality rates were largely influenced by the effects of migration, and of natural disasters. In general, excess mortality was higher in the east, even though the relative shortfall in the rice crop was worst in the western districts of Bengal. Eastern districts were relatively densely populated, were closest to the Burma war zone, and normally ran grain deficits in pre-famine times. These districts also were subject to the boat denial policy, and had a relatively high proportion of jute production instead of rice. Workers in the east were more likely to receive monetary wages than payment in kind with a portion of the harvest, a common practice in the western districts. When prices rose sharply, their wages failed to follow suit; this drop in real wages left them less able to purchase food. The following table, derived from Arup Maharatna (1992), shows trends in excess mortality for 1943–44 as compared to prior non-famine years. Death rate is total number of deaths in a year (mid-year population) from all causes, per 1000.All death rates are with respect to the population in 1941.Percentages for 1943–44 are of *excess deaths* (that is, those attributable to the famine, over and above the normal incidence)as compared to rates from 1937–41.

Table 1.4

Cause-specific death rates during pre-famine and famine periods; relative importance of different causes of death during famine: Bengal

Cause of death	Pre-famine 1937–41	1943		1944	
	Rate	Rate	%	Rate	%
Cholera	0.73	3.60	23.88	0.82	0.99
Smallpox	0.21	0.37	1.30	2.34	23.69
Fever	6.14	7.56	11.83	6.22	0.91
Malaria	6.29	11.46	43.06	12.71	71.41
Dysentery/diarrhoea	0.88	1.58	5.83	1.08	2.27
All other	5.21	7.2	14.11	5.57	0.74
All causes	19.46	31.77	100.00	28.75	100.00

Overall, the table shows the dominance of malaria as the cause of death throughout the famine, accounting for roughly 43%of the excess deaths in 1943 and 71% in 1944. Cholera was a major source of famine-caused deaths in 1943 (24%) but dropped to a negligible percentage (1%) the next year. Smallpox deaths were almost a mirror image: they made up a small percentage of excess deaths in 1943 (1%) but jumped in 1944 (24%). Finally, the sharp jump in the death rate from "All other" causes in 1943 is almost certainly due to deaths from pure starvation, which were negligible in 1944.

Though excess mortality due to malarial deaths peaked in December 1943, rates persistently remained high throughout the following year. Scarce supplies of quinine (the most common malaria medication) were very frequently diverted to the black market. Advanced anti-malarial drugs such as mepacrine (Atabrine) were distributed almost solely to the military and to «priority classes»; DDT (then relatively new and considered «miraculous») and pyrethrum were sprayed only around military installations. Paris Green was used as an insecticide in some other areas.This unequal distribution of anti-malarial measures may explain a lower incidence of malarial deaths in population centres, where the greatest cause of death was «all other» (probably migrants dying from starvation).

Deaths from dysentery and diarrhoea peaked in December 1943, the same month as for malaria. Cholera deaths peaked in October 1943 but receded dramatically in the following year, brought under control by a vaccination program overseen by military medical workers. A similar smallpox vaccine campaign started later and was pursued less effectively; smallpox deaths peaked in April 1944. "Starvation" was generally not listed as a cause of death at the time; many deaths by starvation may have been listed under the "all other" category. Here the death rates, rather than per cents, reveal the peak in 1943.

The two waves – starvation and disease – also interacted and amplified one another, increasing the excess mortality. Widespread starvation and malnutrition first compromised immune systems, and reduced resistance to disease led to death by opportunistic infections. Second, the social disruption and dismal conditions caused by a cascading breakdown of social systems brought mass migration, overcrowding, poor sanitation, poor water quality and waste disposal, increased vermin, and unburied dead. All of these factors are closely associated with the increased spread of infectious disease.

Social disruption

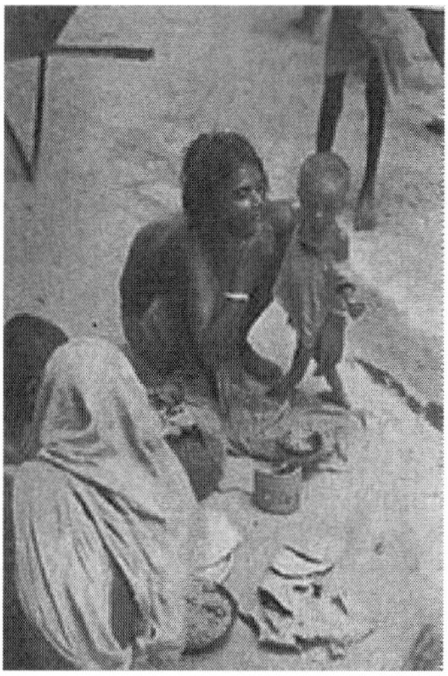

A family on the sidewalk in Calcutta during the Bengal famine of 1943

Despite the organised and sometimes violent civil unrest immediately before the famine, but surprisingly there was no organised rioting when the famine took hold instead it was chaotic. However, the crisis overwhelmed the provision of health care and key supplies: food relief and medical rehabilitation were supplied too late, whilst medical facilities across the province were utterly insufficient for the task at hand.A long-standing system of rural patronage, in which peasants relied on large landowners to supply subsistence in times of crisis, collapsed as patrons exhausted their own resources and abandoned the peasants.

Families also disintegrated, with cases of abandonment, child-selling, prostitution, and sexual exploitation. Lines of small children begging stretched for miles outside cities; at night, children could be heard "crying bitterly and coughing terribly ... in the pouring monsoon rain ... stark naked, homeless, motherless, fatherless and friendless. Their sole possession was an empty tin".A schoolteacher in Mahisadal witnessed «children picking and eating undigested grains out of a beggar›s diarrheal discharge». Author Freda Bedi wrote that it was «not just the problem of rice and the availability of rice. It was the problem of society in fragments».

Population displacement

The famine fell hardest on the rural poor. As the distress continued, families adopted increasingly desperate means for survival. First, they reduced their food intake and began to sell jewellery, ornaments, and smaller items of personal property. As expenses for food or burials became more urgent, the items sold became larger and less replaceable. Eventually, families disintegrated; men sold their small farms and left home to look for work or to join the army, and women and children became homeless migrants, often travelling to Calcutta or another large city in search of organised relief:

Husbands deserted wives and wives husbands; elderly dependents were left behind in the villages; babies and young children were sometimes abandoned in extreme cases. According to a survey carried out in Calcutta during the latter half of 1943, some breaking up of the family had occurred in about half the destitute population which reached the city.

In Calcutta, evidence of the famine was "... mainly in the form of masses of rural destitutes trekking into the city and dying on the streets". Estimates of the number of the sick who flocked to Calcutta ranged between 100,000 and 150,000. Once they left their rural villages in search of food, their outlook for survival was grim: "Many died by the roadside – witness the skulls and bones which were to be seen there in the months following the famine."

Sanitation and undisposed dead

Image of Midnapore famine victim from Chittaprosad's *Hungry Bengal*, five thousand corpses of which were burned by Indian police. The caption read «His name was KshetramohanNaik.»

The disruption of the core elements of society brought a catastrophic breakdown of sanitary conditions and hygiene standards. Large scale migration resulted in the abandonment of the facilities and sale of the utensils and necessary for washing clothes or preparation of food. Many people drank contaminated rainwater from streets and open spaces where others had urinated or defecated. Particularly in the early months of the crisis, conditions did not improve for those under medical care:

Conditions in certain famine hospitals at this time ... were indescribably bad ... Visitors were horrified by the state of the wards and patients, the ubiquitous filth, and the lack of adequate care and treatment ... [In hospitals all across Bengal, the] condition of patients was usually appalling, a large proportion suffering from acute emaciation, with 'famine diarrhoea' ... Sanitary conditions in nearly all temporary indoor institutions were very bad to start with ...

The desperate condition of the healthcare worsened appreciably until the army, under Viscount Wavell, took over the provision of relief supplies in October 1943. At that time medical resources were made far more available.

Disposal of corpses soon became a problem for the government and the public, as numbers outsourced the cremation houses, burial grounds, and those collecting and disposing of the dead. Corpses lay scattered throughout the pavements and streets of Calcutta. In only two days of August 1943, at least 120 were removed from public thorough fares. In the countryside bodies were often disposed of in rivers and water supplies. As one survivor explained, "We couldn't bury them or anything. No one had the strength to perform rites. People would tie a rope around the necks and drag them over to a ditch."Corpses were also left to rot and putrefy in open spaces. The bodies were picked over by vultures and dragged away by jackals. Sometimes this happened while the victim was still living. The sight of corpses beside canals, ravaged by dogs and jackals, was common; during a seven-mile boat ride in Midnapore in November 1943, a journalist counted at least five hundred such sets of skeletal remains. The weekly newspaper *Biplabi* commented in November 1943 on the levels of putrefaction, contamination, and vermin infestation:

Bengal is a vast cremation ground, a meeting place for ghosts and evil spirits, a land so overrun by dogs, jackals and vultures that it makes one wonder whether the Bengalis are really alive or have become ghosts from some distant epoch.

By the summer of 1943, many districts of Bengal, especially in the countryside, had taken on the look of "a vast charnel house".

Cloth famine

Mother with child on a Calcutta street. Bengal famine 1943

As a further consequence of the crisis, a "cloth famine" left the poorest in Bengal clothed in scraps or naked through the winter. The British military consumed nearly all the textiles produced in India by purchasing Indian-made boots, parachutes, uniforms, blankets, and other goods at heavily discounted rates. India produced 600,000 miles of cotton fabric during the war, from which it made two million parachutes and 415 million items of military clothing. It exported 177 million yards of cotton in 1938–1939 and 819 million in 1942–1943.The country's production of silk, wool and leather was also used up by the military.

The small proportion of left over material was purchased by speculators for sale to civilians, subject to similarly steep inflation; in May 1943 prices were 425% higher than in August 1939. With the supply of cloth crowded out by commitments to Britain and price levels affected by profiteering, those not among the "priority classes" faced increasingly dire scarcity. Swami Sambudhanand, President of the Ramakrishna Mission in Bombay, stated in July 1943:

There have been reports regarding the robbing of graveyards for clothes, disrobing of men and women in out of way places for clothes ... and minor rioting here and there. Stray news has also come that women have committed suicide for want of cloth ... Thousands of men and women ... cannot go out to attend their usual work outside for want of a piece of cloth to wrap round their loins.

Many women "took to staying inside a room all day long, emerging only when it was their turn to wear the single fragment of cloth shared with female relatives".

Exploitation of women and children

One of the gruesome effects of famine is that it escalated the exploitation of women; the sale of women and girls, for example, tends to increase. The sexual exploitation of poor, rural, lower-caste and tribal women by the *jotedars* had been difficult to escape even before the crisis.In the wake of the cyclone and later famine, many women lost or sold all their possessions, and lost a male guardian due to abandonment or death. Those who migrated to Calcutta frequently had only begging or prostitution available as strategies for survival; often regular meals were the only payment. Tarakchandra Das suggests that a large proportion of the girls aged 15 and younger who migrated to Calcutta during the famine disappeared into brothels;in late 1943, entire boatloads of girls for sale were reported in ports of East Bengal.Girls were also prostituted to soldiers, with boys acting as pimps.Families sent their young girls to wealthy landowners overnight in exchange for very small amounts of money or rice, or sold them outright into prostitution; girls were sometimes enticed with sweet treats and kidnapped by pimps. Very often, these girls lived in constant fear of injury or death, but the brothels were their sole means of survival, or they were unable to escape.Women who had been sexually exploited could not later expect any social acceptance or a return to their home or family.Bina Agarwal writes that such women became permanent outcastes in a society that highly values female chastity, rejected by both their birth family and husband›s family.

An unknown number of children, some tens of thousands, were orphaned. Many others were abandoned, sometimes by the roadside or at orphanages, or sold for as much as two *maunds* (one *maund* was roughly equal to 37 kilograms (82 lb)),or as little as one *seer* (1 kilogram (2.2 lb))of unhusked rice, or for trifling amounts of cash. Sometimes they were purchased as household servants, where they would «grow up as little better than domestic slaves». They were also purchased by sexual predators. Altogether, according to Greenough, the victimisation and exploitation of these women and children was an immense social cost of the famine.

Relief efforts

Orphans who survived the famine

Besides the relatively prompt but inadequate provision of humanitarian aid for the cyclone-stricken areas around Midnapore beginning in October 1942, the response of both the Bengal Provincial Government and the Government of India was slow. A "non-trivial" yet "pitifully inadequate" amount of aid began to be distributed from private charitable organisationsin the early months of 1943 and increased through time, mainly in Calcutta but to a limited extent in the countryside. In April, more government relief began to flow to the outlying areas, but these efforts were restricted in scope and largely misdirected, with most of the cash and grain supplies flowing to the relatively wealthy landowners and urban middle-class (and typically Hindu) *bhadraloks*. This initial period of relief included three forms of aid: agricultural loans (cash for the purchase of paddy seed, plough cattle, and maintenance expenses),grain given as gratuitous relief, and "test works" that offered food and perhaps a small amount of money in exchange for strenuous work. The "test" aspect arose because there was an assumption that if relatively large numbers of people took the offer, that indicated that famine conditions were prevalent. Agricultural loans offered no assistance to the large numbers of rural poor who had little or no land. Grain relief was divided between cheap grain shops and the open market, with far more going to the markets. Supplying grain to the markets was intended to lower grain prices, but in practice gave little help to the rural poor, instead placing them into direct purchasing competition with wealthier Bengalis at greatly inflated prices. Thus from the beginning of the crisis until around August 1943, private charity was the principal form of relief available to the very poor.

According to Paul Greenough, the Provincial Government of Bengal delayed its relief efforts primarily due to their incapability as they had no idea how to deal with a provincial rice market damaged by the interaction of man-made shocks, contrary to the far more familiar case of localised shortage due to natural disaster. Moreover, the urban middle-class were their overriding concern, not the rural poor. They were also expecting the Government of India to rescue Bengal by bringing food in from outside the province (350,000 tons had been promised but not delivered). And finally, they had long stood by a public propaganda campaign declaring "sufficiency" in Bengal's rice supply, and were afraid that speaking of scarcity rather than sufficiency would lead to increased hoarding and speculation.

There was also rampant corruption and nepotism in the distribution of government aid; often as much as half of the goods disappeared into the black market or into the hands of friends or relatives. Despite a long-established and detailed Famine Code that would have triggered a sizable increase in aid, and a statement privately circulated by the government in June 1943 that a state of famine might need to be formally declared, this declaration never happened.

Since government relief efforts were initially narrowed, a considerably large and diverse number of private groups and voluntary workers attempted on meeting the alarming needs caused by deprivation. Communists, socialists, wealthy merchants, women's' groups, Hindus, Muslims, private citizens from distant Karachi and Indian expatriates from as far away as east Africa aided in relief efforts or sent donations of money, food and cloth. Markedly diverse political groups, including pro-war allies of the Raj and anti-war nationalists, each set up separate relief funds or aid groups. Though the efforts of these diverse groups were sometimes marred by Hindu and Muslim communalism, with bitter accusations and counter-accusations of unfair treatment and favouritism,collectively they provided substantial aid.

Grain started to flow to buyers in Calcutta after the inter-provincial trade barriers were abolished in May 1943, but on 17 July a flood of the Damodar River in Midnapore breached major rail lines, severely crippling import by rail. As the depth and scope of the famine became unmistakable, the Provincial Government began setting up gruel kitchens in August 1943; the gruel, which often provided barely a survival-level caloric intake, was sometimes unfit for consumption – mouldy or contaminated with dirt, sand, and gravel. Unfamiliar and indigestible grains were often substituted for rice, causing intestinal distress that frequently resulted in death among the weakest. Nevertheless, food distributed from government gruel kitchens immediately became the main source of aid for the rural poor.

The rails had been repaired in August and pressure from the Government of India brought substantial supplies into Calcutta during September, Linlithgow's final month as Viceroy. However, a second problem emerged: the Civil Supplies Department of Bengal was undermanned and under-equipped to distribute the supplies, and the resulting transportation bottleneck left very large piles of grain accumulating in the open air in several locations, including Calcutta's Botanical Garden.Field Marshal Archibald Wavell replaced Linlithgow that October, within two weeks he had requested military support for the transport and distribution of crucial supplies. This assistance was delivered promptly, including «a full division of... 15,000 [British] soldiers...military lorries and the Royal Air Force" and distribution to even the most distant rural areas began on a large scale.In particular, grain was imported from the Punjab, and medical resources were made far more available. Rank-and-file soldiers, who had sometimes disobeyed orders to feed the destitute from their rations,were held in esteem by Bengalis for the efficiency of their work in distributing relief. That December, the "largest [rice] paddy crop ever seen" in Bengal was harvested. According to Greenough, large amounts of land previously used for other crops had been switched to rice production. The price of rice began to fall. Survivors of the famine and epidemics gathered the harvest themselves,though in some villages there were no survivors capable of doing the work. Wavell went on to make several other key policy steps, including promising that aid from other provinces would continue to feed the Bengal countryside, setting up a minimum rations scheme, and (after considerable effort) prevailing upon Great Britain to increase international imports. He has been widely praised for his decisive and effective response to the crisis. All official food relief work ended in December 1943 and January 1944.

Economic and political effects

The famine's aftermath highly accelerated pre-existing socioeconomic processes leading to poverty and income inequality, severely disrupted the important elements of Bengal's economy and social fabric, and ruined millions of families. The crisis overwhelmed and impoverished large segments of the economy. A key source of impoverishment was the widespread coping strategy of selling assets, including land. In 1943 alone in one village in east Bengal, for example, 54 out of a total of 168 families sold all or part of their landholdings; among these, 39 (or very nearly 3 out of 4) did so as a coping strategy in reaction to the scarcity of food. As the famine wore on across Bengal, nearly 1.6 million families – roughly one-quarter of all landholders – sold or mortgaged their paddy lands in whole or in part. Some did so to profit from skyrocketing prices, but many others were trying to save themselves from crisis-driven distress. A total of 260,000 families sold all their landholdings outright, thus falling from the status of landholders to that of labourers. The table below illustrates that land transfers increased significantly in each of four successive years. When compared to the base period of 1940–41, the 1941–42 increase was 504%, 1942–43 was 665%, 1943–44 was 1,057% and the increase of 1944–45 compared to 1940–41 was 872%:

1.5 Land alienation in Bengal, 1940–41 to 1944–45: number of sales of occupancy holdings

1940–41	1941–42	1942–43	1943–44	1944–45
141,000	711,000	938,000	1,491,000	1,230,000

This fall into lower income groups happened across a number of occupations. In absolute numbers, the hardest hit by post-famine impoverishment were women and landless agricultural labourers. In relative terms, those engaged in rural trade, fishing and transport (boatmen and bullock cart drivers) suffered the most. In absolute numbers, agricultural labourers faced the highest rates of destitution and mortality.

The "panicky responses" of the colonial state as it controlled the distribution of medical and food supplies during the fall of Burma had profound political ramifications. "It was soon obvious to the bureaucrats in New Delhi and the provinces, as well as the GHQ (India)," wrote Sanjoy Bhattacharya, "that the disruption caused by these short-term policies – and the political capital being made out of their effects – would necessarily lead to a situation where major constitutional concessions, leading to the dissolution of the Raj, would be unavoidable."Similarly, nationwide opposition to the boat denial policy, as typified by Mahatma Gandhi's vehement editorials, helped strengthen the Indian independence movement. The denial of boats alarmed the public; the resulting dispute was one point that helped to shape the "Quit India" movement of 1942 and harden the War Cabinet's response. An Indian National Congress (INC) resolution sharply decrying the destruction of boats and seizure of homes was considered treasonous by Churchill's War Cabinet, and was instrumental in the later arrest of the INC's top leadership. Public thought in India, shaped by impulses such as media coverage and charity efforts, converged into a set of closely-related conclusions: the famine had been a national injustice, preventing any recurrence was a national imperative, and the human tragedy left in its wake was as Jawaharlal Nehru said «...the final judgment on British rule in India».According to historian Benjamin R. Siegel:

...at a national level, famine had transformed India's political landscape, underscoring the need for self-rule to Indian citizens far away from its epicenter. Photographs and journalism and the affective bonds of charity tied Indians inextricably to Bengal and made their suffering its own; a provincial [famine] was turned, in the midst of war, into a national case against imperial rule.

Media Coverage and Other Depictions

Main article: Media coverage of the Bengal famine of 1943

The *People's War*, an organ of the Communist Party of India, published graphic photos of the famine by Sunil Janah.

Calcutta's two leading English-language newspapers were *The Statesman* (at the time British-owned) and Amrita Bazar Patrika (edited by independence campaigner TusharKanti Ghosh).In the early months of the famine, the government applied pressure on newspapers to "calm public fears about the food supply" and follow the official stance that there was no rice shortage. This effort had some success; *The Statesman* published editorials asserting that the famine was due solely to speculation and hoarding, while «berating local traders and producers, and praising ministerial efforts».News of the famine was also subject to strict war-time censorship – even use of the word «famine» was prohibited – leading *The Statesman* later to remark that the UK government «seems virtually to have withheld from the British public knowledge that there was famine in Bengal at all».

Beginning in mid-July 1943 and more so in August, however, these two newspapers began publishing detailed and increasingly critical accounts of the depth and scope of the famine, its impact on society, and the nature of British, Hindu, and Muslim political responses. A turning point in news coverage came in late August 1943, when the editor of *The Statesman*, Ian Stephens, solicited and published a series of graphic photos of the victims. These made world headlinesand marked the beginning of domestic and international consciousness of the famine.The next morning, "in Delhi second-hand copies of the paper were selling at several times the news-stand price,» and soon "in Washington the State Department circulated them among policy makers». In Britain, *The Guardian* called the situation «horrible beyond description».The images had a profound effect and marked «for many, the beginning of the end of colonial rule». Stephens› decision to publish them and to adopt a defiant editorial stance won accolades from many (including the Famine Inquiry Commission),and has been described as «a singular act of journalistic courage and conscientiousness, without which many more lives would have surely been lost». The publication of the images, along with Stephens› editorials, not only helped to bring the famine to an end by driving the British government to supply adequate relief to the victims, but also inspired Amartya Sen›s influential contention that the presence of a free press prevents famines in democratic countries.The photographs also spurred *Amrita Bazar Patrika* and the Indian Communist Party's organ, *People's War*, to publish similar images; the latter would make photographer Sunil Janah famous.Women journalists who covered the famine included Freda Bedi reporting for the *Punjab Times*,and Vasudha Chakravarti and Kalyani Bhattacharjee, who wrote from a nationalist perspective.

The famine has been portrayed in novels, films and art. The novel *AshaniSanket* by BibhutibhushanBandyopadhyay is a fictional account of a young doctor and his wife in rural Bengal during the famine. It was adapted into a film of the same name (*Distant Thunder*) by director Satyajit Ray in 1973. The film is listed in *The New York Times Guide to the Best 1,000 Movies Ever Made*.Also well-known are the novel *So Many Hungers!* (1947) by Bhabani Bhattacharya and the 1980 film *AkalerShandhaney* by Mrinal Sen. Ella Sen's collection of stories based on reality, *Darkening Days: Being a Narrative of Famine-Stricken Bengal* recounts horrific events from a woman›s point of view.

A contemporary sketchbook of iconic scenes of famine victims, *Hungry Bengal: a tour through Midnapur District in November, 1943* by Chittaprosad, was immediately banned by the British and 5,000 copies were seized and destroyed. One copy was hidden by Chittaprosad's family and is now in the possession of the Delhi Art Gallery. Another artist famed for his sketches of the famine was Zainul Abedin.

Historiography

Controversy about the causes of the famine has continued in the decades. Attempting to determine culpability, research and analysis has covered complex issues such as the impacts of natural forces, market failures, failed policies or even malfeasance by governmental institutions, and war profiteering or other unscrupulous acts by private business. The questionable accuracy of much of the available contemporary statistical and anecdotal data is a complicating factor,as is the fact that the analyses and their conclusions are political and politicised.

The degree of crop shortfall in late 1942 and its impact in 1943 has dominated the historiography of the famine. The issue reflects a larger debate between two perspectives: one emphasises the importance of food availability decline (FAD) as a cause for famine, and another focuses on the failure of exchange entitlements (FEE). The FAD explanation blames famine on crop failures brought on principally by crises such as drought, flood, or man-made devastation from war. The FEE account agrees that such external factors are in some cases important, but holds that famine is primarily the interaction between pre-existing "structural vulnerability" (such as poverty) and a shock event (such as war or political interference in markets) that disrupts the economic market for food. When these interact, some groups within society can become unable to purchase or acquire food even though sufficient supplies are available.

Both the FAD and FEE perspectives would agree that Bengal experienced at least some grain shortage in 1943 due to the loss of imports from Burma, damage from the cyclone, and brown-spot infestation. However, FEE analyses do not consider shortage the main factor, while FAD-oriented scholars such Peter Bowbrick hold that a sharp drop in the food supply was the pivotal determining factor. S.Y. Padmanabhan and later Mark Tauger, in particular, argue that the impact of brown-spot disease was vastly underestimated, both during the famine and in later analyses.The signs of crop infestation by the fungus are subtle; given the social and administrative conditions at the time, local officials would very likely have overlooked them.

Academic consensus generally follows the FEE account, describing the Bengal famine of 1943 as an "entitlements famine" as formulated by Amartya Sen. On this view, the prelude to the famine was generalised war-time inflation, and the problem was exacerbated by prioritised distribution and abortive attempts at price control, but the death blow was devastating leaps in the inflation rate due to heavy speculative buying and panic-driven hoarding.This in turn caused a fatal decline in the real wages of landless agricultural workers, transforming what should have been a local shortage into a horrific famine.

More recent analyses often stress on political factors. Discussions of the government's role split into two broad camps: those which suggest that the government unwittingly caused or was unable to respond to the crisis,and those which assert that the government wilfully caused or ignored the plight of starving Indians. The former see the problem as a series of avoidable war-time policy failures and "panicky responses" from a government that was spectacularly inept, overwhelmedand in disarray; the latter as a conscious miscarriage of justice by the "ruling colonial elite" who abandoned the poor of Bengal.

British Prime Minister Winston Churchill in 1943

Sen, for example, does not deny that British mis-government contributed to the crisis, but sees the policy failure as a complete misunderstanding of the cause of the famine. This misunderstanding led to a wholly misguided emphasis on measuring non-existent food shortages rather than addressing the very real and devastating inflation-driven imbalances in exchange entitlements. In stark contrast, though Cormac Ó Gráda notes that the exchange entitlements view of this famine is generally accepted,he lends greater weight to the importance of a crop shortfall than does Sen, and goes on to largely reject Sen's emphasis on hoarding and speculation. He does not stop there, however, but emphasises a "lack of political will" and the pressure of wartime priorities that drove the British government and the provincial government of Bengal to make fateful decisions: the "denial policies", the use of heavy shipping for war supplies rather than food, the refusal to officially declare a state of famine, and the Balkanisation of grain markets through inter-provincial trade barriers.On this view, these policies were designed to serve British military goals at the expense of Indian interests,reflecting the War Cabinet›s willingness to «supply the Army›s needs and let the Indian people starve if necessary».Far from being accidental, these dislocations were fully recognised beforehand as fatal for identifiable Indian groups whose economic activities did not directly, actively, or adequately advance British military goals.The policies may have met their intended wartime goals, but only at the cost of large-scale dislocations in the domestic economy. The British government, this argument maintains, thus bears moral responsibility for the rural deaths.Auriol Law-Smith›s discussion of contributing causes of the famine also lays blame on the British Government of India, primarily emphasising Viceroy Linlithgow›s lack of political will to «infringe provincial autonomy» by using his authority to remove interprovincial barriers, which would have ensured the free movement of life-saving grain.

A related argument, present since the days of the famine but expressed at length by Madhusree Mukerjee, accuses key figures in the British government (particularly Prime Minister Winston Churchill) of genuine antipathy toward Indians and Indian independence – an antipathy arising mainly from a desire to protect imperialist privilege, but tinged also with racist undertones. This is attributed to British anger over widespread Bengali nationalist sentiment and the perceived treachery of the violent Quit India uprising.

For its part, the report of the Famine Commission – its members appointed in 1944 by the British Government of India and chaired by Sir John Woodhead, a former Indian Civil Service (ICS) official in Bengal – absolved the British government from all major blame. It acknowledge some failures in its price controls and transportation efforts and laid additional responsibility at the feet of unavoidable fate, but reserved its broadest and most forceful finger-pointing for local politicians in the (largely Muslim)provincial Government of Bengal: As it stated, "...after considering all the circumstances, we cannot avoid the conclusion that it lay in the power of the Government of Bengal, by bold, resolute and well-conceived measures at the right time to have largely prevented the tragedy of the famine as it actually took place". For example, the position of the Famine Inquiry Commission with respect to charges that prioritised distribution aggravated the famine is that the Government of Bengal's lack of control over supplies was the more serious matter. Some sources allege that the Famine Commission deliberately declined to blame the UK or was even designed to do so;however, Bowbrick defends the report's overall accuracy, stating it was undertaken without any preconceptions and twice describing it as excellent. Meanwhile, he repeatedly and rather forcefully favors its analyses over Sen's.British accusations that Indian officials were responsible began as early as 1943, as an editorial in *The Statesman* on 5 October noted disapprovingly.

Paul Greenough stands somewhat apart from other analysts by emphasising a pattern of victimization. In his account, Bengal was at base susceptible to famine because of population pressures and market inefficiencies, and these were exacerbated by a dire combination of war, political strife, and natural causes. Above all else, however, direct blame should be laid on a series of government interventions that disrupted the wholesale rice market. Once the crisis began, morbidity rates were driven by a series of cultural decisions, as dependents were abandoned by their providers at every level of society: male heads of peasant households abandoned weaker family members; landholders abandoned the various forms of patronage that (according to Greenough) had traditionally been maintained, and the government abandoned the rural poor. These abandoned groups had been socially and politically selected for death.

A final line of blaming holds that major industrialists either caused or at least significantly heightened the famine through speculation, war profiteering, hoarding, and corruption – "unscrupulous, heartless grain traders forcing up prices based on false rumors".Working from an assumption that the Bengal famine claimed 1.5 million lives, the Famine Inquiry Commission made a "gruesome calculation" that "nearly a thousand rupees [£88 in 1944; equivalent to £3,807or $1,278 in 2018] of profits were accrued per death".As the Famine Inquiry Commission put it, "a large part of the community lived in plenty while others starved ... corruption was widespread throughout the province and in many classes of society".

The Hungryalist Movement: Raising the voices through poetry

Post independence, Bengal was caught between a turbulent and chaotic phase. The gradual inflow of the displaced people was taking a major toll on the social fabric of Bengal, with a rise in poverty, hunger and death. The milieu was rearing a young brigade among the middle-class Bengalis who went all out against the government not with their weapons, but with their blunt blow of pen and paper mingling into poetry. Thus gave rise to avant-garde Hungryalist movement in Bengali.

The Hungryalist or the hungry generation movement was a literary movement in Bengali that was launched in 1961, by a group of young Bengali poets. It was spearheaded by the famous Hungryalist quartet — Malay Roychoudhury, Samir Roychoudhury, Shakti Chattopadhyay and Debi Roy. They had coined Hungryalism from the word 'Hungry' used by Geoffrey Chaucer in his poetic line "in the sowre hungry tyme". The central theme of the movement was Oswald Spengler's idea of History, that an ailing culture feeds on cultural elements brought from outside. These writers felt that Bengali culture had reached its zenith and was now living on alien food.

Impact in Bengal's literature

The movement shook the roots of the Bengali literary and cultural establishment in India. The poets significantly brought about a change in the language and vocabulary used in Indian literature. The Hungryalists wanted to disturb the reader's mind that was filled with preconceived colonial ideas. That's why it became one of most controversial movement during its time and faced huge criticism from the upper class of the society and government. The movement was joined by other young poets like Utpal Kumar Basu, BinoyMajumdar, Sandipan Chattopadhyay, Basudeb Dasgupta, Falguni Roy, TridibMitra and many more. Their poetry spoke the displaced people and also contained huge resentment towards the government as well as profanity.

The strangulation of Government

On September 2, 1964, 11 of the Hungry poets were charged with arrest warrants. The charges included obscenity in literature and subversive conspiracy against the state. The court case went on for years, which drew attention worldwide. Poets like Octavio Paz, Ernesto Cardenal and Beat poets like Allen Ginsberg visited Malay Roychoudhury. The Hungryalist movement also influenced Hindi, Marathi, Assamese, Telugu & Urdu literature. The government's decision to crack a whip on the poets slowly saw them leaving the movement out of fear. Later the ultra-leftist student movement by the Naxalites also brought about the end of the Hungry generation. It is still remembered as one of the greatest literary movements in the world.

Some hidden facts about The Hungryalist Movement

MaitreyeeBhattacharjee Chowdhury's ***The Hungryalists-The Poets Who Sparked a Revolution*** is the never-before-told true story of the Hungry Generation (or 'the Hungryalists')-a group of barnstorming, anti-establishment poets, writers and artists in Bengal in the 1960s. Braving social boycott, ridicule and arrests, the Hungryalists changed the literary landscape of Bengal (and many South Asian countries) forever. Along the way, they also influenced iconic poets, such as Allen Ginsberg, who struck up a lifelong friendship with the Hungryalists.

Read on for ten things you didn't know about the Hungryalist movement in India:

The Uprooted Generation that Fired a Literary Revolution

"The feeling of being uprooted was everywhere. Political leaders decided that the second phase of the five-year planning needed to see the growth of heavy industries. The land required for such industries necessitated the evacuation of farmers. Devoid of their ancestral land and in the absence of a proper rehabilitation plan, those evicted wandered aimlessly around the cities—refugees by another name....In a few years, Naxalbari would become a reality, but not yet. Like an infant Kali with bohemian fantasies, Calcutta and its literature sprouted a new tongue—that of the Hungry Generation. Malay, like Samir and many others, found himself at the helm of this madness, and poetry seemed to lick his body and soul in strange colours."

Allen Ginsberg's Meeting with the Hungryalists

"Samir had recalled to Malay: He approached our table, where Sunil, Shakti, Utpal and I sat, with no hesitation whatsoever. There was no awkwardness in talking to people he hadn't ever met. None of us had seen such sahibs before, with torn clothes, cheap rubber chappals and a jhola. We were quite curious. At that time, we were not aware of how well known a poet he was back in the US. But I remember his eyes—they were kind and curious. He sat there with us, braving the most suspicious of an entire cadre of wary and sceptical Bengalis, shorn of all their niceties—they were the fiercest lot of Bengali poets—but, somehow, he had managed to disarm us all. He made us listen to him and tried to genuinely learn from us whatever it was that he'd wanted to learn, or thought we had to offer."

The Hungryalists' Forays into Anti-Establishment Publishing

"The need to present poetry differently from how it had been presented and perceived so far resulted in a number of experiments. Subimal decided to edit and publish a periodical called Pratidwandi. The name signified revolutionaries. The Hungryalists were enthusiastic about writing for it. It would be the first time a collective like this was being published, showcasing anti-establishment writing. Subimal wanted it to be released at Khalasitola on PoilaBoishakh (the Bengali New Year). The cover was a woodcut print, designed by poet Sandipan, who was a firebrand Hungryalist himself. He had done it very cleverly; it looked like a vase but was actually a boy and a girl in an intimate pose."

The Clash Between the Bangla Literary Establishment and the Hungryalists

"By this time, it was a bit obvious that the Hungryalists had been attracting some negative vibes, especially from writers and poets who had already established themselves in the Bangla literary scene. The poetry that the Hungryalists advocated was looked down upon since it boldly and openly spoke of sensuality. In this regard, the differences, especially between the Krittibhash poets and the Hungryalists, soon became apparent. It was obvious that the subjects on which the Hungry poets had begun writing shocked a lot of people. Poetry, which was considered a more refined form of expression till now, was being experimented with and being written in street dialects and on tabooed subjects."

The Varying Attitudes of the Beats and the Hungryalists Towards Religion in India

"Many of the Hungryalists, with their sharp knowledge of Hindu scriptures, had been challenging temple elders on the different rituals and modes of worship. This came as a shock to many, in a country where religion was very much a part of everyday life—a matter of pride and culture even. On the other hand, Ginsberg was evidently quite taken with religion in India and sought out sadhus and holy men wherever he went in the country. While this might have been because he was in search of a guru, he seemed to be fascinated, in equal measure, by the sheer variety that religion opened for him in India—from Kali worship to Buddhism. But like the Beats, the Hungryalists came together in denouncing the politics of war, which merged with their larger world view."

The Hungryalists Turned the 'Starving Artist' Stereotype into a Harsh Reality

"Some of the poets used the public loos in railway stations because they had no homes. Writing poems during the day and finding empty offices of Marwari businessmen to sleep in at Burrabazar became a fine juggle that they soon mastered. Marwari businessmen usually left mattresses in the verandas of their offices for their outstation customers. The poets cleverly made use of these whenever they found them lying empty. On most days, they would pool in money, put all their resources together and eat in small, shady restaurants in Shyambazar or in College Street. The job of being a poet had become a severe adventure."

Malay Roy Chaudhury's Protests Against his Own Poetry-Within his Own Poetry

"As Malay's poems were devoid of lyricism, when read out, they sounded like staccato sentences. The language was colloquial. But during the process of publishing, he realized that there were certain poems he was no longer comfortable with as they were clearly influenced by the poets of the thirties, forties and fifties. He was seized by the desire to make it apparent that he no longer considered them to be adequate. So he stamped 'cancelled' on ten of his poems in the collection. When the book was published, other poets considered this decision to be a sacrilege. They thought the act was defiant and done intentionally to generate shock. The fact that a poet might want to exercise his choice did not convince them."

The Hungryalists and the Beats Revolutionized the Very Language of Poetry

"Like other Hungryalists, and even the Beats, Malay knew for certain that obscenity was a construct. His poem proved, beyond a point, what the Hungryalists had been trying to say for a while— that language and its refinement wasn't the aim of Hungryalist writings. Their emotions were raw, many a time unrefined, and it was all right for the world to see that. In that the poem itself was as much a cry against the conspirators of language as it was the longing for a woman."

The Arrests of the Leaders of the Hungryalists for 'Perversion and Obscenity'

"In a sudden and swift move, the establishment had accused the movement of not only attempting to promote perversion and obscenity in society but also of hurting sentiments that warranted up to two years of rigorous imprisonment. Arrest warrants were issued against Malay, Samir, Debi, Pradip, Subhash, Subimal, Utpal, Saileshwar, Ramananda, Basudeb and Shubho. Samir's strained voice added, 'You are the main accused, Malay, because of the obscenity they found in your poem "Prachanda Baidyutik Chhutar"."

The Outpouring of International Support Protesting the Arrests of the Hungryalists

"But before they left for Kathmandu, a letter arrived from New York. It was from a poet friend, Carol Berge, who had written to tell Malay about a poetry reading that had been arranged in honour of the Hungryalists at St Marks Church in New York. Reading the letter was an emotional moment for Malay as he basked in the warmth and support of poets who had never met him but had found a common link through poetry. This was a streak of hope in the midst of all that was going on, and it reinforced the larger connection with poetry despite differences in geography and individuality of the poets."

Root causes of poverty and amalgamation of hunger and poverty with special reference to the book "Poor Economics" by Abhijit V. Banerjee and Ester Duflo

Reconciliation

Global poverty is to be viewed as a combined outcome of many discrete problems,and not as a single nebulous problem. We should focus on addressing these individual problems. Many economists debate the extent to which poverty traps may exist. The authors define a poverty trap as: "There will be a poverty trap whenever the scope for growing income or wealth at a very fast rate is limited for those who have too little to invest, but expands dramatically for those who can invest a bit more. On the other hand, if the potential for fast growth is high among the poor, and then tapers off as one gets richer, there is no poverty trap." Anti-poverty approaches should be based on research and data such as randomized controlled trials. Lazy thinking (being constrained by ideology, ignorance and inertia) is the enemy of progress.

Numerous Hungry people with limited resources

While lack of food can still be a problem in certain aspects, but in modern society it is not a considered to be a contributor to the general persistence of poverty. A major exception is malnutrition in utero and during childhood, which correlates with lifetime education attained and future wages earned. In these cases anti-malnutrition programs like iodine supplementation for pregnant mothers and deworming pills for youth have demonstrated substantial value. Malnutrition in adults living in extreme poverty has a much less direct effect on wages, but one exception is anemia (extreme iron deficiency). The poor, like everyone, has a basic human tendency of being skeptical to others telling them what to eat. Also like everyone, the poor prioritize spending that make life less boring (TV, cell phones, cultural events, tastier but nutritionally low-value food, etc). Finally, nutrition education is challenging because the personal experience feedback loop takes a long time — investment in childhood nutrition doesn't financially pay off until years later in adult life.

The nexus of malnutrition and poverty

The extreme poor are affected by health poverty traps. Children who are sick go to school less and earn less over the course of their lifetimes. While the poor spend considerable time and money on healthcare, it is most often on expensive cures rather than cheap preventative care. In addition to the lack of education about healthcare, has a negative impact of creating a tendency to over medicate and thus propagate drug-resistant pathogens. People view doctors who prescribe medication or give a shot as effective doctors. There are a number of reasons why preventative care is under-utilized: Facilities may be closed, biological concepts of vaccination and preventative treatments require education, and people tend to procrastinate (time inconsistency). Research suggests that substantial gains in outcomes from a health program can be made simply by making the "right" decision the default decision (for example, only requiring additional action if participants wish to opt out). Also it can be worth it to provide additional incentives to overcome procrastination (for example free food as a reward for getting the next shot in a vaccination sequence).

Education to all: A Myth

The lack of schools cannot be defined as the cause of low educational attainment. Enrollment rates are improving, but current challenges include poor quality of education and students dropping out of school. In poor countries, it is observed that the education outcomes from private schools are significantly better, as the absenteeism among the teachers is much lower. One-on-one tutoring and summer camps have also demonstrated significant education quality gains. Rich countries strategies to keep students in school by mandatorily making the students attend school up to a specific age. Developing countries don't have the capacity to enforce this, thus other interventions are required. Unconditional cash transfers have been shown to increase enrollment rates (interestingly at the same rate as conditional cash transfers). Also, explaining to parents that income returns increase proportionally with each additional year of education — for students from backgrounds similar to their own — has been useful in preventing parents from pooling resources behind one child while their remaining children discontinue further education. As with educational functioning everywhere, parent, teacher, and student perceptions of what the student is capable of achieving strongly correlate with the learning outcomes.

Low education, least awareness, higher population

Family planning is considered a central component to the development strategy by most of the Governments of poor countries While there are clear environmental downsides to large families, there is no research consensus that children in large families have a worse quality of life than children in smaller families. However, parents living in poverty may be incentivized to have a large family as their children's future adult wages represent an important lifeline once the parents are no longer able to work. Husbands, family and the culture of a community have a pivotal role to play on a woman's ability to procure contraceptives. Families are defined by specific social roles, which can be slow to evolve and are resistant to change. The authors posit that instead of attempting to address high fertility by changing family dynamics, high fertility rates may be more easily addressed by the creation of stronger social safety nets, which could reduce incentive for parents to have large families. Strangely, the lower community fertility rate is not correlated with the availability of contraceptives. Contraceptives and family planning programs have been shown, however, to help teenagers delay pregnancy. Teenage pregnancy rates are very high in developing countries and (in addition to health risks) directly impact their ability to stay in school, get more education and earn more over the course of their lives. Unfortunately, teenagers may have the greatest unmet need for for family planning and contraceptives because many countries prohibit teenage access to family planning without consent of their parents. In order to control and avoid HIV, there have been numerous programs promoting abstinence as strategy but it failed to have any significant effect on teenage pregnancy rates, possibly because many poor teens get pregnant (and contract HIV) from older men — a decision that is likely economic need-based. Programs that educate teenagers about getting HIV from older men and programs that provide financial assistance to stay in school have shown reduced pregnancy rates. The authors conclude that teenagers, like adults, make conscious decisions about when to get pregnant. A pregnancy from an older man may provide financial stability down the road. If there are opportunities to stay in school, teenagers are less incentivized to get pregnant in pursuit of maximizing their economic potential.

Financial Risk Managers

The impact of Regional economic disruptions on the extreme poor is not up to the same extent that they impact the general economy because the extreme poor are already living on the absolute margin. However in consideration to their level of exposure, local economic disruptions such as a family's unexpected health emergency can set the poor back substantially — especially if their ability to earn more wealth depends on current wealth (having daily credit available to buy and sell goods, for example). Despite of the poor relying on wildly fluctuating sources of income like non-irrigated farms and day labor work, they limit exposure to any one risk by diversifying their income portfolio (agricultural earnings, seeking work in urban areas, micro-enterprise, tenant farming, large family, etc). However, this has the effect of preventing occupational specialization (and skill-building). The poor's constant financial risk has resultant psychological effect i.e. high stress, which has been shown to impair focus and reduce productivity. In an order to reduce the risk, micro-finance institutions have experimented with offering health insurance, but the extreme poor may not be willing to pay the required premiums at the cost of tying up cash. Subsidized insurance from governments could reduce risk exposure while also dropping premiums to a level that the poor are willing (and able) to pay.

The cobweb of credit and loan

The poor levied with high interest rates, with regard to determining the creditworthiness of the poor is considered time-consuming (and expensive), creditors have a lot of difficulty recovering capital after default, nevertheless net returns per account are still relatively low despite of extremely high interest rates. It is also for these reasons that the poor rarely have access to banks and instead most often have to rely on money lenders, their community, or microcredit for a loan. Despite of microcredit offering better interest rates than money lenders, but its rigid borrower repayment structure means less flexibility with future cash. Microcredit has reached hundreds of millions, and while its effect has not been revolutionarily transformative, it has demonstrated positive gains. Enterprises that receive microcredit loans rarely grow beyond one employee. There is currently a funding gap for medium-scale enterprises that are still too small for banks but too large for money lenders.

Savings

The poor are aware of the challenge of keeping cash on-hand, and also the importance of saving money. Having difficulty saving money isn't necessarily a symptom of frivolous spending — rather, for those living on the absolute margins, by definition there are always necessities that they lack. The poor use non-traditional ways to save including partially building a house brick-by-brick, savings clubs, and storing extra money with money lenders. While the poor generally aren't considered profitable for banks (the per-account administrative costs are too high for the low deposit amount), research reveals that the poor are able to save more than they otherwise would when they have access to affordable savings accounts. However, establishing savings for the poor isn't just a technical challenge of creating affordable savings products. For the extreme poor and poor everywhere, it is hard to stay motivated to save when it seems that improving one's life will be impossible. Thus, for many who may not have encountered examples of how long-term savings can change one's life, it can make more sense to them to prioritize purchases that will improve life today. The perception that opportunity exists is paramount to success in savings.

The middle classes' opportunities

In spite of the marginal returns (return per dollar of investment) of microenterprise investments having the ability to exceed 50%, yet the overall returns are quite small as the businesses themselves are very small. In many cases these businesses do not last more than five years and rarely grow beyond one employee. Additionally, this means that the necessary funds to purchase a technology that might significantly expand production or business can be acquired by decades of steady reinvestment. Knowing this, it can be almost impossible for the extreme poor to believe they'll be able to change their circumstances and therefore attempt to venture down the path of growth. Many of the extreme poor encounter the path to prosperity through their children acquiring a salaried government job — not via a small business. The long term stability these jobs afford offers the mental bandwidth and economic opportunity for long term planning and growth towards middle class. Industrial growth has been observed to disproportionately benefit the wages of the poor. Urban areas offer better job and education opportunities, but general reliance on the social safety net of one's community means a move away can be difficult and risky. Better urban planning policy including public housing and improved safety nets can incentivize such moves and subsequent growth in wages of the poor.

Summery of the discussion

It's often assumed that radical political reforms are required to solve large problems. However, replacing a political institution can be very difficult or impossible. Instead, the authors believe that significant policy improvements can be made in the margins of existing institutions, which may in turn lead to better overall institutions, especially so if the new policies result in small wins and increase trust in existing governments. Greater trust in a government allows it to focus less on voters' immediate approval and more on long-term gains. Small, non-controversial modifications can have big effects. Government's role is to solve the problems that the market is incapable to solve. While there are many cases of inefficiency, inadequacy, or corruption, even mediocre policies (such as imperfect, rigged elections or no elections at all) can catalyse significant progress. Still, whenever possible lazy thinking in policy design (being constrained by ignorance, inertia or ideology) should be avoided. In conclusion, the authors report five key findings. "The poor often lack critical pieces of information and believe things that are not true... The poor bear too much responsibility for their own lives (for the wealthy, the right decisions are often automatic)... There are good reasons that some markets are missing for the poor, or that they face unfavorable prices in them... Poor countries are not doomed to failure because they are poor, or because they have had an unfortunate history... Expectations about what people are able or unable to do all too often end up turning into self-fulfilling prophecies."

3 CHAPTER

CITATION ACROSS THE GLOBE

Review of Literature

Table 2.1: Hunger : A Bleak Scenario

Title	Source	Author/authors	Impact point
Altered baseline brain activities before food intake in obese men: a resting state fMRI study.	Neuroscience Letters. 2015; 584: 156 161	Zhang Bin; Tian DeRun; Yu ChunShui; Zhang Jing; Tian Xiao; Deneen, K M von; Zang YuFeng; Walter, M; Liu, Y J	Obesity as a chronic disease has become a global epidemic. However, why obese individuals eat more still remains unclear. Recent functional neuroimaging studies have found abnormal brain activations in obese people. In the present study, we used resting state functional MRI to observe spontaneous blood-oxygen-level dependent (BOLD) signal fluctuations during both hunger and satiety states in 20 lean and 20 obese men.
Cross-cultural validity of the Intuitive Eating Scale-2. Psychometric evaluation in a sample of the general French population.	Appetite . 2015; 84: 34 42	Camilleri, G M; Mejean, C; Bellisle, F; Andreeva, V A; Sautron, V; Hercberg, S; Peneau, S	Intuitive eating is an adaptive dietary behavior that emphasizes eating in response to physiological hunger and satiety cues. The Intuitive Eating Scale-2 (IES-2) measures such attitudes and behaviors. The aim of the present study was to adapt the IES-2 to the French context and to test its psychometric properties in 335 women and 297 men participating in the NutriNet-Sante study. We evaluated the construct validity of the IES-2 by testing hypotheses with regard to its factor structure, relationships with scores of the revised 21-item Three Factor Eating Questionnaire and the Center for Epidemiologic Studies Depression scale, and differences between "a priori" relevant subgroups.

Politicizing hospitality: the emergency food assistance landscape in Thessaloniki.	Hospitality and Society. 2014; 4(3): 249 274	Kravva, V	The ethnographic focus of this article is on municipal strategies for the 'immediate management of poverty' (gia tin amesi antimetopisi tis ftohias) in the city of Thessaloniki, in northern Greece. These involve several food assistance practices: two daily soup kitchens, two social markets and a food bank. Diefthinsi Koinonikis Politikis/The Department of Social Policy also called Pronoia/Welfare sustains a dormitory for homeless people, clothing banks and provides health services to poor, unemployed people, who have no health insurance (aporoi, anergoi, anasfalistoi).
Assessing progress made toward shared agricultural transformation objectives in Mozambique.	IFPRI Discussion Papers. 2014; (1370): vii + 29 pp	Benson, T; Mogues, T; Woldeyohannes, S	What has been the recent performance of the agricultural sector in Mozambique and the progress made thus far toward achieving the objectives established under the Comprehensive Africa Agriculture Development Programme (CAADP) initiative for Mozambique that began in late-2011? CAADP promotes agriculture-driven economic growth to reduce hunger, malnutrition, and poverty across Africa.
"Too fat to be an orphan": the moral semiotics of food aid in Botswana.	Cultural Anthropology. 2014; 29(4): 626 647	Dahl, B	The iconography of the African AIDS orphan, captured in National Geographic-style images of half-starved toddlers with distended bellies, inspires humanitarian aid for the continent. In Botswana, stereotypes underlying both foreign-funded and governmental programs for orphaned children - which imply that orphans are underfed and underloved - initially resonated with Tswana people's anxieties that neglect by overburdened kin results in parentless children going hungry.
Legally entitled to a full stomach.	D + C, Development and Cooperation. 2012; 39(3): 117 119	Baski, B	This paper looks at the provisions of a Right to Food Bill that is about to be passed in India, also discussing the controversial issues and ethical questions raised by this bill. The paper asks whether this bill will be the end of malnutrition and starvation in India.
Hunger in the Horn of Africa.	D + C, Developme nt and Cooperation. 2012; 39(3): 110 111	Oneko, J	This paper looks at the recent and ongoing food emergency in East Africa where, despite early warnings, the crisis hit full force before the world community reacted. It is indicated that the enormity of the disaster reveals the need for long-term interventions. The contribution of drought and poor governance to the disaster is discussed.

Green renaissance, not revolution.	D + C, Development and Cooperation. 2012; 39(3): 107 109	Bass, H H	As Africa's food problems are becoming more acute, the calls for a green revolution are increasing. It is argued that such ideas are often based on misconceptions - that much African land is not being utilized; that an increase in food production will, by itself, overcome hunger; and that leasing land to investors offers net benefits. It is suggested that a green renaissance with a focus on food security would make more sense.
Changing role of Anganwadi workers, a study conducted in Vadodara district.	Healthline, Journal of Indian Association of Preventive and Social Medicine. 2012; 3(1): 41 44	Gaurav Desai; Niraj Pandit; Diwakar Sharma	The Integrated Child Development Scheme (ICDS), was initiated nearly 35 years ago, in October 1975, in response to the evident problems of persistent hunger and malnutrition especially among children. The role of Anganwadi Worker (AWW) MahilaMandal and Village Health Committee. To understand this changing roles, the present study was initiated with the objective to study the changing role of Anganwadi worker in present scenario. METHOD- The study was a cross sectional study.
Nutrition information in crisis situations.	Nutrition Information in Crisis Situations. 2012; (24): 70 pp		This report summarizes outcome indicators related to population nutrition of emergency-affected groups (i.e., refugees, internally displaced population, resident population) in crisis-stricken countries like Djibouti, Ethiopia, Kenya, Somalia, Chad, Democratic Republic of Congo, Rwanda, Afghanistan, Pakistan, Philippines, and Guatemala. Details of the specific humanitarian situation in each country are discussed, focusing on public nutrition data and mortality rates.
Famine foods of Brazil's seasonal dry forests: ethnobotanical and nutritional aspects.	Economic Botany. 2012; 66(1): 22 34	Nascimento, V T do; Vasconcelos, M A da S; Maciel, M I S; Albuquerque, U P	Famine foods are used by people in times of food scarcity. Although the northeast of Brazil regularly faces periods of drought that cause a shortage of traditional foods, the use of famine foods is insufficiently recorded. The purpose of this study was to record the knowledge of famine food in two rural Caatinga communities, Carao and Cachoeira, in the states of Pernambuco and Paraiba, respectively, and to determine the chemical composition of the major famine foods utilized.
Development, acceptability, and nutritional characteristics of a low-cost, shelf-stable supplementary food product for vulnerable groups in Kenya.	Food and Nutrition Bulletin. 2012; 33(1): 43 52	Kunyanga, C; Imungi, J; Okoth, M; Vadivel, V; Biesalski, H K	Background. Food-based approaches have been advocated as the best strategies to curb hunger and malnutrition in developing countries. The use of low-cost, locally available, nutritious foods in the development of supplementary foods has been recommended.

A fine-grained analysis of eating behavior in women with bulimia nervosa.	International Journal of Eating Disorders. 2012; 45(3): 400 406	Mitchell, J E; Karr, T M; Peat, C; Wonderlich, S; Crosby, R D; Engel, S; Simonich, H	In the current study,they were interested in developing a typology of eating in patients with bulimia nervosa (BN) based on the size of the eating episode, whether the episode was followed by self-induced vomiting, and the degree of loss of control (LOC) self-reported by participants.
Climate change and floods in Yemen: impacts on food security and options for adaptation.	IFPRI Discussion Papers. 2011; (1139): viii + 48 pp	Wiebelt, M; Breisinger, C; Ecker, O; Al Riffai, P; Robertson, R; Thiele, R	This paper uses both a global and local perspective to assess the impacts of climate change on the Yemeni economy, agriculture, and household income and food security. The major impact channels of climate change are through changing world food prices as a result of global food scarcities, long-term local yield changes as a result of temperature and rainfall variations, and damages and losses of cropland, fruit trees, livestock, and infrastructure as a result of natural disasters such as recurrent storms and floods.
Making agriculture pro-nutrition opportunities in Tanzania.	IFPRI Discussion Papers. 2011; (1124): viii + 60 pp	Ecker, O; Mabiso, A; Kennedy, A; Diao, X S	Extreme poverty and chronic malnutrition are widespread in rural Tanzania, where smallholder agriculture dominates the economy.
Was the global food crisis really a crisis? Simulations versus self-reporting.	IFPRI Discussion Papers. 2011; (1087): vi + 65 pp	Headey, D	Estimates by the U.N. Food and Agriculture Organization (FAO), the U.S. Department of Agriculture (USDA), and the World Bank concerning the welfare impact of the 2007/08 global food crisis conclude that between 75 million and 160 million people were thrown into hunger or poverty. However, these simulation-based approaches suffer from inherent deficiencies as well as insufficient coverage of the largest developing countries, especially China and India. This paper therefore assesses the usefulness of an alternative to simulation-based approaches, self-reported food insecurity data from the Gallup World Poll (GWP), a survey conducted before, during, and after the 2007/08 crisis.
Are our babies hungry? Food insecurity among infants in urban clinics.	Clinical Pediatrics. 2012; 51(3): 238 243	Burkhardt, M C; Beck, A F; Kahn, R S; Klein, M D	Background. Food insecurity (FI) is common, but studies in families with infants are rare. Objectives. To determine prevalence of FI, assess the effect public benefits have on FI, assess strategies to stretch nutritional resources (eg, using generic formula), and investigate FI's relationship to anthropometric measurements. Methods. A cross-sectional survey was completed. FI was classified using the US Department of Agriculture's 6-item indicator set.

The Social Ecology of Migration in Agriculture: The Micro Level Reality	ACTA Scientific Agriculture.	Anannya Chakraborty, S K Acharya, Biswajit Goswami, Chethana. K	Rural-urban migration implies geo-spatial movement of population from the countryside/ rural areas into the cities, often the metropolitan cities of a country. Nearly 29% of rural male migrants had migrated due to employment related reasons and 91% of the rural female migrants due to marriage. The majority of migrants move because of economic reasons. The inadequate irrigation facilities, lack of employment opportunities in rural non-household manufacturing activities and decline in the average size of operational holdings as the major 'push' factors; and increase in rural literacy and expansion of non-household manufacturing and construction activities in urban areas as the leading 'pull' factor in rural-urban migration. This factors implies socio-cultural, economical and ecological dynamics of a performing both rural and urban areas. neighbouring rural areas to meet non-local demand. The present study is conducted in Devarahalli, Chikkanahalli, under Chikkanahalli gram panchayat in Tumkur district of Karnataka state. 60 peoples were selected randomly and the data were selected through pilot study, structured interview and focused group interview. 19 independent variables were studied against 8 dependent variables. The study reveals that the availability of assets and better technology channels in rural areas would play a crucial role in stemming unnecessary ruralurban migration.
Impoverishment and Vulnerability of The Farmers: A Harsh Reality	Indian Research Journal of Extension Education	Anannya Chakraborty, S K Acharya.	A survey of FAO reveals that nearly 75 per cent of food insecure and vulnerable people are directly or indirectly related to agriculture. Another report of FAO has revealed that about 78 per cent of the farmers are ready to quit agriculture. This is really alarming as well as distressing while we have made a call for complete digitization to create ICT driven modern India. Poverty can be measured in terms of income. In other way the other forms of poverty are educational starvation, cultural deprivation and social depletion. Silence can be perceived in terms of inability of a person to raise voices against discrimination, both social and economic atrocities as well as a decision to go silent whenever it needs to utter voices .On this preamble the present study was conducted in Beraberi GP under Nadia district of WB. 150 respondents were selected purposively and they were interviewed thoroughly with a structure interview schedule. In this study 21 independent variables were selected against a dependent variable poverty(Y). Here in this study it has been revealed that the most important aspects of impoverishment of the farming community are cropping intensity, communication variables, livestock possession and stress perception.

Agricultural export subsidies and domestic support reform under the WTO system: what does it mean for welfare in West Africa?	World Economy. 2011; 34(12): 2044 2062	Nuetah, J A; Zuo Ting; Xian Xin	This paper analyses the potential welfare impacts West Africa tends to encounter from liberalising agricultural export subsidies and domestic support in developed and developing countries according to the reform proposals of the Doha Development Agenda. We base our analysis on the July framework scenario which considers a tiered formula that adopts a Harbinson Approach plus sensitive and special product provisions.
Impact of climate change on food security.	International Journal of Agriculture Environment and Biotechnology. 2011; 4(2): 125 127	Minaxi, R P; Acharya, K O; Santosh Nawale	The impact of climate change will be particularly significant on smallholder and subsistence agriculture. Livelihood systems predominantly in low latitudes will be affected by major changes due to climate change. The farming system will be directly affected by changing weather patterns, sea level rise, and the increase in frequency and intensity of extreme events. The productivity of livestock and fisheries systems will also be affected, as well as potential income from collecting activities in forests (Christoph Bals et al, 2008).
Global hunger phenomenon and seeking for solution.	GIDA Journal of Food. 2012; 37(1): 39 45	Eksi, A; Isci, A	Food insecurity is the biggest issue of our century. In fact, one out of every seven people is undernourished. Worse still, the number of starving people is increasing although globally enough food is produced to feed the entire world. The major cause of hunger is regional differences in food production and consumption. Countries with a high prevalence of undernourishment have lower income levels. It would be difficult to correlate the number of starving people to the fluctuations in food prices.
Special Issue: Frugality and food in contemporary and historical perspective.	Food and Foodways. 2012; 20(1): 1 92		This special issue contains articles exploring the relationships between frugality and food in the past and present context. The topics discussed include: the impact of over-indebtedness on contemporary foodways in low-income families in the UK; frugality as a way of life for French rural workers during the second half of the nineteenth century; hunger, collective memory, and the meaning of soul food in Gee's Bend, Alabama, USA, during a time of struggle for survival; and food frugality in relation to both the complexities of climate change and conceptions of food as a right.

Hunger and Voice: Marginalization and Impoverishment towards a Chaotic Social Ecology	Hunger and Voice: Marginalization and Impoverishment towards a Chaotic Social Ecology	Anannya Chakraborty, SK Acharya	After 72 years of independence India is still suffering from the malicious problem of and poverty. The farmers,who are providing food to our 1.31 billion population are in empty stomach and bare hands. 21.7 per cent ofit's people are below the poverty line. According to National Crime Record Bureau, in the year 2015 the total numbers of farmers who have committed suicide is 5670. A survey of FAO reveals that nearly 75 per cent of food insecure and vulnerable people are directly or indirectly related to agriculture. the present study was conducted in Beraberi GP under Nadia district of WB. 150 respondents were selected purposively and they were interviewed thoroughly with a structuredinterview schedule. In this study 21 independent variables were selected against a dependent variable poverty (Y). Here in this study it has been revealed that the most important aspects of impoverishment of the farming community are cropping intensity, communication variables, livestock possession and stress perception.
Special Section: Food security.	D + C, Development and Cooperation. 2012; 39(3): 98 119		This section, consisting of six papers and two interviews, focuses on the problem of food insecurity in Africa and other developing countries, its causes, and the measures that can be implemented to improve agricultural productivity and food security.
Food insecurity is associated with attitudes towards exclusive breastfeeding among women in urban Kenya.	Maternal and Child Nutrition. 2012; 8(2): 199 214	Webb Girard, A; Cherobon, A; Mbugua, S; Kamau Mbuthia, E; Amin, A; Sellen, D W	This study aimed to document whether food insecurity was associated with beliefs and attitudes towards exclusive breastfeeding (EBF) among urban Kenyan women. We conducted structured interviews with 75 human immunodeficiency virus (HIV)-affected and 75 HIV-status unknown, low-income women who were either pregnant or with a child <=24 months and residing in Nakuru, Kenya to generate categorical and open-ended responses on knowledge, attitudes and beliefs towards EBF and food insecurity. We facilitated six focus group discussions (FGD) with HIV-affected and HIV-status unknown mothers (n=50 women) to assess barriers and facilitators to EBF. Of 148 women with complete interview data, 77% were moderately or severely food insecure (FIS).

An assessment of the effects of the 2002 food crisis on children's health in Malawi.	Journal of African Economies. 2012; 21(1): 124 165	Hartwig, R; Grimm, M	The food crisis encountered in 2002 in Malawi was arguably one of the worst in the recent history of the country. The World Food Programme estimated that between 2.1 and 3.2 million people were threatened by starvation. Despite this assumed severity, not much research on the actual consequences of the crisis has been carried out so far. In order to fill this gap, this paper aims to identify the effects of the 2002 food crisis on the health status of the very young children exposed to it.
Social scientific reflections on hunger, appetite, satiety.	Industrie Alimentari. 2012; 51(520): 34 36	Morini, P	A sociological discourse on different aspects of diet is presented. This paper discusses hunger, appetite and satiety and how man perceives these concepts. Some physiological aspects of satiety and appetite are also tackled.
NERICA: a hope for fighting hunger and poverty in Africa.	Molecular Pl ant Breeding. 2011; 2011: No. 11	Nassirou, T Y; He YuQing	NERICA (new rice for Africa), a new promising African upland rice species, is getting into the limelight in West-Africa, it has been developed through crossing African rice species (known for resistant to disease and drought) and Asian rice species (for its high yield potential) with the assistance from Japan, UNDP and other organizations. Its varieties are being hailed as a "miracle crop" that can bring Africa its long-promised green revolution in rice that is why a powerful coalition of governments, research institutes, private seed companies and donors are leading a major effort to spread NERICA seeds to all the continent's rice fields.
Ensuring future food and nutritional security in the context of a warming world with more limited natural resources.	Acta Horticulturae. 2011; (916): 47 58	Keatinge, J D H; Easdown, W J; Hughes, J d'A; Yang, R Y; Symonds, R	Malnutrition is worst in sub-Saharan Africa and South Asia, but over the next 40 years food production in these regions will need to increase more than double to meet the needs of population growth. In both regions smallholder farmers will continue to provide the bulk of food production as climate change brings more extremes of drought and flooding, making crop production more risky.
Aquaculture, fisheries, poverty and food security.	WorldFish Center Working Paper. 2011; (2011 65): 60 pp	Allison, E H	Fisheries and aquaculture play important roles in providing food and income in many developing countries, either as a stand-alone activity or in association with crop agriculture and livestock rearing. The aim of this paper is to identify how these contributions of fisheries and aquaculture to poverty reduction and food security can be enhanced while also addressing the need for a sustainability transition in over-exploited and over-capitalized capture fisheries, and for improved environmental performance and distributive justice in a rapidly growing aquaculture sector.

D-town farm: African American resistance to food insecurity and the transformation of Detroit.	Environmental Practice. 2011; 13(4): 406 417	White, M M	This article analyzes community building and political agency through an investigation of the Detroit Black Community Food Security Network (DBCFSN). By using farming as a strategy of resistance against the structural factors that have left much of Detroit in a condition of food insecurity, DBCFSN not only meets citizens' needs for fresh produce, but also builds community by transforming the social, economic, and physical environment
Roles and capacity of duty bearers in the realization of the human right to adequate food in Uganda.	African Journal of Food, Agriculture, Nutrition and Development. 2011; 11(7): 5493 5509	Rukundo, P M; Kikafunda, J K; Oshaug, A	The right to adequate food recognised under international law provides a strong foundation for eradicating hunger and malnutrition in all nations. Uganda ratified the International Covenant on Economic, Social, and Cultural Rights (ICESCR) in 1987 and thereby committed itself to ensure the realization of the right to adequate food recognised under Article 11 of the Covenant. This study analysed the roles and capacity of duty bearers in the realization of the right to adequate food in Uganda. Structured interviews were held with purposefully selected duty bearers from 11 districts in the country between February and July 2007.
Hunger and nutritional poverty in Germany. quantitative and qualitative empirical insights.	Critical Public Health. 2011; 21(4). 417 428	Pfeiffer, S; Ritter, T; Hirseland, A	Despite increasing social inequality, hunger and nutritional poverty are not regarded as phenomena of German reallty; Germany's debate on eating patterns is largely dominated by the issue of obesity. The article challenges this view and shows by means of empirical approaches that hunger and nutritional poverty tend to be underestimated in a supposedly affluent society.
Special Issue: Food and public health.	Critical Public Health. 2011; 21(4): 385 508		This special issue contains 9 papers presented at the British Sociological Association's Food Study Group second international conference on Food, Society, and Public Health, held in London in 2010. The conference focused on food security and sustainability, an area of considerable import for public health.
Assessing food-shelves' ability to distribute healthy foods to foodshelf clients.	Journal of Hunger and Environmental Nutrition. 2011; 6(1): 10 26	Rochester, J S; Nanney, M S; Story, M	Food shelves provide a vital service of distributing food to food insecure families, but the resources food shelves have to source and distribute healthy food and the extent to which this emergency food is healthy have not been evaluated. All member food shelves of the Emergency Food shelf Network (n=58), a food bank in Minnesota, were sent a survey that assessed resources and barriers related to organizational policies, infrastructure, and access to healthy food that food shelves face.

In longitudinal data from the survey of program dynamics, 16.9% of the U.S. population was exposed to household food insecurity in a 5-year period.	Journal of Hunger and Environmental Nutrition. 2010; 5(3): 380 398	Wilde, P E; Nord, M; Zager, R E	This study provides US national-level estimates of exposure to and persistence of food insecurity in a period of 5 consecutive years. Based on longitudinal data from the Survey of Program Dynamics (SPD) (n=12,185), 16.9% of the US population was exposed to household food insecurity in at least 1 year and 1.04% experienced persistent insecurity for 5 years from 1998 to 2002.
Investigation of the hunger-obesity paradigm among shelter-based homeless women living in Minnesota.	Journal of Hunger and Environmental Nutrition. 2010; 5(3): 339 359	Richards, R; Smith, C	A 264-item survey evaluated how two shelter environments influenced homeless women's health (n=259). Measured heights and weights determined obesity prevalence. Nonparametric tests evaluated body mass index (BMI) and dietary intake data. Principle components analysis reduced the data and correlations were made to BMI and dietary intake. Almost 80% had a BMI>=25 and 67% were food insecure. Median milk intake was significantly different between shelters (1.0 [S1] vs 0.5 [S2], P<0.05). Most servings were from the fats/sugars group (median servings: 19.6 [S1]; 15.1 [S2], P=0.73). Obesity is a public health problem among homeless women, with shelter environments limiting the modification of associated risk factors.
Fighting hunger through innovation: evaluation of a food bank's social enterprise venture.	Journal of Hunger and Environmental Nutrition. 2010; 5(1): 56 69	Popielarski, J A; Cotugna, N	Innovative solutions are required to overcome increasing demands on food bank resources. A social enterprise venture in the form of an agency-run grocery store called "The Market" was developed by the Food Bank of Delaware (FBD) with the goal of bringing food to the citizens of an inner-city neighborhood. The project began with a pilot store and a feasibility study evaluated the challenges and successes. The evaluation generated recommendations for development of a permanent store and future research. The Market is one model of social enterprise in the area of nutrition and food security.
How can climate change and the development of bioenergy alter the long-term outlook for food and agriculture?	Looking ahead in world food and agriculture: perspectives to 2050. 2011; 95 155	Fischer, G	This chapter presents an integrated agro-ecological and socio-economic spatial global assessment of the interlinkages among emerging biofuel developments, food security and climate change. Its purpose is to quantify the extent to which climate change and expansion of biofuel production may alter the long-term outlook for food, agriculture and resource availability, based on work by FAO and using the International Institute for Applied Systems Analysis (IIASA) modelling framework and models to analyse the world food and agriculture system spatially and to evaluate the impacts and implications of agricultural policies. The modelling framework has recently been extended and adapted to incorporate biofuel development issues.

GM foods and their impact on food security.	Higiene Alimentar. 2011; 25(200/201): 40 44	Nunes, S M; Carneiro, A A	Genetically modified (GM) food is a creation of genetic engineering and biotechnology with the aim of improving the food production, nutritional quality, shelf life, etc. Several impacts on food safety were reported on the production of GM foods. Discussions were conducted between companies possessing the genetically modified organisms (GMOs), governments and organizations representing the interests of social communities. This study aims to develop a critical analysis of GM foods with regards to food safety. The approach will be made according to several factors that directly affect food transgenics: their use to combat hunger in the world, the benefits, risks, food security, focus on biosafety and ethical aspects, public perception and consumer law.
Intellectual property, agriculture and global food security: the privatization of crop diversity.	Intellectual property, agriculture and global food security: the privatization of crop diversity. 2011; xxi + 232 pp	Chiarolla, C	The book focuses on international governance of crop diversity and agricultural innovation. It highlights the implications that the future control of food, including access to agricultural resources and technologies, might have for global food security.
Accounting for hunger: the right to food in the era of globalisation.	Accounting for hunger: the right to food in the era of globalisation. 2011; x + 278 pp	Schutter, O De; Cordes, K Y	The challenge of global hunger is now high on the agenda of governments and international policy-makers. This work contributes to addressing that challenge by looking at the obstacles which stand in the way of implementing a right to food in the era of globalisation. The book describes the current situation of global hunger; it considers how it relates both to the development of food systems and to the merger of the food and energy markets; and it explains how the right to food contributes to identifying solutions at the domestic and international levels. The right to food, it argues, can only be realised if governance improves at the domestic level, and if the international environment enables governments to adopt appropriate policies, for which they require a certain policy space.

Governance for a resilient food system.	Governance for a resilient food system. 2011; 30 pp	Evans, A	Today, the world produces enough to feed all seven billion of its inhabitants - but nearly a billion people still go without. This paper is about why this global scandal continues, and what can be done to solve it. Its central argument is that access to food is as important as how much food is produced - and that in a world of food price volatility, climate change and other kinds of shocks and stresses, the challenge of building resilience in the food system takes on overwhelming importance. Section One of the paper looks at what needs to happen within developing countries, focusing, in particular, on a massive scale-up in provision of social protection systems that target the poorest and most vulnerable people.

Table no. 2.2: Poverty :The Lexicon of Agony

Title	Source	Author/authors	Impact point
Creating a policy environment for pro-poor agricultural extension: the who? what? and how?	ODI Natural Resource Perspectives. 2002; (80): 4 pp.	Farrington, J; Christoplos, I; Kidd, A; Beckman, M; Cromwell, E	This paper examines the target group of, the need for, and the ways in which to deliver pro-poor agricultural extension. It is concluded that for agricultural extension to address poverty effectively, it must: address poverty from a livelihoods perspective within a wider, poverty-focused policy environment; contribute to enhanced 'voice' among the poor; and seize opportunistically on possibilities for enhancing incomes or reducing vulnerability, based on sharper alertness to new opportunities..
Uncommon ground: the 'poverty of history' in common property discourse.	Development and Change. 2004; 35(3): 407 433	Johnson, C	It would not be a great exaggeration to say that scholars of environmental conservation and conflict have re-discovered the institutional foundations of social and economic life. At the heart of this 'renaissance' is the belief that property and property relations have a strong bearing on how people use, manage and abuse natural resource systems, and that institutional arrangements based on the creation and management of common property can have positive impacts on resource use and conservation.

Labour force participation and the influence of having CVD on income poverty of older workers.	International Journal of Cardiology. 2012; 156(1): 80 83	Schofield, D J; Callander, E J; Shrestha, R N; Percival, R; Kelly, S J; Passey, M E	Background: In addition to being the leading cause of death, cardiovascular disease (CVD) also impacts upon the ability of individuals to function normally in everyday activities, which is likely to affect individuals' employment. This paper will quantify the relationship between labour force participation, CVD and being in poverty.
A study of the Healthy Growth Charter in socially disadvantaged children.	Annali dell' Istituto Superiore di Sanita. 2011; 47(4): 417 423	Musicco, M; Pulvirenti, R A; Gainotti, S; Petrini, C; Riccio, C; Silvestrini, B	The Healthy Growth Charter is an educational and health-promoting project aimed at an active involvement of the primary school children in health surveillance and protection. Being duly acquainted of the matter, scholars are asked to fill by their own a questionnaire on height, weight, sport attendance and other items of medical and demographic interest.
Utilization of non-timber forest products (NTFPs) for economic development in Nigeria.	Nature and Faune. 2011; 25(2): 68 71	Israel, B T	Non-timber forest products (NTFPs) contribute immensely to food security, poverty alleviation, economic development, and household and national income generation among many other benefits. This paper gives a synopsis of ntfps in Nigeria, their diversity and diverse uses, with specific examples of the economic potentials of Moringa oleifera, Lonchocarpus cyanescens, Vitellaria paradoxa, Acacia senegal and Dacryodes edulis. The paper highlights the challenges facing the economic utilization of ntfps in Nigeria and solutions were suggested.
Persistent poverty in rural China: where, why, and how to escape?	World Development Oxford. 2012; 40(4): 784 795	Glauben, T; Herzfeld, T; Rozelle, S; Wang XiaoBing	Using rural household panel data from three Chinese provinces, this paper identifies determinants of long-term poverty and tests the duration dependence on the probability to leave poverty. Special emphasis is given to the selection of the poverty line and inter-regional differences across provinces. Results suggest that the majority of population seems to be only temporary poor.
Psychological disorders among inhabit residing in poor social district of Gorgan, Iran.	Journal of Gorgan University of Medical Sciences. 2011; 13(3): 87 93	Ghaffari, E; Shahi, A S; Ozouni, D R; Rostami, R	Background and Objective: Normal psychological behaviour is one of the general concept of health. This study was carried out to evaluate psychological disorders among inhabitants residing in a poor social district of Gorgan, Iran. Materials and Methods: This cross-sectional study was done

			on 376 subjects (188 males and 188 females) using a simple random sampling method. Data were gathered using the general health questionnaire (GHQ-28).
Taking action on the social determinants of health: improving health access for the urban poor in Mongolia.	International Journal for Equity in Health. 2012; 11(15): (20 March 2012)	Lhamsuren, K; Choijiljav, T; Budbazar, E; Vanchinkhuu, S; Blanc, D C; Grundy, J	In recent years, the country of Mongolia (population 2.8 million) has experienced rapid social changes associated with economic growth, persisting socio-economic inequities and internal migration. In order to improve health access for the urban poor, the Ministry of Health developed a "Reaching Every District" strategy (RED strategy) to deliver an integrated package of key health and social services. The aim of this article is to present findings of an assessment of the implementation of the RED strategy, and, on the basis of this assessment, articulate lessons learned for equitable urban health planning.
SWOT analysis on rural poverty alleviation of Bijie Experimental Zone in the new period.	Guizhou Agricultural Sciences. 2011; (12): 262 267	Sun Qiu; Cui Wei; Zhou Pi-Dong; Zhang Pei	The strengthens, weaknesses, opportunities and threats of rural poverty alleviation of Bijie Experimental Zone in the new period were analyzed by SWOT model to provide references for promoting the reform and development. It was pointed out that in order to achieve virtuous circle of "industrial development, ecological improvement and income increase of farmers", Bijie Experimental Zone needs to adhere to development-oriented poverty reduction policy, place emphasis on reducing poor population, increasing income of poor people and enhancing their capacity, take infrastructure construction, development of social programs, industrial development and environment protection as priorities, and focus on the key poverty-targeted counties (townships), minority people concentrated areas, rock-desertification stricken region.
An appraisal of urban poverty and measures for its alleviation in Makurdi Metropolis, Benue State, Nigeria.	Agricultural Journal. 2012; 7(1): 65 69	Okwoche, V A; Obinne, C P O; Otijele, A E	The study was undertaken to appraise urban poverty and measures for its alleviation in Makurdi metropolis. Double-stage random sampling was used to select the sample size of 120 respondents. Frequency distribution and percentages were used to de

			scribe the data. The study identified poverty of income and access as the main causes of urban poverty. The results of the study indicated that about 52% of the respondents earned an average income of<200,000.00 naira pa, 59% do not own personal houses and about 56% send their wards to public institutions. Only about 54% have access to pipe borne water; 51% go to public hospitals and about 71% have access to public power supply that is erratic. The study concludes that accessibility of functional public facilities stand out as the key determinants of employment outcome and income earnings. It recommended that the extent to which urban poverty can be alleviated depends on household's access to functional public facilities and subsequent increased income.
Microfinance and livelihood development in poor coastal communities in Eastern Sri Lanka.	Tropical Agricultural Research. 2011; 22(3): 330 336	Thilepan, M; Thiruchelvam, S	Various organizations are involved in micro-financial programmes in Sri Lanka. There is limited knowledge on the impact of microfinance on the livelihood improvements among poor coastal communities who were affected by Tsunami in Eastern Sri Lanka. This study investigated the effectiveness of microfinance support for coastal communities' livelihood development in Trincomalee.
'Tourism poverty' in affluent societies: voices from inner-city London.	Tourism Management. 2012; 33(4): 951 960	Sedgley, D; Pritchard, A; Morgan, N	There is now a significant body of work analysing the multifaceted connections between tourism and poverty in less developed economies. Far fewer studies discuss the relationships between tourism and poverty in the world's affluent societies and most of these concentrate on social tourism and on the benefits of these holidays for deprived and marginalised groups. This paper provides an insight into the experiences of families unable to afford any form of paid holiday away from home.
Evaluating the long-term impact of antipoverty interventions in Bangladesh: an overview.	IFPRI Discussion Papers. 2011; (1077): vi + 23 pp	Quisumbing, A R; Baulch, B; Kumar, N	This paper provides an overview of a research project that assessed the long-term impact of three antipoverty interventions in Bangladesh - the introduction of new agricultural technologies, educational transfers, and microfinance - on monetary and non-monetary measures of well-being. This paper begins by setting out the conceptual framework, methodology, and empirical methods used for the evaluation of long-term impacts.

The wealth and gender distribution of rural services in Ethiopia: a public expenditure benefit incidence analysis.	IFPRI Discussion Papers. 2011; (1057): vii + 62 pp	Mogues, T; Petracco, C; Randriamamonjy, J	Over the past several years, the Ethiopian government has committed a substantial portion of the public budget to expanding public services and infrastructure in rural areas. This paper assesses who exactly is benefiting from this public spending. To do so, this paper undertakes a public expenditure benefit incidence analysis across gender and wealth groups of three public services/programs in rural Ethiopia: (1) selected components of the Food Security Program (FSP), (2) drinking water supply, and (3) agricultural extension services. The analysis uses data at the individual, household, kebele (a subdistrict administrative unit), and district level in Ethiopia.
Environmental services and poverty alleviation: either, or, or both?	Discussion Paper Series Environmental Economy and Policy Research, University of Cambridge. 2009; (46.2009): 23 pp	Groom, B; Palmer, C	Payments for environmental services (PES) schemes in developing countries face trade-offs between environmental and development objectives. This tension is inherent in cost effective direct PES since, by their very nature, they limit transfers to recipients. However, where recipients of PES are subject to market constraints (e.g. credit rationing, input constraints etc.), we show that indirect payments which relax constraints can be cost effective and achieve both environmental and poverty alleviation objectives.
Relaxing rural constraints: a 'win-win' policy for poverty and environment in China?	Discussion Paper Series Environmental Economy and Policy Research, University of Cambridge. 2008; (30.2008): 55 pp	Groom, B; Grosjean, P; Kontoleon, A; Swanson, T; Zhang ShiQiu	The link between local institutional and market failures, rural poverty and environmental degradation suggests a 'win-win' policy intervention: solve local 'constraints' and achieve both poverty alleviation and environmental goals.
Free trade in agriculture and global poverty.	World Economy. 2011; 34(12): 2019 2043	Bussolo, M; Hoyos, R de; Medvedev, D	This paper assesses the potential impacts of the removal of agriculture trade distortions using a newly developed data set and methodological approach for evaluating the global poverty and inequality effects of policy reforms. It finds that liberalisation of agriculture and food could increase global extreme poverty by 0.2 per cent and lower moderate poverty by 0.3 per cent.

Negotiating trade-offs: choices about ecosystem services for poverty alleviation.	Economic and Political Weekly. 2012; 47(9): 67 75	Vira, B; Adams, B; Chetan Agarwal; Shrinivas Badiger; Hope, R A; Jagdish Krishnaswamy; Kumar, C	Exploring the prospects of the ecosystem services approach for natural resource management and poverty alleviation in India, this paper points out that it is vital to have an understanding of the political economy of negotiations over natural resource use. An appreciation of the synergies and trade-offs between ecosystem services is equally important to develop better strategies for pro-poor ecosystem management. If the distributional outcomes associated with alternative options for natural resource management are neglected, there is a risk that such interventions may fail because of resistance from those who are excluded or those who stand to lose.
Impact of national special programme for food security on poverty alleviation among women in Oyo State, Nigeria.	nternational Journal of Agriculture Environment and Biotechnology. 2011; 4(2): 139 151	Ayoade, A R; Ogunwale, A B; Adewale, J G	The study investigated the impact of National Special Programme for Food Security (NSPFS) on poverty alleviation among women in the three project sites of Oyo State, Nigeria. Purposive sampling technique was used to select 166 participants from the sites of the programme while Systematic sampling technique was used to select 110 non-participants from the list of the farmers provided by the Agricultural Development Programme (ADP) officials in Oyo State. A structured interview schedule was used to gather information from the respondents. Data collected were collated and subjected to both descriptive and inferential statistics such as frequency counts, percentages, Tobit regression estimates, analysis of variance (ANOVA) and t-test.
Livestock-keeping systems and poverty.	Sustainable livestock management for poverty alleviation and food security. 2012; 43 52	Hooft, K E van't; Wollen, T S; Bhandari, B P	This chapter discusses: (i) livestock-keeping systems relevant to the poor; (ii) the role of livestock as part of integrated agricultural systems and households' strategy of risk minimization and diversification; (iii) the link between livestock and the most vulnerable groups, including women; (iv) the influence of livestock on environment and climate change; (v) the risk and advantages of low-input livestock keeping for human health; and (vi) the potential of smallholder mixed farming and pastoralism for the achievement of the Millennium Development Goals (MDGs) and biodiversity conservation.

Sustainable livestock management for poverty alleviation and food security.	Sustainable livestock management for poverty alleviation and food security. 2012; x + 194 pp	Hooft, K E van't; Wollen, T S; Bhandari, B P	This practical learning guide, composed of 12 chapters, focuses on supporting livestock within the smallholder mixed farming systems. It provides practical ways to understand and improve smallholder animal husbandry under a variety of agricultural and ecological conditions based on the Endogenous Livestock Development (ELD) approach.
Enhancing agricultural productivity and rural incomes through sustainable use of natural resources in the Semi Arid Tropics.	Journal of the Science of Food and Agriculture. 2012; 92(5): 1054 1063	Wani, S P; Yin DiXin; Li Zhong; Dar, W D; Girish Chander	In dryland tropics, integrated watershed management approach enabled farmers to diversify the systems along with increasing agricultural productivity through increased water availability, while conserving the natural resource base. Household incomes increased substantially, leading to improved living and building the resilience of the community and natural resources.
A strategic framework for strengthening rural incomes in developing countries.	Agricultural policies for poverty reduction. 2012; 23 59	Brooks, J	This chapter examines the role of agricultural policies in raising rural incomes in developing countries. The underlying premise is that policies need to be effective given current economic structures, yet anticipate and facilitate the transition to structures that are capable of generating higher incomes. This means improving the productivity and competitiveness of smallholder farmers, who dominate developing country agriculture, while widening opportunities outside the sector as the economy diversifies.
Agricultural policies for poverty reduction.	Agricultural policies for poverty reduction. 2012; 191 pp	Brooks, J	This volume (consisting of 5 chapters, in addition to an introduction) is concerned with the role of agricultural policies in developing countries, with a focus on their potential for raising the incomes of agricultural and rural households and thereby reducing poverty. The welfare impacts of agricultural policies and commodity prices, the stabilization policies after the 2007-08 food crisis, and the use of input subsidies, are addressed.

Table no. 2.3: Voice: The interlink of hunger and poverty

Title	source	Author/authors	Impact point
Acute sleep deprivation enhances the brain's response to hedonic food stimuli: an fMRI study.	Journal of Clinical Endocrinology and Metabolism. 2012; 97(3): E443 E447	Benedict, C; Brooks, S J; O'Daly, O G; Almen, M S; Morell, A; Aberg, K; Gingnell, M; Schultes, B; Hallschmid, M; Broman, J E; Larsson, E M; Schioth, H B	There is growing recognition that a large number of individuals living in Western society are chronically sleep deprived. Sleep deprivation is associated with an increase in food consumption and appetite. However, the brain regions that are most susceptible to sleep deprivation-induced changes when processing food stimuli are unknown.
Altered social cohesion and adverse psychological experiences with chronic food insecurity in the non-market economy and complex households of Burkina Faso.	Social Science and Medicine. 2012; 74(3): 444 451	Nanama, S; Frongillo, E A	Food insecurity negatively impacts outcomes in adults and children including parenting practices, child development, educational achievement, school performance, diet, and nutritional status. Ethnographic and quantitative research suggests that food insecurity affects well-being not only through the lack food, poor diet, and hunger, but also through social and psychological consequences that are closely linked to it.
What is eating you? Stress and the drive to eat.	Appetite . 2012; 58(2): 717 721	Groesz, L M; McCoy, S; Carl, J; Saslow, L; Stewart, J; Adler, N; Laraia, B; Epel, E	Non-human animal studies demonstrate relationships between stress and selective intake of palatable food. In humans, exposure to laboratory stressors and self-reported stress are associated with greater food intake. Large studies have yet to examine chronic stress exposure and eating behavior. The current study assessed the relationship between stress (perceived and chronic), drive to eat, and reported food frequency intake (nutritious food vs. palatable non-nutritious food) in women ranging from normal weight to obese (N=457). Greater reported stress, both exposure and perception, was associated with indices of greater drive to eat - including feelings of disinhibited eating, binge eating, hunger, and more ineffective attempts to control eating (rigid restraint; r's from .11 to .36, p's <.05).
Effects of starvation stress on body shape, chemical composition and blood physiological of Cyprinus carpio.	Journal of Shanghai Ocean University. 2011; 20(6): 814 819	Feng GongNeng; Yang WenPing; Wang AiMin; Li Jing; Xu HangFeng; Han GuangMing; Xu Pao	The aim of the present study was to determine the effect of starvation on body appearance, muscle Composition and blood physiological of the carp (Cyprinus carpio) at the water temperature 18-20 degrees C in the cycle and flow water aquarium. The experiment was

			designed for four groups with three replication, which were 0 day (control), 10 days, 20 days, 30 days group. The results showed: With the carp hunger time extending, there was a trend to decline for the obese relative, intestine and body length ratio, live index of the carp, but the effects was not significantly (P>0.05); there was a trend to increase for contents of moisture, crude ash in the body muscle of the carp, but a trend to decline for the crude fat in the body muscle of the carp, and the differences was not significantly (P>0.05), while body contents of crude protein was remarkable difference (P<0.05); With the carp hunger time extending, red blood cell quantity and hemoglobin value dropped significantly (P<0.05), and white blood cell quantity decreased not significantly (P>0.05), blood glucose levels raised first and then went down. The experiment illustrated that the carp body appearance had changed, and with the prolongatin of starvation, the muscle fat was decompose to exploit first, the muscle protein then to be used, at the same time, the body physiological of the carp was changed to be fit for the starvation condition.
Using translational research to enhance farmers' voice: a case study of the potential introduction of GM cassava in Kenya's coast.	Agriculture and Human Values. 2014; 31(4): 673 681	Valdivia, C; Danda, M K; Sheikh, D; James, H S Jr; Gathaara, V; Mbure, G; Murithi, F; Folk, W	Genetically modified (GM) cassava is currently being developed to address problems of diseases that threaten the food security of farmers in developing countries. The technologies are aimed at smallholder farmers, in hopes of reducing the vulnerability of cassava production to these diseases. In this paper we examine barriers to farmers' voice in the development of GM cassava.
Choice and voice: creating a community of practice in KwaZulu-Natal, South Africa.	Agriculture and Human Values. 2014; 31(4): 665 672	Hendrickson, M K; Gilles, J L; Meyers, W H; Schneeberger, K C; Folk, W R	The development and utility of genetically modified (GM) crops for smallholders around the world is controversial. Critical questions include what traits and crops are to be developed; how they can be adapted to smallholders' ecological, social and economic contexts; which dissemination channels should be used to reach smallholders; and which policy environments will enable the greatest benefits for smallholders and the rural poor. A key question is how the voices of smallholders who have experience with or desire to use GM technologies enter the larger debate.

Silenced voices, vital arguments: smallholder farmers in the Mexican GM maize controversy.	Agriculture and Human Values. 2014; 31(4): 655 663	Carro Ripalda, S; Astier, M	Smallholder producers are the collective most likely to be affected by the introduction of GMOs globally, yet the least included in public debates and consultation about the development, implementation or regulation of this agricultural biotechnology.
Special Symposium: Smallholder farmer voice.	Agriculture and Human Values. 2014; 31(4): 637 681	James, H S Jr; Sulemana, I	This collection of five essays brings attention to the issue of farmer voice. These essays highlight specific barriers and challenges to farmer voice and provide insights into efforts intended to make the voice of smallholder farmers heard, especially regarding the development of new agricultural technologies, particularly GM crops.
Mobile phones: the next step towards healthcare delivery in rural India?	PLoS ONE. 2014; 9(8): e104895	DeSouza, S I; Rashmi, M R; Vasanthi, A P; Joseph, S M; Rashmi Rodrigues	Background: Given the ubiquity of mobile phones, their use to support healthcare in the Indian context is inevitable. It is however necessary to assess end-user perceptions regarding mobile health interventions especially in the rural Indian context prior to its use in healthcare. This would contextualize the use of mobile phone communication for health to 70% of the country's population that resides in rural India.
Towards gender equitable schooling: insights from rural teachers' voice in the local context.	Studies of Tribes and Tribals. 2014; 12(1): 103 111	Morojele, P	This paper foregrounds local teachers' views to understand how we could address gender inequalities in schools. It asks: How do teachers' constructions of gender shape gendered social relations? What are the implications of these on gender equitable schooling? The paper draws from semi-structured interviews with 12 teachers in three South African primary schools. The findings denote how essentialist teachers' constructions, which polarised children into masculine and feminine beinngs, had the likelihood to compromise the quality of children's schooling experience. The existing dominant (and cultural) discourses of gender were found to inform how teachers socialised girls and boys into inequitable gender relations.

A study of sustainable development of Self Help Groups: a review of literature.	Review of Research Journal. 2014; 3(6): ROR 666	Barbole, A N; More, S M	In India, Self Help Groups are now supported by government policy and NGO intervention. These are being seen as a magic wand for all the development of poor. The concept of micro finance was introduced in India in 1992 for overcoming the existing constraints and providing adequate credit to the poor in general and the weaker section in particular by following a simple procedure. Self Help Groups are important vehicles for credit delivery to self-employed persons particularly women in rural and semi urban areas. 6
From 'sustainable rural communities' to 'social sustainability': giving voice to diversity in Mangakahia Valley, New Zealand.	Journal of Rural Studies. 2000; 16(4): 433 446	Scott, K; Park, J; Cocklin, C	This paper maintains that, despite widespread interest in the notion of sustainability, little progress has been made towards an understanding of its social dimensions. Nonetheless, the concept of 'sustainable rural communities' is embedded in popular, policy and academic discourses, where the needs of 'rural communities' are usually equated with those of farm families. This ethnographic research in Northland, New Zealand illustrates the diverse interests to be found within 'rural communities'.
A voice in policy and practice: advocacy in the water and sanitation sector.	Waterlines . 2000; 18(4): 8 11	Calaguas, B U	The paper argues that it is now time for all water and sanitation (WATSAN) NGOs to put aside their fears and become fully involved in the development of advocacy work. Ignoring it is a luxury that can no longer be afforded. Issues discussed are: understanding advocacy; challenges to watsan NGOs; current trends in watsan NGO advocacy; exploring the potential; a global network; and eyes on the prize (the global vision of a fairer world; a more sustainable world for children; a clean and healthy world where every person has safe and adequate water and sanitation; and lives in a hygienic environment)..
Consumer exit, voice, and loyalty: indicative, legitimation, and regulatory role in agricultural and food ethics.	Journal of Agricultural and Environmental Ethics. 2000; 12(2): 153 164	Newholm, T	Disputes over agriculture and food production have occurred against a background of disputed authority with regard to governments, experts, and single issue pressure groups. Consumers have intervened in quite significant ways with many altering their buying patterns.

CANEGROWERS - the strong voice of the Australian cane growing industry.	International Sugar Journal. 2001; 103(1233): 425 429	Kerr, B	This paper describes the role and the structure of CANEGROWERS (a membership organization supported by over 94% of 6500 cane farmers in Queensland, Australia) and its links with the research establishments in the country and its affiliation with international bodies..
The farmer's voice in priority setting: a cross-country experiment in eliciting technological preferences.	Economic Development and Cultural Change. 2001; 49(3): 591 610	Pingali, P L; Rozelle, S D; Gerpacio, R V	A study was conducted in 1995-96 in the Philippines, China, Vietnam, and India to elicit and examine the technological preferences of farmers (n=401), using an experimental voting (ranking) method that allows for reasonable quantitative estimates of preferences and the socioeconomic determinants of adoption. In particular, the study sought to discover which of 4 post-Green Revolution technologies currently under research would most likely be adopted by Asian farmers, and why: higher yielding varieties; higher-quality varieties; a labour-saving mechanical stripper-harvester; or a simple, environmet-friendly method of integrated pest management.
Engendering development: Through gender equality in rights, resources, and voice.	Engendering development: through gender equality in rights, resources, and voice. 2001; 364 pp.		Existing and new research from several social science disciplines are brought together to examine the costs of persistent gender disparities to wellbeing and countries' development prospects. How formal and informal institutions shape gender roles and relations and how household decisions and behaviours reproduce those roles, the relationship between economic development and gender equality, and the role of public policy and civic action in promoting gender equality.
Reflections on the "voice of authority" in leisure research and practice.	Loisir et Societe. 2000; 23(1): 237 250	Hutchinson, S L; Samdahl, D M	This paper critically evaluates a previous study that introduced reflective theory to the field of leisure research. Implications for future research are introduced in conclusion.
Strengthening the voice of supporters.	Soccer and Society. 2000; 1(3): 13 16	Smith, C	The rationale for the formation of Supporters Direct, an initiative by the UK government to provide legal and practical support for supporters to form supporters' trusts with a view to being actively involved in the way their clubs are run, is described. Two possible scenarios for the future development of football are hypothesized.

Consumer-oriented functional food development: how well do functional disciplines reflect the 'voice of the consumer'?	Trends in Food Science and Technology, 2002; 13(3): 93 101.	Kleef, E van; Trijp, H C M-van; Luning, P; Jongen, W M F	Food innovation can have its source in either superior understanding of consumer demand (pull) or in superiority at the supply side (science and technology push). However in either case, market success depends on the degree to which the new product reflects unmet consumer needs. The present study provides a framework, which allows to obtain relevant consumer and expert input in the early stages of functional food development.
Timber certification: where is the voice of the biologist?	Conservation Biology. 2001; 15(2): 308 319	Bennett, E L; Cauley, H A; Peters, C M; Donovan, R Z; O' Connor, J M; Putz, F E; Romero, C; Ghazoul, J	This forum begins with a reprinted editorial [Bennett; see Conservation Biology (2000) 14 (4) 921-923], which discusses the relationship between timber certification in tropical forests and biodiversity, with an emphasis on wildlife, and uses Forest Stewardship Council (FSC) certification as an example to argue that certification standards should be expanded to include the effects of logging.
Voice of the people.	Health Club Management. 2002; (August): 36 37	Cooney, A	This paper reports on the findings of a survey of public attitudes towards health and fitness clubs in the UK. The survey, which was conducted on 20-24 June 2002, involved a sample of 1982 adults aged 15 years or over. The findings suggest that as the health and fitness industry matures, operators will need to pay as much attention to membership retention as they do to acquiring new members. Plenty of potential members are still out there, but accessing them will depend on the right facilities and the right deals being offered..
Improving basic health service delivery in low-income countries: 'voice' to the poor.	Social Science and Medicine. 2002; 54(11): 1685 1690	Santosh Mehrotra; Jarrett, S W	This paper basically shows the economic vulnerability of lower income groups and their struggle for existence.
How the poor can have a voice in government policy.	Finance and Development. 2000; 37(4): 22 25	Robb, C M	This paper describes participatory poverty assessment, a tool that allows policymakers to consult the poor directly in formulating poverty reduction strategies. The benefits of a participatory approach are discussed.

4 CHAPTER

SOCIAL ECOLOGY OF STUDY

Research setting refers to the detailed information of an area where the study was conducted. The study area generally comprises of a particular geographical area viz. a state, district, block, or gram panchayat area selected according to the convenience of the investigator who possess adequate knowledge regarding the location, communication facility etc. of the locality so that he can easily approach each and every corner of the area for data collection. Besides the investigators must also have the basic knowledge about the socio-demographic background of the people so as to have an easy understanding of their knowledge, attitude and behavior. Thus, a study on Monsoon impact analysis certainly requires the information about the impact of monsoon on the economic, social and cultural aspects of people.

The present study was taken up at Ghoragachha villages of Rautari gram panchayat in Haringhata block in West Bengal. A brief description of the state and district in general and the block and village in particular is given below.

AREA OF STUDY

The present study has conducted in two separate socio-ecological strata. One study has conducted among the farm families and the other one has conducted on the non farm families. Here we have only selected women at the age group of 15-60 as respondents.

The farm locality is Beraberi gram panchayat, Habra II Block, North 24 pgs and the non farm locality is Bilkanda gram panchayat, Barrackpur 2 block, North 24 pgs.

PROFILE OF THE STATE WEST BENGAL

West Bengal was created as a one of the constituent state of the Indian union among the 29'h states in India on I5" August, I947 as the result of partition of the undivided

British Indian province of Bengal into West Bengal. West Bengal is situated in the north eastern part of India and lies between 21 37′-27 10′ north latitude and 85 51′-89 53′ east longitude.

It is flanked by the Bay of Bengal on the South, Sikkim on the north, Assam on the east and Bihar, Jharkhand and Orissa on the west. It also has common borders with the three neighbouring countries viz. Nepal and Bhutan in the north and Bangladesh in the east. The tropic of cancer runs across the middle of the state and passes through four districts viz. Bankura, Burdwan, Nadia and Purulia. The state has a total geographical area of 88,752 sq. km. representing only 2.7% of the total area of India. It is territorially divided into 19 districts i.e. Coochbehar, Darjeeling, Jalpaiguri, Uttar Dinajpur, DakshinDinajpur, Malda, Murshidabad Hoogly Howrah, Nadia, Birbhum Burdwm Bankura, Purulia, 24 Par2ana(N) 24 Pargana(S) PurbaMidnapur, PashimMidnapur and Kolkata. There are more than 341 Community Development bocks in West Bengal. There are 3354 Gram Panchayats and more than 42000 villages in West Bengal.

In 1947, when India became independent Bengal was partitioned between India lndias share came to be known as West Bengal and Pakistans share was called East Pakistan Later, the state of Cooch Behar, French enclave of Chandranagore and some parts of Bihar were added to West Bengal West Bengal represents the lands that possess a district culture with its indigenous art and crafts and make it an important part of the Indian Union.

Table-3.1. General Information of West Bengal

Total population	91,347,736
Male	46,927,3 89
Female	44,420,347
Population Growth	13.93%
Sex ratio	947
Population density	1029/ kmz
Literacy rate	77.08%
Male	82.67%
Female	71.16%
No of Sub divisions	66
No of blocks	341
No of Gram Panchayats	3354
Inhabited Villages	37945
Area (Sq. Km)	88752
% of Electrified Vtllaves	83.6

Source: Office of the Register General and Census Commissioner 2011 Ministry of Home Affairs, GOI.

CLIMATE OF THE STATE

In terms of the standard climate types, tropical climate viz. humid, sub humid, semi arid and arid occurs nvcr different regions of the stale. lls climate is very much influenced by monsoon rains which, in turn arc of two types (1) Seuth-West monsoon during the rainy season and (ll) North-East inensnon during thc winter season. Rainfall in West Bengal averages about W50 mm, ranging from 2500 to 3500 mm in Dooars and Tarai regions to I200 to 1800 mm in the plains. The maximum and minimum temperatures recorded during March to October are 30-40 C and l6-26 C respectively in the plains while in lhc hills these anc IS-20 C.

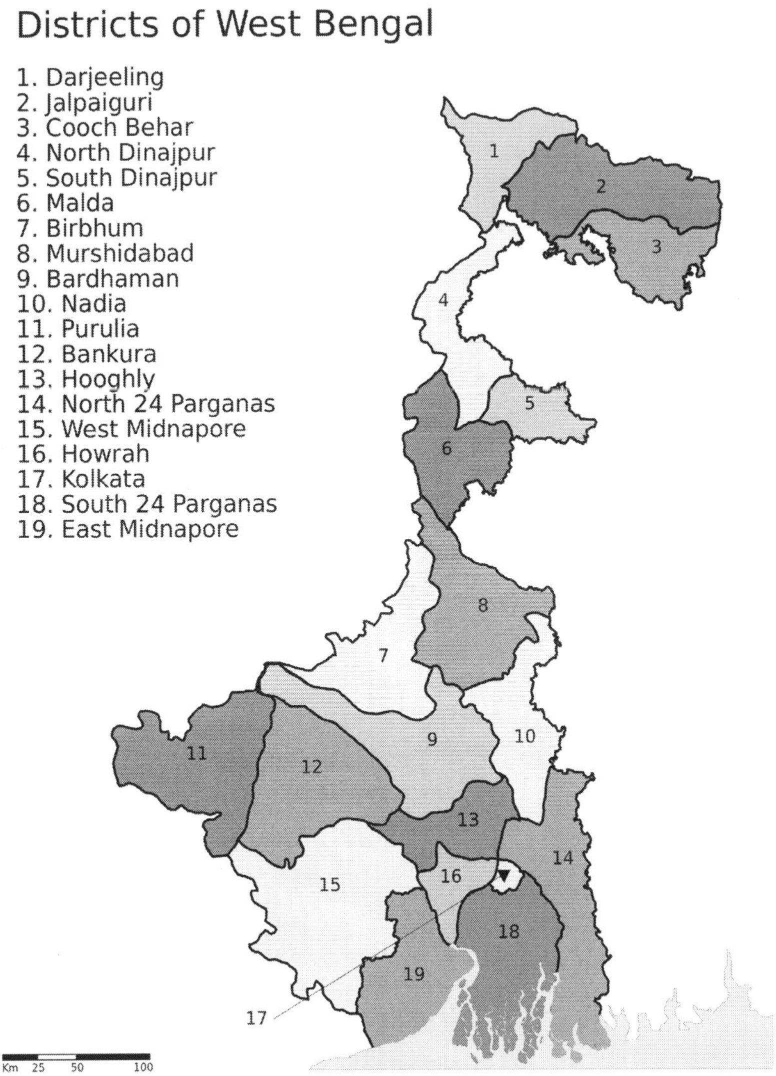

http://census.gov.in/2011-prov results/prov data_products_wb.html

PROFILE OF NORTH 24 PGS

North 24 Parganas (Pron: pɔrgɔnɔs) (abv. 24 PGS(N)) or sometimes North Twenty Four Parganas is a district in southern West Bengal, of eastern India. North 24 Parganas extends in the tropical zone from latitude 22°11′6″ north to 23°15′2″ north and from longitude 88°20′ east to 89°5′ east. It is bordered to Nadia by north, to Bangladesh (Khulna Division) by north and east, to South 24 Parganas and Kolkata by south and to Kolkata, Howrah and Hoogly by west. Barasat is the district headquarters of North 24 Parganas. North 24 Parganas is West Bengal's most populous district[2] and also the most populated district in the whole of India. It is the tenth-largest district in the State by area.

According to Ptolemy's Treatise on geography, written in the 2nd Century A.D., the ancient land of Gangaridi was stretched between the rivers Bhagirathi-Hoogly (lower Ganges) and Padma-Meghna. The modern-day 24 Parganas was the southern and the south-eastern territory of that legendary kingdom.

Archaeological excavation at Berachampa village in Deganga PS proves that though the area was not directly attached to the rule of the Guptas, yet it could not shun their cultural influence. Xuanzang (c. 629-685) visited 30 Buddhist Biharas and 100 Hindu Temples in India and some of these were in the Greater 24 Parganas region.

The district was not a part of Shashanka's unified Bengali empire known as Gauda, but it is assumed that the district which was the south-west frontier territory of ancient Bengal, was comprised in under the rule of Dharmapala (estimated c. 770-810). The Pala rule was not quite strong in this part, as no excavation uncovered any of Buddhist Pala antiquities but many Hindu Sena sculptures.

In the middle of the 16th century, Portuguese pirates began to invade and plunder many of the waterways and prosperous human settlements in the lower delta region. People left these places out of the fear of being murdered, raped, or captured to be sold as slaves. The Basirhat sub-division of North 24 Parganas suffered most from these torments.

Shrihari (Shridhar), a Kayastha, was an influential officer in the service of Daud Khan Karrani. On the fall of Daud he fled away with the government treasure in his custody. He then set up a kingdom for himself in the marshy land to the extreme south of Khulna district (1574) and assumed the title of Maharaja. Pratapaditya succeeded to the kingship in 1574. The baharistan and the travel diary of Abdul Latif and the contemporary European writers, all testify to the personal ability of Pratapaditya, his political pre-eminence, material resources. His territories covered the greater part of what is now included in the greater Jessore, Khulna and Barisal districts. He established his capital at Dhumghat, a strategic position at the confluence of the Jamuna and Ichhamati before it was shifted to Ishwaripur.

Maharaja Pratapaditya, soon became one of the 12 feudal lords of Bengal who not only declared their sovereignty from the Mughal Empire in the ruling of

Jessore, Khulna, Barisal and Greater 24 Parganas, but also fought and resisted the Portuguese in the early years of the 17th Century. When he was finally defeated by the Mughals. Pratapaditya lost both the battles of Salka and Magrahat. His fate was sealed and he was compelled to tender submission to Islam Khan at Dhaka. His kingdom was annexed. Probably he died at Benares on his way to Delhi from Dhaka, as a prisoner.[5] of war to the Mughals.

After his death, Bhavanand Majumdar, who had been in the service of Pratapaditya, was given the throne by Raja Man Singh, and he later became the founder of the Nadiya Raj family.[6] Laksmikanta Gangopadhyay better known as Laksmikanta Roy Choudhury, the well known Brahmin scholar who was the son of great saint Kamdev Brahmachari and also a close associate of Raja Basanta Ray, was given tax free jaigir of eight parganas, including the areas in and around Kalikatah as Gurudakshina by Raja Man Singh in 1608.[7][8]

Jashoreshwari Kali Temple (built by Pratapaditya), Chanda Bhairab Mandir at Ishwaripur (built during the Sena period), Five domed Tenga Mosque at Banshipur (Mughal period), two big and four small domed Hammankhana (constructed by Pratapaditya) at Bangshipur, Govinda Dev Temple at Gopalpur (built by Basanta Roy, uncle of Maharaja Pratapaditya in 1593), Jahajghata Port (Khanpur), Kalighat Temple (owned by Laksmikanta Roy Choudhury) bear archaeological evidences of the Bhuiyan and Majumdar kings.

BRITISH RAJ

The territory of Greater 24 Parganas were under the Satgaon (ancient Saptagram, now in Hoogly district) administration during the Mughal era and later it was included in Hoogly chakla (district under post-Mughal Nawabi rule) during the rule of Murshid Quli Khan. In 1757, after the Battle of Plassey, Nawab Mir Jafar conferred the Zamindari of 24 parganas and janglimahals (small administrative units) upon the British East India Company. These were Amirpur, Akbarpur, Balia, Birati, Azimabad, Basandhari, Baridhati, Bagjola, Kalikata, Garh, Hatiagarh, Islampur[disambiguation needed], Dakshin Sagar, Kharijuri, Khaspur, Ikhtiarpur, Magura, Medanmalla, Maida, Manpur, Barasat, Madhyamgram, Muragachha, Pechakuli, Paikan, Rajarhat, Shahpur, Shahnagar, Satal, New Barrackpore, Aharampur and Uttar Pargana. Since then, this entire territory is known as Twentyfour Parganas.

In 1751, the Company assigned John Zephaniah Holwell as zemindar of the District.[9] In 1759, after the Bengali War of 1756-57, the Company assigned it to Lord Clive as a personal Jaghir (zamindari) and after his death it again came under the direct authority of the Company.

In 1793, during the rule of Lord Cornwallis, entire Sunderbans were in Twentyfour Parganas. In 1802, some parganas on the western banks of river Hoogly were included into it. These parganas were in Nadia earlier. In 1814, a separate collectorate was established in Twenty-four Parganas. In 1817, Falta and Baranagar and in 1820, some portions of Nadia's Balanda and Anwarpur were encompassed to it. In 1824, portions of Barasat, Khulna and Bakhargunge (now in Bangladesh) were also included to it. In 1824, the district headquarters was shifted from Kolkata to Baruipur, but in 1828, it was removed to Alipore. In 1834, the district was split into two districts – Alipore and Barasat, but later these were united again.

In 1905, some portion of this district around the Sunderbans was detached and linked to Khulna and Barishal. These parts remained in Bangladesh territories where Jessore's Bangaon was joined to Twentyfour Pargana after the 1947 partition.

AFTER INDEPENDENCE

In 1980, an administrative reform committee under the chairmanship of Dr. Ashok Mitra suggested to split the district into two and as per the recommendation of the committee in 1983, 1 March two districts – North 24 Parganas (24 PGS N) and South 24 Parganas (24 PGS S) were created. The North 24 Parganas which was included to the Presidency Division has been formed with 5 sub-divisions of the Greater 24 Parganas namely Barasat (Headquarters), Barrackpore, Basirhat, Bangaon and Bidhannagar (a satellite township of Kolkata, popularly known as Salt Lake).

GEOGRAPHY

The district lies within the Ganges-Brahmaputra delta. The river Ganges flows along the western border of the district. There are many other rivers, which include the Ichhamati, Jamuna, and Bidyadhari.

LOCATION

Latitude: 23 15'North - 22 11' North

Longitude: 89 5'East - 88 20' East

LAND

The type of soil varies widely from alluvial to clay loam.

GROUNDWATER ARSENIC CONTAMINATION

North 24 Parganas is one of the nine severely arsenic affected district in West Bengal. On the basis of updated survey conducted by School of Environmental Studies (SOES), Jadavpur University, out of total 22 administrative blocks in 22, 21 and 16 blocks arsenic above the 10 µg/L (WHO Recommended value of arsenic in drinking water), 50 µg/L (Indian standard value of arsenic in drinking water) and 300 µg/L (the concentration predicting overt arsenical skin lesions) was noted respectively.

The maximum arsenic contamination level found in this district is 2830µg/L in the Baduria block.[citation needed]

CLIMATE

The climate is tropical, like the rest of the Gangetic West Bengal. It is also characterised by the Monsoon, which lasts from early June to mid September. The weather remains dry during the winter (mid November to mid February) and humid during summer.[citation needed]

Temperature ranges from 41 C in May and 10 C in January while relative humidity ranges between 50% in March & 90% in July. The average annual rainfall is 1,579mm.

ECONOMY

This section does not cite any sources. Please help improve this section by adding citations to reliable sources. Unsourced material may be challenged and removed.

Find sources: "North 24 Parganas district" – news newspapers books scholar JSTOR (October 2016) (Learn how and when to remove this template message)

People are mainly engaged in farming, fishing and other agricultural activities. The average size of agricultural landholdings is about 3.2 Bighas. North 24 Parganas is one of the less economically backward districts of West Bengal, but there is chronic poverty in the southern half of the District (the Sundarbans area)

Table-3.2. General Information of North 24 pgs

Area	**4094 Sq. Km.**
Population	**10,009,781** as per 2011 Census.
No. of Subdivision	**5**
Blocks/Munici-palities/ Canton-ment Board	Nos. of Blocks: 22, Nos. of Municipalities: 27, Cantonment Board : 1.
Languages	Main Language: Bengali with other regional languages. Link Language English & Hindi.
Transport	Road Transport : Bus, Taxi, Truck, Auto, Rickshaw, Toto, Cycle. Railway : Surface/ Circular Railway, Metro railway, Airways : international and Domestic. Water ways : Ferry services.
Ferry Ghats	Naihati, Jagatdal, Shyamnagar, Garulia, Ichhapur, Halisahar, Barrackpore, Titagarh, Khardah, Panihati, Ariadaha, Baranagar.
Railway Stations	**On Sealdah - Ranaghat main line:** Dum Dum- Belghoria - Agarpara - Sodepur - Khardaha - Titagarh - Barrackpore - Palta - Ichapur - Shyamnagar - Kankinara - Jagatdal - Naihati - Halisahar - Kanchrapara - (to Ranaghat via Kalyani). **Naihati -Bandel line (Burdwan/ Katoa via Jubilee Bridge):** Naihati, Garifa, Hooghly Ghat, Bandel. **Naihati - Budge Budge line:** Naihati - Dum Dum, Bidhan Nagar, Sir Gurudas Halt station to Budge Budge via Ballygunj. **Sealdah - Dankuni line:** Dum Dum - Baranagar - Dakshineswar - Bally Halt (To Dankuni / Burdwan / Howrah/Tarkeshwar/ Bandel). **Sealdah - Bongaon line:** Dum Dum - Dum Dum Cant. - Birati - Durganagar - New Barrackpore - Madhyamgram to Barasat / Bongaon/ Basirhat.

Circular Rail	**Naihati - Princep Ghat :** Naihati - Dum Dum, Patipukur, Tala to B.B.D.Bag, Princep Ghat.
Big Rly, Stations	Naihati Junction, Dum Dum Junction, Sealdah, Howrah, Barasat, Bangaon
Major Bus Depots	Belghoria- CSTC, South Bengal State Transport Corporation, Nilgunj- CSTC Depot, Barrackpore Court terminus, Naihati.
Common Vehicles	Bus, Taxi, Trucks, Auto, Two wheelers, Rikshaw & Cycle/Van.
Airports	Nataji Subhas Chandra Bose International Airport at DumDum, Kolkata connecting major national and International flights.
Auditoriums	Barrackpore - Sukanta Sadan, Khardaha- Rabindra Bhavan, Panihati - Lok Sanaskriti Bhawan, Kamarhati - Nazrul Mancha, Naihati - Aikyatan, Titagarh - Rabindra Bhavan etc.
Cinema Halls	Debasree, Atindra, Rathindra, Jayanti, Sonali, Anannya, Prafulla, Charuchitra, Laxmi (more than 50)
Visiting Places	Ramprasad's Vita-Halisahar, Dakshineswar Kali temple, Adyapith, Sarada Math, Gandhi Ghat/ Gandhi Museum-Barrackpore, Bankim Institute-Naihati, Panihati-Mahotsab Tala Ghat, Mangal Pandey Uddyan, Maha Milan Math- Dakshineswar, Baro Mandir Ghat- Panihati, Khardah - Lakshmi Narayan Mandir, Rasmandir- Khardah, Gobinda Home- Panihati, Ramakrishna Mission Boys Home - Rahara, Annapurna mandir - Barrackpore.
Important Festivals	Durga Puja, Kali Puja, Id, Dol Jatra, Ras mela, Bengali New years day, Baisakhi & Charak, X-mas day, Chat Puja, Pongal. Various other religious and cultural functions are held.
Nearest Sea Ports	Kolkata, Haldia
Adjacent to	Kolkata
Major Fairs	Kalpataru Utsav, Book Fairs held at various places, Flower shows, Ras mela at Khardaha, Gopastami fair- Sodepur, Rath yatra, Panihati Chira mela, .
Climate	Tropical, Barrackpore is warm and humid most of the year around. However, the winter months, say, from November to February are cool, sunny, dry and very pleasant. Monsoon hits generally in the second week of June and lasts till the end of September. During this period it rains regularly. The relative humidity is around 50%. Light woolen clothing is necessary during winter. Summer: 24-40 degree C Winter: 9-26 degree C Monsoons: June to September Rainfall: 175 Cms

Source: Office of the Register General and Census Commissioner 2011 Ministry of Home Affairs, GOI.

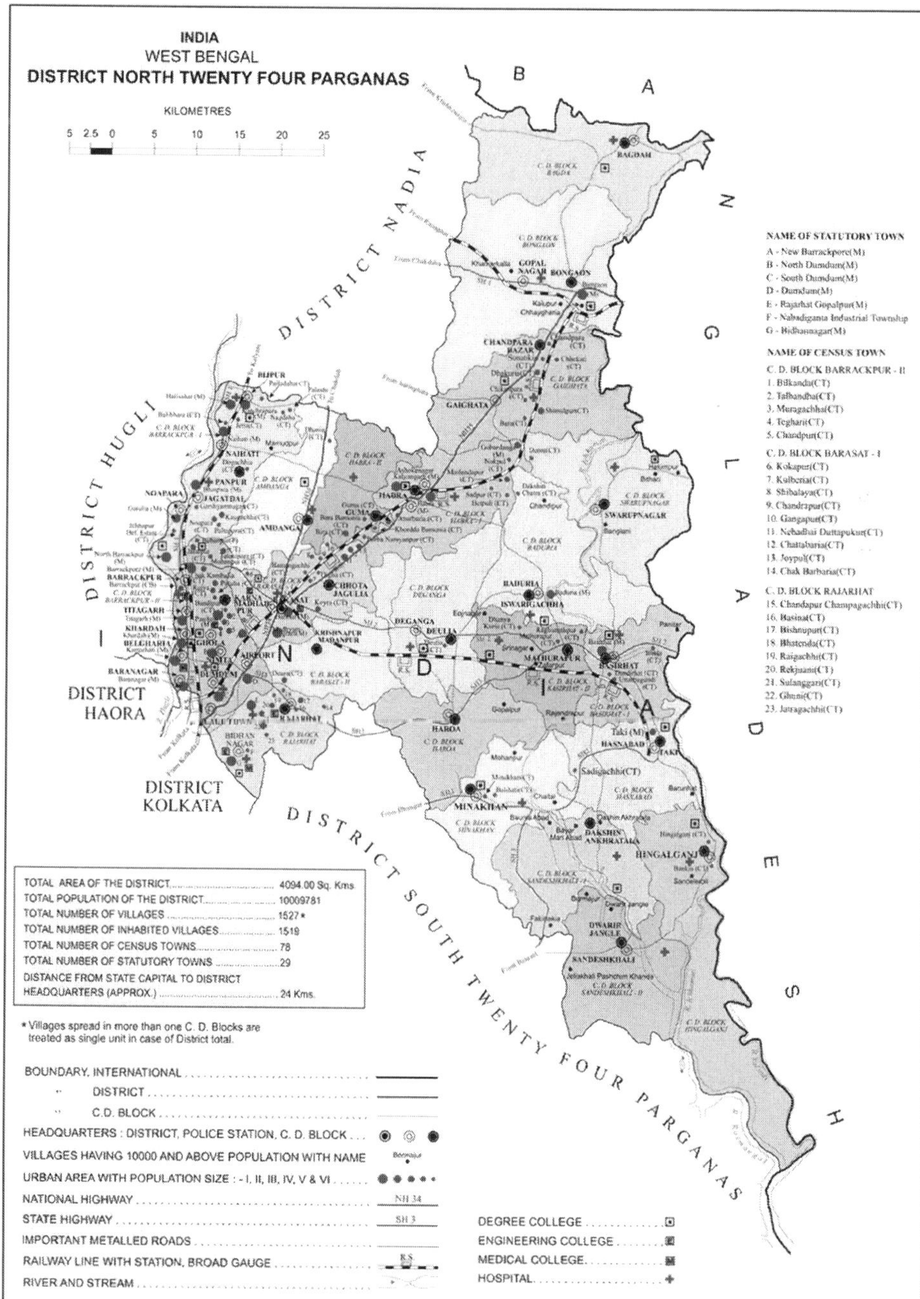

Source- Office of the District Magistrate, New Administrative Building

Barasat, North 24-Parganas, Kolkata-700 124, India

PROFILE OF HABRA II BLOCK

Habra II CD Block is bounded by Haringhata CD Block in Nadia district in the north, Habra I CD Block in the east, Barasat I CD Block in the south and Amdanga CD Block in the west.

Gram panchayats of Habra II block/ panchayat samiti are: Banspole, Bhurkunda, Guma I, Rajibpur Bira, Beraberi, Dighara Malikberia, Guma II and Srikrishnapur.

Habra II CD Block is part of the North Hooghly Flat, one of the three physiographic regions in the district located in the lower Ganges Delta. It is a raised alluvium area along the Hooghly, which forms the western boundary of the district.

Habra II CD Block has an area of 112.67 km2. It has 1 panchayat samity, 8 gram panchayats, 106 gram sansads (village councils), 78 mouzas and 78 inhabited villages.[4]Ashoknagar police station serves this block.[5] Headquarters of this CD Block is at Guma.

DEMOGRAPHICS

POPULATION

As per 2011 Census of India Habra II CD Block had a total population of 176,490, of which 140,675 were rural and 35,185 were urban. There were 90,466 (51%) males and 86,024 (49%) females. Population below 6 years was 19,756. Scheduled Castes numbered 31,665 (17.94%) and Scheduled Tribes numbered 4,075 (2.31%).

As per 2001 census, Habra II block has a total population of 149,803 out of which 77,014 were males and 72,709 were females. Population of the rural areas in the block was 134,322 and that of the urban areas was 15,471.

Census towns in Habra II CD Block were (2011 census figures in brackets): Guma (12,025), Bara Bamonia (7,193), Khorddabamonia (5,856) and Bira (10,741).

Large villages in Habra II CD Block were (2011 census figures in brackets): Dighra (4,257) and Tajpur (4,092).

North 24 Parganas district is densely populated, mainly because of the influx of refugees from East Pakistan (later Bangladesh). With a density of population of 2,182 per km2 in 1971, it was 3rd in terms of density per km2 in West Bengal after Kolkata and Howrah, and 20th in India.[9] According to the District Human Development Report: North 24 Parganas, "High density is also explained partly by the rapid growth of urbanization in the district. In 1991, the percentage of urban population in the district has been 51.23."

LITERACY

As per the 2011 census, the total number of literates in Habra II CD Block was 127,039 (81.05% of the population over 6 years) out of which males numbered 68,148

(84.77% of the male population over 6 years) and females numbered 58,891 (77.14% of the female population over 6 years). The gender disparity (the difference between female and male literacy rates) was 7.64%.

See also – List of West Bengal districts ranked by literacy rate

Literacy in CD Blocks of North 24 Parganas district

LANGUAGE AND RELIGION

As per census definition, mother-tongue is the language spoken in childhood by the person's mother to the person. As a mother-tongue, Bengali has decreased its share from 91.3% of the population of North Twenty Four Parganas district in 1961 to 87.9% in 2001, Hindi has increased its share from 5.0% in 1961 to 8.3% in 2001, Urdu has increased its share from 2.0% in 1961 to 2.5% in 2001. Other mother-tongues spoken in 2001 were Odiya (0.4%), Telugu (0.2%), Santhali (0.1%), Nepali (0.1%), Bhojpuri (0.1%) and Sadan/ Sadri (0.1%).

Religion in Habra II CD Block

Hindu	50.85%
Muslim	48.76%
Others	0.39%

In the 2011 census Hindus numbered 89,747 and formed 50.85% of the population in Habra II CD Block. Muslims numbered 86,052 and formed 48.76% of the population. Others numbered 691 and formed 0.39% of the population.

In 1981 Hindus numbered 79,050 and formed 80.30% of the population and Muslims numbered 18,883 and formed 19.20% of the population in Habra I CD Block. In 1981 Hindus numbered 60,268 and formed 67.12% of the population and Muslims numbered 29,206 and formed 32.53% of the population in Habra II CD Block. In 1991 Hindus numbered 180,690 and formed 63.80% of the population and Muslims numbered 101,571 and formed 35.84% of the population in Habra I and Habra II CD Blocks taken together. (In 1981 and 1991 census was conducted as per jurisdiction of the police station.)

In the 2011 census, Hindus numbered 7,352,769 and formed 73.46% of the population in North 24 Parganas district. Muslims numbered 2,584,684 and formed 25.82% of the population.[21] In West Bengal Hindus numbered 64,385,546 and formed 70.53% of the population. Muslims numbered 24,654,825 and formed 27.01% of the population.

See also - North 24 Parganas: minority concentrated district

RURAL POVERTY

31.75% of households in Habra II CD Block lived below poverty line in 2001, against an average of 29.28% in North 24 Parganas district.

ECONOMY AND LIVELIHOOD

Cultivators (15.50%)

Agricultural labourers (30.97%)

Household industries (4.96%)

Other Workers (48.57%)

In Habra II CD Block in 2011, amongst the class of total workers, cultivators numbered 10,001 and formed 15.50% of the total workers, agricultural labourers numbered 19,978 and formed 30.97%, household industry workers numbered 3,199 and formed 4.96% and other workers numbered 31,329and formed 48.57%. Total workers numbered 64,507 and formed 36.55% of the total population, and non-workers numbered 111,983 and formed 63.45% of the population.

In more than 30 percent of the villages in North 24 Parganas, agriculture or household industry is no longer the major source of livelihood for the main workers there. The CD Blocks in the district can be classified as belonging to three categories: border areas, Sundarbans area and other rural areas. The percentage of other workers in the other rural areas category is considerably higher than those in the border areas and Sundarbans area.

> **Note:** In the census records a person is considered a cultivator, if the person is engaged in cultivation/ supervision of land owned by self/government/ institution. When a person who works on another person's land for wages in cash or kind or share, is regarded as an agricultural labourer. Household industry is defined as an industry conducted by one or more members of the family within the household or village, and one that does not qualify for registration as a factory under the Factories Act. Other workers are persons engaged in some economic activity other than cultivators, agricultural labourers and household workers. It includes factory, mining, plantation, transport and office workers, those engaged in business and commerce, teachers, entertainment artistes and so on.

INFRASTRUCTURE

There are 76 inhabited villages in Habra II CD Block. 100% villages have power supply and drinking water supply. 13 villages (17.11%) have post offices. 75 villages (98.68%) have telephones (including landlines, public call offices and mobile phones). 49 villages (64.47%) have a pucca approach road and 13 villages (17.11%) have transport communication (includes bus service, rail facility and navigable waterways). 10 villages (13.16%) have agricultural credit societies and 8 villages (10.53%) have banks.

AGRICULTURE

The North 24 Parganas district Human Development Report opines that in spite of agricultural productivity in North 24 Parganas district being rather impressive

81.84% of rural population suffered from shortage of food. With a high urbanisation of 54.3% in 2001, the land use pattern in the district is changing quite fast and the area under cultivation is declining. However, agriculture is still the major source of livelihood in the rural areas of the district.

From 1977 on wards major land reforms took place in West Bengal. Land in excess of land ceiling was acquired and distributed amongst the peasants.[29] Following land reforms land ownership pattern has undergone transformation. In 2010-11, persons engaged in agriculture in Habra II CD Block could be classified as follows: bargadars 815 (2.76%), patta (document) holders 2,730 (9.25%), small farmers (possessing land between 1 and 2 hectares) 915 (3.10%), marginal farmers (possessing land up to 1 hectare) 12,255 (41.53%) and agricultural labourers 12,792 (43.35%).

Habra II CD Block had 120 fertiliser depots, 30 seed stores and 34 fair price shops in 2010-11.

In 2010-11, Habra II CD Block produced 15,117 tonnes of Aman paddy, the main winter crop from 6,054 hectares, 11,066 tonnes of Boro paddy (spring crop) from 3,424 hectares, 554 tonnes of Aus paddy (summer crop) from 215 hectares, 154 tonnes of wheat from 98 hectares, 34,873 tonnes of jute from 1,932 hectares and 8,143 tonnes of potatoes from 329 hectares. It also produced pulses and oilseeds.

In 2010-11, the total area irrigated in Habra II CD Block was 955 hectares, out of which 631 hectares were irrigated by deep tube well and 324 hectares by shallow tube well.

Pisciculture

In 2010-11, the net area under effective pisciculture in Habra II CD Block was 824.84 hectares. 18,675 persons were engaged in the profession. Approximate annual production was 24,745.2 quintals.

Banking

In 2010-11, Habra II CD Block had offices of 6 commercial banks and 2 gramin banks.

Transport

Sealdah-Hasnabad-Bangaon-

Ranaghat line

In 2010-11, Habra II CD Block had 10 originating/ terminating bus routes.

NH 112 (old numbering NH 35) (also known as Jessore Road) passes through the Guma area of this CD Block.

There are stations like - Guma railway station and Bira railway station on the Sealdah-Bangaon line.

EDUCATION

In 2010-11, Habra II CD Block had 75 primary schools with 8,883 students, 8 high schools with 6,091 students and 10 higher secondary schools with 13,664 students. Habra II CD Block had 239 institutions for special and non-formal education with 10,229 students.

There are 12 Secondary Schools in this community development block affiliated to West Bengal Board of Secondary Education.

- Dighara Haradayal Vidyapith
- Tangra Adarsha Sikshaniketan
- Rajibpore A.V. High School
- Rajibpore Girls High School
- Bira Ballavpara High School
- Bira Ballavpara Santipipasha Ballav High School
- Guma Rabindra Vidyapith
- Nazrul Balika Vidyalaya
- Suria Moulana Abul Kalam Azad High School
- Natni High School
- Mena Sarojini Pahar High School
- Sendanga Chaturdash Pally High School

As per the 2011 census, in Habra II CD Block, amongst the 76 inhabited villages, 2 villages did not have a school, 18 villages had more than 1 primary school, 19 villages had at least 1 primary and 1 middle school and 13 villages had at least 1 middle and 1 secondary school.

Healthcare

In 2011, Habra II CD Block had 1 block primary health centre and 2 primary health centres, with total 23 beds and 4 doctors (excluding private bodies). It had 27 family welfare subcentres. 1,413 patients were treated indoors and 71,591 patients were treated outdoor in the hospitals, health centres and subcentres of the CD Block.

Sabdalpur Rural Hospital at Sabdalpur with 30 beds functions as the main medical facility in Habra II CD Block. There are primary health centres at Bira Ballabhpur (Pukurkona PHC with 10 beds) and Sendanga (Pumlia PHC with 6 beds).

Habra II block is one of the areas where ground water is affected by arsenic contamination.

PROFILE OF BARRACKPORE II BLOCK

Barrackpore II CD Block is bounded by Barrackpore I CD Block in the north, Amdanga and Barasat I CD Blocks in the east and Chinsurah Mogra CD Block in

Hooghly district, across the Hooghly River, in the west. The Barrackpore industrial belt with municipalities such as New Barrackpur, Panihati, Khardaha, Titagarh and Barrackpore is in the south and west.

Barrackpore II CD Block is part of the North Hooghly Flat, one of the three physiographic regions in the district located in the lower Ganges Delta. It is a raised alluvium area along the Hooghly, which forms the western boundary of the district.

Barrackpore II CD Block has an area of 40.74 km2. It has 1 panchayat samity, 6 gram panchayats, 104 gram sansads (village councils), 21 mouzas and 21 inhabited villages, as per the District Statistical Handbook: North 24 Parganas. Khardaha police station serves this block. Headquarters of this CD Block is at Karna Madhabpur.

Gram panchayats of Barrackpore II block/ panchayat samiti are: Bandipur, Bilkanda–II, Patulia, Bilkanda–I, Mohanpur and Sewli Telenipara.

DEMOGRAPHICS

Population

As per 2011 Census of India Barrackpur II CD Block had a total population of 217,171, of which 51,874 were rural and 165,297 were urban. There were 111,282 (51%) males and 105,889 (49%) females. Population below 6 years was 20,953. Scheduled Castes numbered 53,987 (24.86%) and Scheduled Tribes numbered 4,906 (2.26%).

As per 2001 census, Barrackpore II block has a total population of 158,778 out of which 82,573 were males and 76,205 were females.

There are several census towns in Barrackpore II CD Block (2011 census figures in brackets): Babanpur (P) (8,942), Jafarpur (19,062), Mohanpur (9,096), Telenipara (17,781), Chak Kanthalia (11,108), Ruiya (17,661), Patulia (16,979), Bandipur (8,115), Talbandha (17,802), Bilkanda (6,081), Muragachha (13,249), Teghari (8,491) and Chandpur (10,930).

Large villages in Barrackpore II CD Block (2011 census figures in brackets): Chhota Kanthalia (4,052), Surjyapur (5,714), Dopere (5,276), Dhangadihila (5,827) and Iswaripur (5,155).

North 24 Parganas district is densely populated, mainly because of the influx of refugees from East Pakistan (later Bangladesh). With a density of population of 2,182 per km2 in 1971, it was 3rd in terms of density per km2 in West Bengal after Kolkata and Howrah, and 20th in India.[9] According to the District Human Development Report: North 24 Parganas, "High density is also explained partly by the rapid growth of urbanization in the district. In 1991, the percentage of urban population in the district has been 51.23."

LITERACY

As per the 2011 census, the total number of literates in Barrackpore II CD Block was 165,866 (84.53% of the population over 6 years) out of which males numbered

88,801 (88.33% of the male population over 6 years) and females numbered 77,065 (80.54% of the female population over 6 years). The gender disparity (the difference between female and male literacy rates) was 7.79%.

See also – List of West Bengal districts ranked by literacy rate

Literacy in CD Blocks of North 24 Parganas district

LANGUAGE AND RELIGION

As per census definition, mother-tongue is the language spoken in childhood by the person's mother to the person. As a mother-tongue, Bengali has decreased its share from 91.3% of the population of North Twenty Four Parganas district in 1961 to 87.9% in 2001, Hindi has increased its share from 5.0% in 1961 to 8.3% in 2001, Urdu has increased its share from 2.0% in 1961 to 2.5% in 2001. Other mother-tongues spoken in 2001 were Odiya (0.4%), Telugu (0.2%), Santhali (0.1%), Nepali (0.1%), Bhojpuri (0.1%) and Sadan/ Sadri (0.1%).[20] Around 35% of the population in Barrackpore (Lok Sabha constituency) is Hindi-speaking.

Religion in Barrackpur II CD Block

Hindu	77.70%
Muslim	21.46%
Others	0.84%

In the 2011 census Hindus numbered 168,756 and formed 77.70% of the population in Barrackpur II CD Block. Muslims numbered 46,594 and formed 21.46% of the population. Others numbered 1,821 and formed 0.84% of the population.

In 1981 Hindus numbered 133,647 and formed 79.18% of the population and Muslims numbered 33,444 and formed 20.56% of the population in Barrackpur I CD Block. In 1981 Hindus numbered 113,755 and formed 76.76% of the population and Muslims numbered 34,003 and formed 22.94% of the population in Barrackpur II CD Block. In 1991 Hindus numbered 39,880 and formed 86.10% of the population and Muslims numbered 61,462 and formed 13.27% of the population in Barrackpur I and Barrackpur II CD Blocks taken together. (In 1981 and 1991 census was conducted as per jurisdiction of the police station.)

In the 2011 census, Hindus numbered 7,352,769 and formed 73.46% of the population in North 24 Parganas district. Muslims numbered 2,584,684 and formed 25.82% of the population.[22] In West Bengal Hindus numbered 64,385,546 and formed 70.53% of the population. Muslims numbered 24,654,825 and formed 27.01% of the population.

See also - North 24 Parganas: minority concentrated district

RURAL POVERTY

16.78% of households in Barrackpore II CD Block lived below poverty line in 2001, against an average of 29.28% in North 24 Parganas district.

- Economy
- Livelihood
- Livelihood
- in Barrackpore II CD Block
- Cultivators (1.89%)
- Agricultural labourers (1.91%)
- Household industries (3.88%)
- Other Workers (92.32%)

In Barrackpore II CD Block in 2011, amongst the class of total workers, cultivators numbered 1,444 and formed 1.89% of the total workers, agricultural labourers numbered 1,461 and formed 1.91%, household industry workers numbered 2,963 and formed 3.88% and other workers numbered 70,541 and formed 92.32%. Total workers numbered 76,409 and formed 35.18% of the total population, and non-workers numbered 140,762 and formed 64.82% of the population.

In more than 30 percent of the villages in North 24 Parganas, agriculture or household industry is no longer the major source of livelihood for the main workers there. The CD Blocks in the district can be classified as belonging to three categories: border areas, Sundarbans area and other rural areas. The percentage of other workers in the other rural areas category is considerably higher than those in the border areas and Sundarbans area.

> **Note:** In the census records a person is considered a cultivator, if the person is engaged in cultivation/ supervision of land owned by self/government/ institution. When a person who works on another person's land for wages in cash or kind or share, is regarded as an agricultural labourer. Household industry is defined as an industry conducted by one or more members of the family within the household or village, and one that does not qualify for registration as a factory under the Factories Act. Other workers are persons engaged in some economic activity other than cultivators, agricultural labourers and household workers. It includes factory, mining, plantation, transport and office workers, those engaged in business and commerce, teachers, entertainment artistes and so on.

INFRASTRUCTURE

There are 15 inhabited villages in Barrackpore II CD Block, as per the District . 100% villages have power supply and drinking water supply. 2 villages (13.33%) have post offices. 15 villages (100%) have telephones (including landlines, public call offices and mobile phones). 15 villages (100%) have a pucca approach road and 7 villages (46.67%) have transport communication (includes bus service, rail facility and navigable waterways). 1 village (6.67%) has a bank.

AGRICULTURE

The North 24 Parganas district Human Development Report opines that in spite of agricultural productivity in North 24 Parganas district being rather impressive 81.84% of rural population suffered from shortage of food. With a high urbanisation of 54.3% in 2001, the land use pattern in the district is changing quite fast and the area under cultivation is declining. However, agriculture is still the major source of livelihood in the rural areas of the district.

From 1977 on wards major land reforms took place in West Bengal. Land in excess of land ceiling was acquired and distributed amongst the peasants.[30] Following land reforms land ownership pattern has undergone transformation. In 2010-11, persons engaged in agriculture in Barrackpore II CD Block could be classified as follows: bargadars 22 (0.48%), patta (document) holders 632 (13.92%), small farmers (possessing land between 1 and 2 hectares) 180 (3.97%), marginal farmers (possessing land up to 1 hectare) 2,170 (47.81%) and agricultural labourers 1,535 (33.82%).

Barrackpore II CD Block had 10 fertiliser depots, 2 seed stores and 24 fair price shops in 2010-11.

In 2010-11, Barrackpore II CD Block produced 322 tonnes of Aman paddy, the main winter crop from 122 hectares, 1,515 tonnes of Boro paddy (spring crop) from 514 hectares, 513 tonnes of jute from 30 hectares, 287 tonnes of potatoes from 15 hectares and 81 tonnes of sugar cane from 1 hectare. It also produced oilseeds.

In 2010-11, the total area irrigated in Barrackpore II CD Block was 209 hectares, of which 100 hectares were irrigated by river lift irrigation and 109 hectares by deep tube well.

PISCICULTURE

In 2010-11, the net area under effective pisciculture in Barrackpore II CD Block was 294.71 hectares. 5,785 persons were engaged in the profession. Approximate annual production was 8,841.3 quintals.[31]

BANKING

In 2010-11, Barrackpore II CD Block had offices of 12 commercial banks and 1 gramin bank.

TRANSPORT

In 2010-11, Barrackpore II CD Block had 2 originating/ terminating bus routes.

The SH 1 (between Shaymbazar and Barrackpore it is also known as Barrackpore Trunk Road), the Sealdah-Ranaghat line and the Sealdah-Bangaon line pass through the adjoining urban area. SH 2 passes through the CD Block.

EDUCATION

In 2010-11, Barrackpore II CD Block had 24 primary schools with 3,091 students, 6 high schools with 3,025 students and 9 higher secondary schools with 10,520 students. Barrackpore II CD Block had 1 general college with 1,782 students and 280 institutions for special and non-formal education with 14,275 students.

As per the 2011 census, in Barrackpore II CD Block, amongst the 15 inhabited villages, all villages had a school, 9 villages had more than 1 primary school, 5 villages had at least 1 primary and 1 middle school and 4 villages had at least 1 middle and 1 secondary school.

HEALTHCARE

In 2011, Barrackpore II CD Block had 1 block primary health centre and 2 primary health centres, with total 15 beds and 3 doctors (excluding private bodies). It had 31 family welfare subcentres. 768 patients were treated indoor and 39,571 patients were treated outdoor in the hospitals, health centres and subcentres of the CD Block.

Table 3.3: Beraberi gram panchayat at a glance

Area	25 sq.kms
Total population	26,253
Male	13651
Female	12602
Heal center	4
BPHC(block primary health centre)	1
Total agricultural area(bigha)	692
Post office	1
Rural bank	1
Anganwari	30
Primary school	12
High school	2
Madhyamik shiksha kendra	2
Sishu ahiksha kendra	5

Table 3.4: Bilkanda gram panchayat at a glance

Area	0.5321sq.kms
Total population	6,081
Male	3,156
Female	2,925
Children	514
Sex ratio	927
Child sex ratio	797
No. of SHG	5
Literacy rate	82.72%
No. of houses	1,532
Bank	1
Post office	1
Police station	1
BPHC(block primary health centre)	1
Primary school	1
Secondary school	1

5

CHAPTER

METHODS AND APPROACHES

The deliberation on the methodology has been made to understand the concept, methods and techniques which are utilized to design the study, collection of information, analysis of data and interpretation of the findings for revelation of truths and formulation of theories. These chapter deals with the method and a procedure used in the study an consist of eight main parts-

(a) Locale of research.

(b) Pilot study.

(c) Sampling Design.

(d) Empirical measurement of the variables.

(e) Preparation of interview Schedule.

(f) Pre-testing of Interview Schedule.

(g) Techniques of Data collection.

(h) Statistical Tools used for Analysis of Data.

A. LOCALE OF RESEARCH

Ghoragachha village of the Haringhata block of Nadia district in West Bengal was randomly selected for the study. The area has been selected for the study because of – a)There is sample scope for collecting relevant data for the present study, b) Acquaintance with the local people as well as local language, c)The concern area was very easily accessible to the researcher in terms of place of residence, d)the area was very easily accessible to the researcher in terms of transportation and e) The closer familiarities of the student researchers with the area, people, officials and local dialects.

B. PILOT STUDY

Before taking up actual field work a pilot study was conducted to understand the area, its people, institution, communication and extension system and the knowledge, perception and attitude of the people towards climate changes concept.

C. SAMPLING DESIGN

Purposive as well as simple random sampling techniques were adopted for the study.

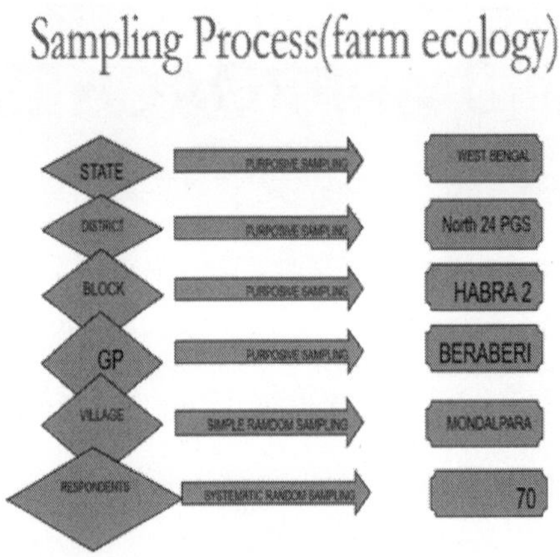

Model 5.1

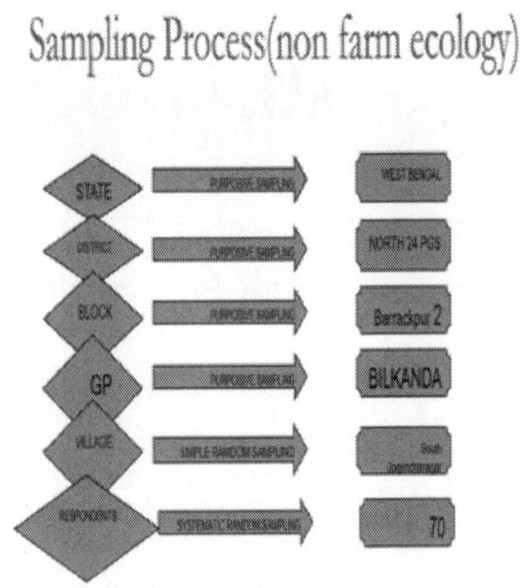

Model 5.2

D. EMPIRICAL MEASUREMENT OF THE VARIABLES

After reviewing various literature related to the field of study and consultation with respected chairman Advisory Committee and other experts, a list of variables was prepared.

INDEPENDENT VARIABLES (FARM SECTOR)

- X1 = age
- X2 = education
- X3 = family size
- X4 = economic motivation
- X5 = risk orientation
- X6 = management orientation
- X7 = stress perception on hunger
- X8 = stress perception on poverty
- X9 = stress perception on voice

- X10 = size of holding
- X11 = cropping intensity
- X12 = livestock count
- X13 = livestock yield
- X14 = pond and fish
- X15 = total crop yield
- X16 = total cost of cultivation
- X17 = No. of fragments
- X18 = communication variables
- X19 = marketed surplus
- X20 = energy consumption
- X21 = BMI

INDEPENDENT VARIABLES (NON FARM SECTOR)

- X1 = age
- X2 = education
- X3 = family size
- X4 = economic motivation
- X5 = risk orientation
- X6 = management orientation
- X7 = stress perception on hunger
- X8 = stress perception on poverty
- X9 = stress perception on voice
- X10 = homestead land
- X11 = occupation(total no. of years engaged with the particular occupation)
- X12 = ancillary income
- X13 = expenditure
- X14 = savings
- X15 = bank account
- X16 = communication variables
- X17 = management variables
- X18 = energy consumption
- X19 = BMI

INDEPENDENT VARIABLES (POOLED SAMPLE)

Independent variables

- X1 = age
- X2 = education
- X3 = family size
- X4 = economic motivation
- X5 = risk orientation
- X6 = management orientation
- X7 = stress perception on hunger
- X8 = stress perception on poverty
- X9 = stress perception on voice
- X10 = Homesteadland
- X11 = Communication variables
- X12 = Energy consumption
- X13 = BMI

DEPENDENT VARIABLES

- Y1 = Hunger
- Y2 = poverty
- Y3 = voice
- Y4 = Cognitive differential.

To compare the parameters and factors of hunger, poverty and voicelessness of both farm and non farm sector, we have done pooled analysis with thirteen independent variables those are common for both the sectors and four dependent variables.

Those 13 common independent variables are:

- X1 = age
- X2 = education
- X3 = family size
- X4 = economic motivation
- X5 = risk orientation
- X6 = management orientation
- X7 = stress perception on hunger
- X8 = stress perception on poverty

- X9 = stress perception on voice
- X10 = Homesteadland
- X11 = Communication variables
- X12 = Energy consumption
- X13 = BMI

Dependent variables are the same as farm and non farm sectors.

MEASUREMENT OF VARIABLES (FARM)

Age (X1)

In all societies, age is one of the most important determinants of social status and social role of the individual. In the present study, age of the respondent was measured on the basis of their chronological age at the time of investigation.

Education (X2)

Education is instrumental is building personality structure and helps in changing one's behavior in social life. Education may be conceptualized as the amount of formal schooling literacy acquired by the responded.

Family Size (X3)

Number of family members of an individual farmers.

Economic motivation (X4)

Asked the respondents a few pertinent question and recorded the responses in a 10 point score.

Risk Orientation (X5)

Asked the respondents a few pertinent question and recorded the responses in a 10 point score.

Management Orientation (X6)

Asked the respondents a few pertinent question and recorded the responses in a 10 point score.

Stress perception on hunger (X7)

Asked the respondents a few pertinent question and recorded the responses in a 10 point score.

Stress perception on poverty (X8)

Asked the respondents a few pertinent question and recorded the responses in a 10 point score.

Stress perception on voice (X9)

Asked the respondents a few pertinent question and recorded the responses in a 10 point score.

Size of Holding (X10)

The amount of land owned by a person is an important parameter to access the economic status of the person in the society.

Cropping Intensity (X11)

It has been conceptualized as the proportion of total annual cropped area to the size of holding expressed in percentage. Its calculated as-

Cropping Intensity = Total annual cropped area/size of holding 100

Livestock count (X12)

Per capita possession of livestock i.e total no. of livestock owned by the family is divided by the family size.

Livestock yield (X13)

Total no. of eggs/day + total amount of milk /day(in lt.) is divided by the family size.

Pond and fish (X14)

Total no. of ponds is divided by the family size

Crop Yield (X15)

Yield of crops in the farmers' field is divided by the family size.

Total cost of cultivation (X16)

Per bigha cost of cultivation is multiplied by the size of holding and then was divided by the family size to bring out per capita cost of cultivation.

Number of Fragments (X17)

Number of fragments in each plot.

Communication variables (X18)

Asked the respondents a few pertinent question and recorded the responses in a 10 point score.

Management variables (X19)

Asked the respondents a few pertinent question and recorded the responses in a 10 point score.

Energy consumption (X20)

Total energy (wood+LPG+electricity) consumed by the family per day is converted into Joule equivalent and then divided by the family size.

BMI (X21)

Body weight of the respondents (in Kgs) is divided by the Height of the respondents (in cm.).

Note: For all the qualitative variables, items were selected through expert rating and finalized by summative rating scales. Pre post application of the trial items have been carried out and the validity of the variables have been estimated from R value.

MEASUREMENT OF INDEPENDENT VARIABLES (NON-FARM)

Age (X1)

In all societies, age is one of the most important determinants of social status and social role of the individual. In the present study, age of the respondent was measured on the basis of their chronological age at the time of investigation.

Education (X2)

Education is instrumental is building personality structure and helps in changing one's behavior in social life. Education may be conceptualized as the amount of formal schooling literacy acquired by the responded.

Family Size (X3)

Number of family members of an individual farmers.

Economic motivation (X4)

Asked the respondents a few pertinent question and recorded the responses in a 10 point score.

Risk Orientation (X5)

Asked the respondents a few pertinent question and recorded the responses in a 10 point score.

Management Orientation (X6)

Asked the respondents a few pertinent question and recorded the responses in a 10 point score.

Stress perception on hunger (X7)

Asked the respondents a few pertinent question and recorded the responses in a 10 point score.

Stress perception on poverty (X8)

Asked the respondents a few pertinent question and recorded the responses in a 10 point score.

Stress perception on voice (X9)

Asked the respondents a few pertinent question and recorded the responses in a 10 point score.

Homestead Land (X10)

Amount of land acquired by the home building and surrounding (in katha). This total amount is divided by the family size.

Occupation (X11)

Number of the years since the respondent is engaged in the present occupation.

Ancillary income (X12)

Extra income other than the main income of the family from is divided by the family size.

Expenditure (X13)

Total family expenditure per month is divided by family size.

Savings (X14)

Per capita Income- per capita expenditure

Bank account (X15)

Asked the respondents whether they have bank account or not (if yes then for how many years they are having it) and recorded the response accordingly. If they haven't bank account then they are scored 1 and if they have then the score given is number of years of having bank account plus 1.

Communication variables (X16)

Asked the respondents a few pertinent question and recorded the responses in a 10 point score.

Management variables (X17)

Asked the respondents a few pertinent question and recorded the responses in a 10 point score.

Energy consumption (X18)

Total energy (wood+LPG+electricity) consumed by the family per day is converted into Joule equivalent and then divided by the family size.

BMI (X19)

Body weight of the respondents (in Kgs) is divided by the Height of the respondents (in cm.).

Note: For all the qualitative variables, items were selected through expert rating and finalized by summative rating scales. Pre post application of the trial items have been carried out and the validity of the variables have been estimated from R value.

MEASUREMENT OF DEPENDENT VARIABLES (FARM AND NON-FARM SECTOR)

Hunger (Y_1)

Total calorie intake of the respondents per day.

Poverty (Y_2)

Per capita family income plus per capita credit (if any) taken.

Voice (Y_3)

Participation in community gatherings, number of times acts as speaker, number of times acts a listener are measured in 10 point score.

Cognitive Differential (Y_4)

Awareness of the respondents towards the development schemes of the government is measured in 10 point score.

E. CONSTRUCTION OF INTERVIEW SCHEDULE.

On the basis of finding pilot study a preliminary schedule was formed with the help of literature and by the assistance of Chairman of Advisory Committee. The interview schedule consisted of three major parts according to the specific objective of the study.

F. PRE-TESTING OF INTERVIEW SCHEDULE.

It's the process of advance testing of the study design after the schedule has been prepared. The object of pretesting is to detect the discrepancies that have emerged and to remove them after necessary modification in the schedule. After conducting pretesting appropriate changes and modification of the interview schedule have been made. The individuals who responded in pretesting have been excluded in the final sample selected for the study.

G. TECHNIQUES OF FIELD DATA COLLECTION.

The respondents were personally interviewed during durga puja vacation and summer vacation. The items were asked in local language, Bengali .The entire were done in the schedule by student investigator himself at the time of interview.

H. STATISTICAL TOOLS FOR ANALYSIS

The statistical methods for analysis and interpretation of raw data were..

(a) Mean.
(b) Standard deviation.
(c) Coefficient of Variance.
(d) Correlation of coefficient.
(e) Step down regression analysis.
(f) Canonical covariate analysis.
(g) Factor analysis
(h) Path analysis
(i) Artificial Neural Network Analysis
(j) Partial correlation
(k) Participatory Rural Appraisal
(l) Multiple stepwise logistic forward regressions

(a) Mean

The mean is the arithmetic average and is the result obtained when the sum of the of value of individual in the data is divided by the number of individuals in the data. Mean is and relatively stable measure of central tendency. The mean reflects and is by every score in the distribution.

When the data are expressed in a frequency distribution (grouped), the mean is calculated using the following formula -

$$X = \frac{\sum_{1=1}^{N} fixi}{N}$$

Where,

x = Mean of the observation.

fi = Frequency ofthe class.

xi = Mid value of the class.

N = Total number of observation

(b) Standard deviation

Standard deviation (SD) of a set of observation is the square root of the arithmetic mean of the squares of the deviations. The deviations being measured from the arithmetic mean of the distributions. lt is commonly denoted by the symbol (Sigma). To measure the average deviation from the standard value of the data standard deviation is used. It is less affected by sampling en-ors and is a more stable measure of dispersion.

The standard deviation of the data grouped in the form of frequency distribution is computed by using the following formula –

$$S.D. = \sqrt{\frac{\sum_{1=1}^{N} fixi}{N} - \left[\frac{\sum_{1=1}^{N} fixi}{N}\right]^2}$$

When,

δ = Standard deviation

N = total No of observation in a particular coll.

X = value of observation in a particular cell

F = Frequency of observation

$\overline{X}$ = mean number of observation

I = any number (e.g. 1, 2, 3) denoting position

(c) Coefficient of Variance

A measure of variation which is independent of the unit of measurement is proved by the coefficient of variation. Being unit free, this is useful For comparison of variability between different populations. The coefficient of variation is standard deviation expressed as percentage of the mean.

Coefficient of variation is measured by using the following formula —

$$C.V. = \frac{S.D.}{Mean} \times 100$$

(d) Coefficient of correlation:

When increase or decrease in one variety is accompanied by an increase or in another variety, the two are said to be correlated and Ute phenomenon is known as correlation. Correlation coefficient (r) is a measure of the relationship between two variables, which are at the interval or rational level of measurement and are linearly related. A Person product-moment "r" is computed by the formula.

$$r_{xy} = \frac{N\sum XY(\sum X)(\sum Y)}{\sqrt{[N\sum X^2 - (\sum X)^2][N\sum Y^2 - (\sum Y)^2]}}$$

When,

X and Y = Original scares in variables X and Y

N = Number ofpaired scores

$\sum XY$ = Each X multiplied by its comzsponding Y. then summed

$\sum X$ = Sum ol'X scores

$\sum X^2$ = Each of X squared, then summed

$(\sum X)^2$ = Sum of X score squared

$\sum Y$ = Sum of Y scores

$\sum Y^2$ = Each of Y squared, than summed

$(\sum Y)^2$ = Sum of Y score squared

The range of correlation coefficient is between -1 to +1. This perfect negative correlation +1 perfect positive correlation. A perfect correlation is, however, seldom achieved. A correlation Coefficient to be acceptable should be statistically significant. Otherwise, we say that no significant relationship exist between the variables.

(e) Step down regression analysis

Stepwise regression is a which provides a means of choosing independent variables that yield the best possible with the fewest independent variables. It permts the user to solve a sequence of one more multiple linear regression problems by stepwise application of the least square method. At each step in the analysis. a variable is added or removed which results in the greatest production in the error sum of squares .

(f) Canonical covariate analysis

In statistics, canonical-covariate analysis (CCA) is a way of making sense of cross-covariance matrices. If we have two vectors X = (Xi, ..Xa) and Y = (Yl, ... Ym) of random variables, and there are variables, then canonical-correlation analysis will find linear combinations of the Xi and Yj whish have maximum correlation with each other. Virtually all of the commonly encountered parametric tests of significance can be treated as special cases of canonical-correlation analysis, which is the general procedure for investigating the relationships between two sets of variables. The method was first introduced by Harold Hotelling in I936.

Given two column vectors X = $(x_1 .....x_a)^1$and Y = $(y_1 ...y_m)^1$ of random variables with finite second moments, one may define the cross-covariance $\sum_{xy} = CN(X.Y)$ to be the n m matrix whose (i.j) entry is the convarience$^{CON\ (r,y)}$. In the covariance $^{CON\ (r,y)}$. In practice, we would estimate the covariance matrix based on sampled data from X and Y (i.e. from a pair of data matrices).

Canonical-correlation analysis seeks vectors a′ and b′ such that the random variables a^1X and b′Y maximize the correlation p = *corr (a′X,b′Y)*. The random variables $U = a'X$ and $V = b'Y$ are the first pair of canonical variables. Then one seeks vectors maximizing the same correlation subject to the constraint that they are to be uncorrelated with the first pair of canonical variables; this gives the second pair of canonical variables. This procedure may be continued up to $^{\min(m,\ n)}$ times.

(g) Factor Analysis

Factor analysis is a very useful and popular method of multivariate research technique, mostly used in social and behavioral sciences. factor analysis seeks to resolve a large set of measured variables in terms of relatively few categories, known as factors. This technique allows the researcher to group variables into factors (based on correlation between variables); the factors so derived may be treated as new variables (often termed as latent variables) and their value derived by summing the values of the original variables, which had been grouped into the factor. The meaning and name of such new variable is subjectively determined by the researcher. Since the factors happen to be linear combinations of data, the coordinates of each observation or variable is measured to obtain what are factor loadings. Such factor loading represent the correlation between the variable and the factor and are usually placed in a matrix of correlations of the variables and the factors. In the Factor Analysis the "Principle Component Method" was followed.

(h) Path Analysis

Path Analysis is a straight forward extension of multiple regression. Its aim is to provide estimations of the magnitude and significance of hypothesized causal connections between sets of variables. Path analysis is a statistical method with a structural model, but not measurement model. In other terms it is called as causal modeling, analysis of co variance structures and latent variable models. The reason for the name is that the techniques allow us to test theoretical propositions about cause and effect without manipulating variables. However , the causal in causal modeling refers to an assumption of the model rather than a property of the output or consequence of the technique. That is, people assume some variables are causally related and the test propositions about them using the technique. If the propositions are supported , it does not prove that the causal assumptions are correct.

(i) Artificial Neural Network Analysis

Artificial neural networks (**ANN**) or **connectionist systems** are computing systems vaguely inspired by the biological neural networks that constitute animal brains. Such systems «learn» to perform tasks by considering examples, generally without being programmed with task-specific rules. For example, in image recognition, they might learn to identify images that contain cats by analyzing example images that have been manually labeled as «cat» or «no cat» and using the results to identify cats in other images. They do this without any prior knowledge of cats, for example, that they have fur, tails, whiskers and cat-like faces. Instead, they automatically generate identifying characteristics from the examples that they process.

An ANN is based on a collection of connected units or nodes called artificial neurons, which loosely model the neurons in a biological brain. Each connection, like the synapses in a biological brain, can transmit a signal to other neurons. An artificial neuron that receives a signal then processes it and can signal neurons connected to it.

In ANN implementations, the "signal" at a connection is a real number, and the output of each neuron is computed by some non-linear function of the sum of its inputs. The connections are called *edges*. Neurons and edges typically have a *weight* that adjusts as learning proceeds. The weight increases or decreases the strength of the signal at a connection. Neurons may have a threshold such that a signal is sent only if the aggregate signal crosses that threshold. Typically, neurons are aggregated into layers. Different layers may perform different transformations on their inputs. Signals travel from the first layer (the input layer), to the last layer (the output layer), possibly after traversing the layers multiple times.

The original goal of the ANN approach was to solve problems in the same way that a human brain would. However, over time, attention moved to performing specific tasks, leading to deviations from biology. ANNs have been used on a variety of tasks, including computer vision, speech recognition, machine translation, social network filtering, playing board and video games, medical diagnosis and even in activities that have traditionally been considered as reserved to humans, like painting.

(j) Partial Correlation

In probability theory and statistics, **partial correlation** measures the degree of association between two random variables, with the effect of a set of controlling random variables removed. If we are interested in finding whether or to what extent there is a numerical relationship between two variables of interest, using their correlation coefficient will give misleading results if there is another, confounding, variable that is numerically related to both variables of interest. This misleading information can be avoided by controlling for the confounding variable, which is done by computing the partial correlation coefficient. This is precisely the motivation for including other right-side variables in a multiple regression; but while multiple regression gives unbiased results for the effect size, it does not give a numerical value of a measure of the strength of the relationship between the two variables of interest.

For example, if we have economic data on the consumption, income, and wealth of various individuals and we wish to see if there is a relationship between consumption and income, failing to control for wealth when computing a correlation coefficient between consumption and income would give a misleading result, since income might be numerically related to wealth which in turn might be numerically related to consumption; a measured correlation between consumption and income might actually be contaminated by these other correlations. The use of a partial correlation avoids this problem.

Like the correlation coefficient, the partial correlation coefficient takes on a value in the range from –1 to 1. The value –1 conveys a perfect negative correlation controlling for some variables (that is, an exact linear relationship in which higher values of one variable are associated with lower values of the other); the value 1 conveys a perfect positive linear relationship, and the value 0 conveys that there is no linear relationship.

The partial correlation coincides with the conditional correlation if the random variables are jointly distributed as the multivariate normal, other elliptical, multivariate hypergeometric, multivariate negative hypergeometric, multinomial or Dirichlet distribution, but not in general otherwise.

(k) Participatory Rural Appraisal

The philosophical roots of participatory rural appraisal techniques can be traced to activist adult education methods such as those of Paulo Freire and the study clubs of the Antagonist Movement. In this view, an actively involved and empowered local population is essential to successful rural community development. Robert Chambers, a key exponent of PRA, argued that the approach owes much to "the Freirian theme, that poor and exploited people can and should be enabled to analyze their own reality."

By the early 1980s, there was growing dissatisfaction among development experts with both the reductionism of formal surveys, and the biases of typical field visits. In 1983, Robert Chambers, a Fellow at the Institute of Development Studies (UK), used the term *rapid rural appraisal* (RRA) to describe techniques that could bring about a "reversal of learning", to learn from rural people directly. Two years later, the first international conference to share experiences relating to RRA was held in Thailand.[8] This was followed by a rapid acceptance of usage of methods that involved rural people in examining their own problems, setting their own goals, and monitoring their own achievements. By the mid 1990s, the term RRA had been replaced by a number of other terms including *participatory rural appraisal* (PRA) and *participatory learning and action* (PLA).

(l) Multiple stepwise logistic forward regressions

Data were analyzed using multiple stepwise logistic forward regressions (PROC LOGISTIC; SAS, 2012) to do the stepwise selection for identifying the prognostic factors for perceived hunger score of respondents in both rural and semi-urban areas. The model used was the logistic regression (LR) model: where, perceived hunger score is the dependent variable; age, educational qualification, family size, ... etc are the predictor variables that 'x' represents; a = intercept of the model and b = the coefficient of predictor variables. LR enters predictor variables in a stepwise manner; it will also fit specified models or perform forward selection of variables. At each step in the stepping process, an attempt was made to remove any insignificant variables from the model before adding a significant variable to the model. A significance level of 0.25 was required to allow a variable to enter into the model, and a significance level of 0.30 was required for a variable to stay in the model. The step selection was based on the maximum likelihood ratio. The response variable perceived hunger score = 1 (when the perceived hunger score is above average) as against perceived hunger score = 0 (when the perceived hunger score is below average) so that the probability of perceived hunger score is modeled. From stepwise logistic regressions, some variables were emerged as the prognostic factors that could influence perceived hunger score significantly. The Hosmer and Lemeshow Goodness-of- Fit test was done for goodness of fit.

6

CHAPTER

EMPIRICAL STUDY: ON FARM REALITY

DESCRIPTIVE STATISTICS (ON FARM)

Table 5.1.1: Descriptive statistics of independent variable with respect to Mean, Standard Deviation of values and Coefficient of variance:

S. No.	Variables	Mean	SD	CV
1	Age (x_1)	38.49	10.44	27.13
2	Education (x_2)	6.30	4.24	67.30
3	Family size (x_3)	3.90	1.19	30.51
4	Economic motivation (x_4)	1.13	0.54	47.58
5	Risk orientation (x_5)	0.52	0.22	42.48
6	Management orientation (x_6)	5.02	2.87	57.16
7	Stress perception on hunger (x_7)	5.25	2.19	41.74
8	Stress perception on poverty (x_8)	5.37	2.77	51.61
9	Stress perception on voice (x_9)	6.07	1.32	21.74
10	Size of holding (x_{10})	2.70	3.49	129.17
11	Cropping intensity (x_{11})	233.65	85.51	36.60
12	Livestock count (x_{12})	1.27	1.04	82.01
13	Livestock yield (x_{13})	1.66	2.17	130.72
14	Pond and fish (x_{14})	0.39	0.19	49.18
15	Total crop yield (x_{15})	6883.98	16565.58	240.64
16	Cost of cultivation (x_{16})	56664.55	116911.83	206.32
17	No. of fragments (x_{17})	4.54	4.25	93.70

S. No.	Variables	Mean	SD	CV
18	Communication variables (x_{18})	1.03	1.48	143.61
19	Marketed surplus (x_{19})	37.71	36.46	96.67
20	Energy consumption (x_{20})	5.75	2.69	46.77
21	BMI (x_{21})	26.42	5.16	19.53

It has been found from the study that the mean Age group (x_1) is 38.49 years with standard deviation 10.44 for the total distribution taken for the study. Coefficient of variation of Age (x_1) is 27.13% which shows a very high level of consistency in its distribution.

The independent variable Education (x_2) of farmer, has been found from the study that the mean 6.30 with standard deviation, 4.24 for the total distribution taken for the study. Coefficient of variation of Education (x_2) is 67.30% which show a medium level of consistency in its distribution.

The independent variable Family size (x_3) of farmer, has been found from the study that the mean 3.90 with standard deviation, 1.19 for the total distribution taken for the study. Coefficient of variation of Family size (x_3) is 30.51% which show a high level of consistency in its distribution.

The independent variable Economic motivation (x_4) of farmer, has been found from the study that the mean 1.13 with standard deviation, 0.54 for the total distribution taken for the study. Coefficient of variation of Economic motivation (x_4) is 47.58% which show a high level of consistency in its distribution.

The independent variable Stress Risk orientation (x_5) of farmer, has been found from the study that the mean 0.52 with standard deviation, 0.22 for the total distribution taken for the study. Coefficient of variation of Risk orientation (x_5) is 42.48% which show a high level of consistency in its distribution.

The independent variable Management orientation (x_6) of farmer, has been found from the study that the mean 5.02 with standard deviation, 2.87 for the total distribution taken for the study. Coefficient of variation of Management orientation (x_6) is 57.16% which show a high level of consistency in its distribution.

The independent variable Stress perception on hunger (x_7) of farmer, has been found from the study that the mean 5.25 with standard deviation, 2.19 for the total distribution taken for the study. Coefficient of variation of Stress perception on hunger (x_7) is 41.74% which show a high level of consistency in its distribution.

The independent variable Stress perception on poverty (x_8) of farmer, has been found from the study that the mean 5.37 with standard deviation, 2.77 for the total distribution taken for the study. Coefficient of variation of Stress perception on poverty (x_8) is 51.61% which show a high level of consistency in its distribution.

The independent variable Stress perception on voice (x_9) of farmer, has been found from the study that the mean 6.07 with standard deviation, 1.32 for the total distribution taken for the study. Coefficient of variation of Stress perception on voice (x_9) is 21.74% which show a very high level of consistency in its distribution.

The independent variable Size of holding (x_{10}) of farmer, has been found from the study that the mean 2.70 with standard deviation, 3.49 for the total distribution taken for the study. Coefficient of variation of Size of holding (x_{10}) is 129.17% which show a very low level of consistency in its distribution.

The independent variable Stress Cropping intensity (x_{11}) of farmer, has been found from the study that the mean 233.65 with standard deviation, 85.51 for the total distribution taken for the study. Coefficient of variation of Cropping intensity (x_{11}) is 36.60% which show a high level of consistency in its distribution.

The independent variable Livestock count (x_{12}) of farmer, has been found from the study that the mean 1.27 with standard deviation, 1.04 for the total distribution taken for the study. Coefficient of variation of Livestock count (x_{12}) is 82.01% which show a low level of consistency in its distribution.

The independent variable Livestock yield (x_{13}) of farmer, has been found from the study that the mean 1.66 with standard deviation, 2.17 for the total distribution taken for the study. Coefficient of variation of Livestock yield (x_{13}) is 130.72% which show a very low level of consistency in its distribution.

The independent variable Pond and fish (x_{14}) of farmer, has been found from the study that the mean 0.39 with standard deviation, 0.19 for the total distribution taken for the study. Coefficient of variation of Pond and fish (x_{14}) is 49.18% which show a high level of consistency in its distribution.

The independent variable Total crop yield (x_{15}) of farmer, has been found from the study that the mean 6883.98 with standard deviation, 16565.58 for the total distribution taken for the study. Coefficient of variation of Total crop yield (x_{15}) is 240.64% which show a very low level of consistency in its distribution.

The independent variable Cost of cultivation (x_{16}) of farmer, has been found from the study that the mean 56664.55 with standard deviation, 116911.83 for the total distribution taken for the study. Coefficient of variation of Cost of cultivation (x_{16}) is 206.32% which show a very low level of consistency in its distribution.

The independent variable No. of fragments (x_{17}) of farmer, has been found from the study that the mean 4.54 with standard deviation, 4.25 for the total distribution taken for the study. Coefficient of variation of No. of fragments (x_{17}) is 93.70% which show a low level of consistency in its distribution.

The independent variable Communication variables (x_{18}) of farmer, has been found from the study that the mean 1.03 with standard deviation, 1.48 for the total distribution taken for the study. Coefficient of variation of Communication variables (x_{18}) is 143.61% which show a very low level of consistency in its distribution.

The independent variable Marketed surplus (x_{19}) of farmer, has been found from the study that the mean 37.71 with standard deviation, 36.46 for the total distribution taken for the study. Coefficient of variation of Marketed surplus (x_{19}) is 96.67% which show a low level of consistency in its distribution.

The independent variable Energy consumption (x_{20}) of farmer, has been found from the study that the mean 5.75 with standard deviation, 2.69 for the total distribution taken for the study. Coefficient of variation of Energy consumption (x_{20}) is 46.77% which show a high level of consistency in its distribution.

The independent variable BMI (x_{21}) of farmer, has been found from the study that the mean 26.42 with standard deviation, 5.16 for the total distribution taken for the study. Coefficient of variation of BMI (x_{21}) is 19.53% which show a very high level of consistency in its distribution.

Table 5.1.2: Coefficient of Correlation (r): Hunger.(y_1)vs.21independent variables($x_{1-}x_{21}$)

Notation	Variables	Coefficient of correlation (r)	Remarks At .05& .01 levels
x_1	age	0.044	
x_2	education	-0.046	
x_3	Family size	0.171	
x_4	Economic motivation	-0.276	**
x_5	Risk orientation	-0.009	
x_6	Management orientation	-0.150	
x_7	Stress perception on hunger	0.127	
x_8	Stress perception on poverty	0.103	
x_9	Stress perception on voice	-0.138	
x_{10}	Size of holding	0.131	
x_{11}	Cropping intensity	-0.230	**
x_{12}	Livestock count	-0.084	
x_{13}	Livestock yield	0.011	
x_{14}	Pond and fish	-0.103	
x_{15}	Total crop yield	-0.100	
x_{16}	Cost of cultivation	-0.013	
x_{17}	No. of fragments	-0.015	
x_{18}	Communication variables	-0.103	
x_{19}	Marketed surplus	-0.183	*
x_{20}	Energy consumption	-0.198	*
x_{21}	BMI	0.994	**

RESULT

The above table presents the coefficient of correlation between Hunger (y_1) and 21 independent variables. It has been found that following variables viz. economic motivation (x_4), cropping intensity (x_{11}),marketed surplus (x_{19}) and energy consumption (x_{20}) and BMI (x_{21}) have recorded significant correlation with the dependent variable.

REVELATION

Economic motivation helps building up label of compliances or resilience with the risk and jerk. Hungry people has to comply with lots of jerk in the forms of inflicting physiological stress, dejection by the society and accelerating physio-social entropy. But eradication all these negative vibes, the farming community if become motivated, charged, positive and striving, many of their problems would get solution. That is how it stands by the significant correlation with the hunger and vice versa. Cropping intensity and marketable surplus are no doubt, two important factors as far the level of hunger is concerned. Because is cropping intensity would be high, then less resources would be required for more output. Subsequently due to higher production, marketable surplus would be higher, and this automatically would positively affect on the status of hunger. Energy indicates the status of the economy. The more is consumption the better is the status. So the farm ecology with the higher energy consumption indicates the better deterrence to inflicting hunger. Here, BMI is also gets significantly correlated with the level of hunger, rightly because a normal BMI is an indicator of good nutrition, similarly an aberrated BMI is an indicator of bad hunger status.

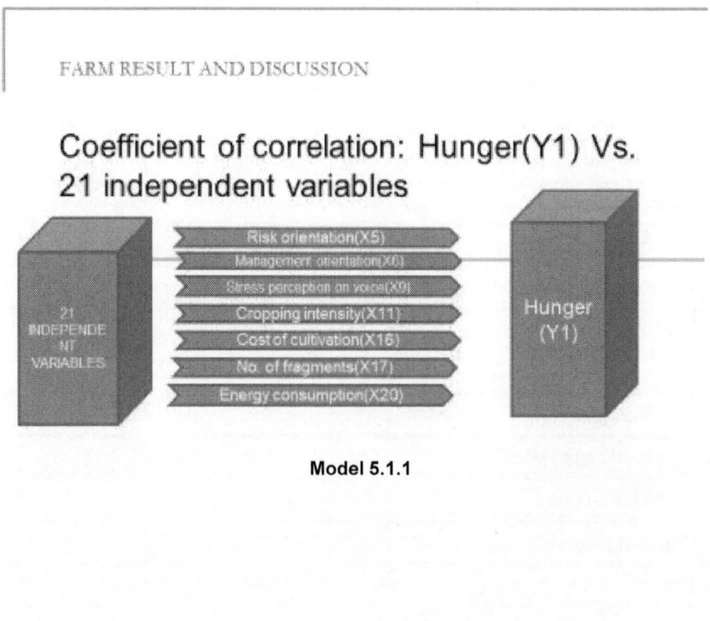

Model 5.1.1

Table 5.1.3: Coefficient of Correlation (r): Poverty (y_2) vs. 21 independent variables (x_{1-} x_{21}).

Notation	Variables	Coefficient of correlation (r)	Remarks At .05& .01 levels
x_1	age	0.035	**
x_2	education	-0.168	
x_3	Family size	0.066	
x_4	Economic motivation	0.150	
x_5	Risk orientation	-0.081	
x_6	Management orientation	-0.090	
x_7	Stress perception on hunger	0.042	
x_8	Stress perception on poverty	0.443	**
x_9	Stress perception on voice	0.051	
x_{10}	Size of holding	-0.177	
x_{11}	Cropping intensity	-0.041	
x_{12}	Livestock count	-0.038	
x_{13}	Livestock yield	0.053	
x_{14}	Pond and fish	-0.062	
x_{15}	Total crop yield	-0.159	
x_{16}	Cost of cultivation	-0.179	*
x_{17}	No. of fragments	0.001	
x_{18}	Communication variables	0.030	
x_{19}	Marketed surplus	-0.130	
x_{20}	Energy consumption	0.040	
x_{21}	BMI	0.068	

RESULT

The above table presents the coefficient of correlation between poverty (y_2) and 21 independent variables. It has been found that following variables viz.age (x_1), stress perception on poverty (x_8), stress perception on hunger (x_7) and cost of cultivation (x_{16}) have recorded significant correlation with the dependent variable.

REVELATION

Age is significantly correlated with poverty. There in our study we have found many of the rural and agrarian old women who are physically unable to do their farming job and some of them are suffering from loneliness as their children has quit agriculture and migrated to the towns and cities. So they are subjected to poverty and living a life of destitute. Hungry people in a family like situation or in a hunger endemic social ecology are through in fear psychosis and constant intimidating surroundings. Chronic hunger and sufferings of the farm women have driven them into a stage of distressing rejection and simmering frustration. So hunger and stress perceptions have been clinically linked with each other.If the income is less rather the people are poor then there would be a psychological stress on the stakeholders about their livelihood and bread earning. The cost of cultivation is negatively correlated with the poverty status because higher cost of cultivation means lower profit and lower profit subsequently creates impact on the poverty status.

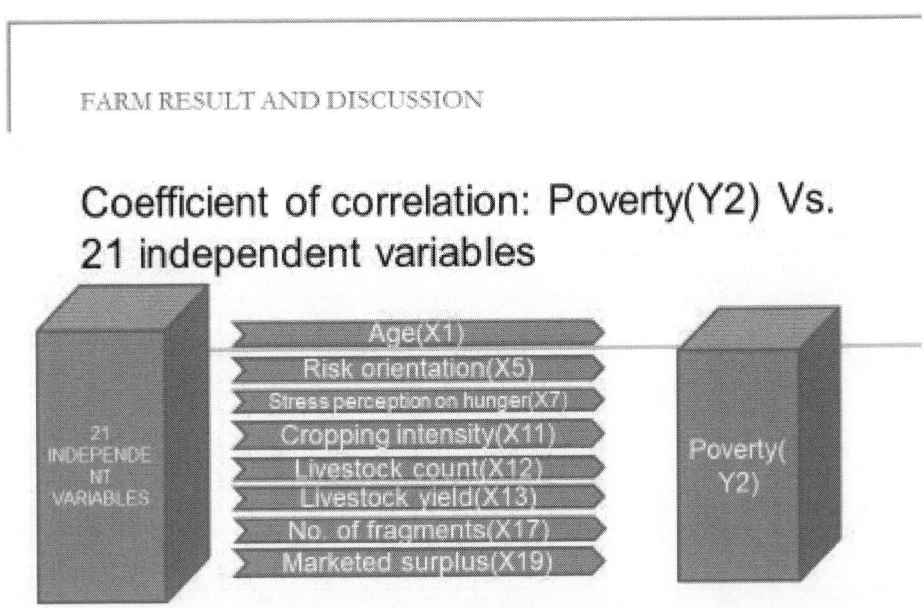

Model 5.1.2

Table 5.1.4: Coefficient of Correlation (r): Voice (y_3) vs. 21 independent variables (x_{1-} x_{21})

Notation	Variables	Coefficient of correlation (r)	Remarks At .05& .01 levels
x_1	age	-0.168	
x_2	education	0.066	
x_3	Family size	0.150	
x_4	Economic motivation	-0.081	
x_5	Risk orientation	-0.081	
x_6	Management orientation	-0.090	
x_7	Stress perception on hunger	0.042	
x_8	Stress perception on poverty	0.443	**
x_9	Stress perception on voice	0.051	
x_{10}	Size of holding	-0.177	
x_{11}	Cropping intensity	-0.041	
x_{12}	Livestock count	-0.038	
x_{13}	Livestock yield	0.053	
x_{14}	Pond and fish	-0.062	
x_{15}	Total crop yield	-0.159	
x_{16}	Cost of cultivation	-0.179	
x_{17}	No. of fragments	0.001	
x_{18}	Communication variables	-0.130	
x_{19}	Marketed surplus	0.040	
x_{20}	Energy consumption	-0.068	
x_{21}	BMI	0.259	**

RESULT

The above table presents the coefficient of correlation between voice (y_3) and 21 independent variables. It has been found that following variables viz.Stress perception on poverty (x_8), and BMI (x_{21}) have recorded significant correlation with the dependent variable.

REVELATION

Poverty status and stress due to lower income make a person frustrated, depressed, dejected and socially aloof. If a person is poor then he or she would be suffering from lower self esteem. An inferiority complex would be develop within him/ her. The same thing is happening in case of these farm women under study. Due to their poor condition, they have developed a kind of withdrawal syndrome with low confidence and morale. They are considering themselves as incapable of giving suggestion, unentitled of being vocal and this condition has push them into the doom of absolute voicelessness. BMI is not directly impactful as far the the level of voice is concerned. It is an important parameter of hunger and poverty status. Therefore as hungry and poor people has got more aberrated BMI, as well as are also suffering from voicelessness, in that way they both (voice and BMI) are linked.

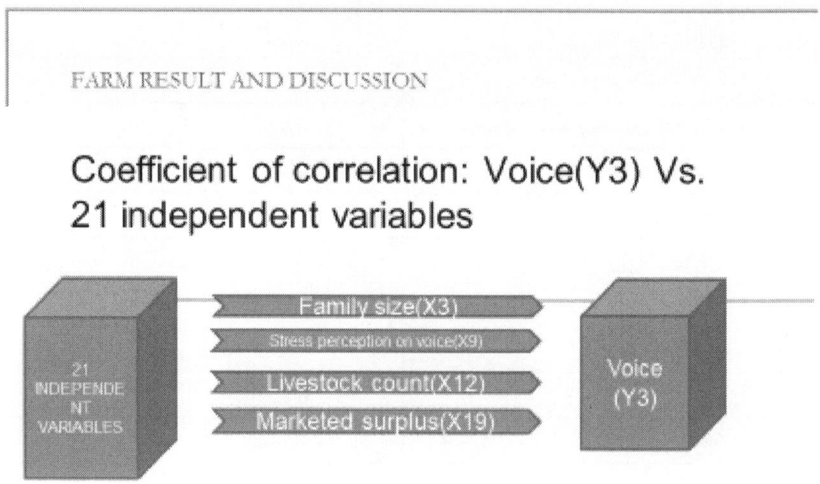

Model 5.1.3

Table 5.1.5: Coefficient of Correlation (r): Cognitive differential (y_4) vs. 21 independent variables ($x_{1-}x_{21}$)

Notation	Variables	Coefficient of correlation (r)	Remarks At .05& .01 levels
x_1	Age	0.099	
x_2	Education	-0.018	
x_3	Family size	-0.135	
x_4	Economic motivation	0.026	
x_5	Risk orientation	-0.057	
x_6	Management orientation	0.275	**
x_7	Stress perception on hunger	0.059	
x_8	Stress perception on poverty	0.206	*

x_9	Stress perception on voice	0.109	
x_{10}	Size of holding	0.315	**
x_{11}	Cropping intensity	0.171	
x_{12}	Livestock count	0.289	**
x_{13}	Livestock yield	0.028	
x_{14}	Pond and fish	0.055	
x_{15}	Total crop yield	0.086	
x_{16}	Cost of cultivation	0.223	*
x_{17}	No. of fragments	-0.176	*
x_{18}	Communication variables	0.193	*
x_{19}	Marketed surplus	-0.068	
x_{20}	Energy consumption	0.040	
x_{21}	BMI	0.110	

RESULT

The Table above presents the coefficient of correlation between voice (y_3) and 21 independent variables. It has been found that following variables viz.management orientation (x_6),stress perception on poverty (x_8), size of holding (x10), livestock count (x_{12}), cost of cultivation (x_{16}) ,No. of fragments (x_{17}) and communication variables (x_{18})have recorded significant correlation with the dependent variable.

REVELATION

Management orientation helps orchestrating time cost and resources and bringing all these on a platform of performance and achievement. The poor farm women mostly are in a fragile reality are not that adept either to manage cost or to save resources. Poor management and poverty are organically linked with each other and sometimes they together build up a resonance of concurrent impoverishment. So, inefficient management of resources and higher cost of cultivation caused lower profit margin which is substantially impactful on poverty status and creates stress. Stress restrain a person from communication and interaction , ultimately ended up with higher cognitive differential. Upscaling of productivity needs optimum investment below which crop will decline and above which there is an incremental impact. Fragmentation has got didactic effect, having both incremental and depletive effect when number of fragments are more, it indicates the cumulative size of holding will be more and basket of choices across the ecosystem within a micro-farming, elasticity would be more. So in the present situation the number of fragments have been a provider for redressing risk out of crop failure or yield losses. The more is the number fragments the more will be the flexibility of redressing the poor and reducing yield losses.Lower the resource like livestock counts, lesser would be their economic potentials. Livestock products like milk and egg additionally can be provided as resources daily or weekly. If these resources are not adequate the rural women would consider themselves as deprived and subject of exploitation as well as deprivation. Higher the resource, higher would be the marketability and this would impact positively on their self confidence and self-esteem. Then they can pull themselves off from the pale, dark and bleak scenario of silence and can cry out the louder for their legitimate rights.

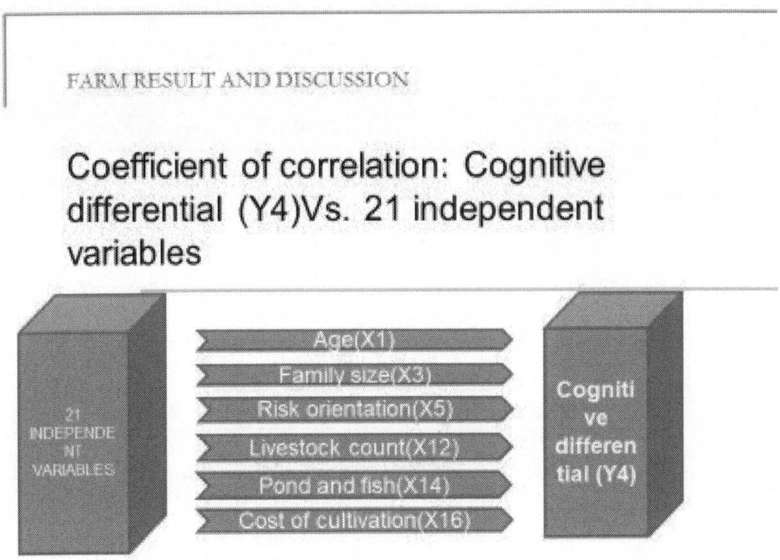

Model 5.1.4

Table 5.1.6- Step down Regression Analysis, Hunger (y_1) VS 21 Causal variables (x_1-x_{21})

Notation	Variables	Beta Value	B Value	Std. Error	T Value
x_1	age	-1.056	-9.424	8.922	-1.056
x_2	education	-0.054	-6.736	21.666	-0.311
x_3	Family size	-0.073	-32.350	68.835	-0.470
x_4	Economic motivation	-0.068	-66.776	175.312	-0.381
x_5	Risk orientation	-0.457	-1089.172	346.538	-3.143
x_6	Management orientation	0.085	15.675	27.913	0.562
x_7	Stress perception on hunger	-0.146	-35.350	34.502	-1.025
x_8	Stress perception on poverty	0.219	41.973	27.647	1.518

x_9	Stress perception on voice	0.090	36.269	56.083	0.647
x_{10}	Size of holding	0.321	48.793	57.272	0.852
x_{11}	Cropping intensity	0.484	3.004	1.078	2.787
x_{12}	Livestock count	-0.489	-249.398	76.649	-3.254
x_{13}	Livestock yield	0.039	9.483	32.694	0.290
x_{14}	Pond and fish	0.031	83.957	437.689	0.192
x_{15}	Total crop yield	-0.832	-0.027	0.017	-1.527
x_{16}	Cost of cultivation	0.370	0.002	0.003	0.660
x_{17}	No. of fragments	-0.078	-9.770	28.743	-0.340
x_{18}	Communication variables	0.014	4.940	46.004	0.107
x_{19}	Marketed surplus	-0.163	-2.370	2.905	-0.816
x_{20}	Energy consumption	-0.127	-25.125	30.471	-0.825
x_{21}	BMI	-0.055	-5.613	15.177	-0.370

R square value 50.6 %

SE 0.47

Table 5.1.7 -Regression Analysis, Hunger (y_1) vs 3 causal variables (x_{11}, x_{12} and x_{15})

S.no.	variables	Beta value	B value	Std. error	t value
1.	Cropping intensity (x_{11})	0.367	2.274	0.734	3.096
2.	Livestock count (x_{12})	-0.473	-241.297	58.466	-4.127
3.	Total crop yield (x_{15})	-0.408	-0.013	0.004	-3.417

R square value 30.3%

SE 0.48

RESULT

Table 5.1.6 presents the multiple regression analysis between exogenous variable Hunger (y_1)vs. 21 causal variables. It has been found that the variable cropping intensity (x_{11}), livestock count (x_{12}) and total crop yield (x_{15}) has contributed to the substantive variance embedded with the consequent variable y1.

The R^2 value being 0.506, it is to infer that 50.60% of variants in the consequent variable has been explained by the combination of these 21 causal variables.

Table 5.1.7 presents the step wise regression and it has been depicted that the 3 causal variables that are cropping intensity (x_{11}), livestock count (x_{12}) and total crop yield (x_{15}) have been retained at the last step.

The R^2 value being 0.303, it is to infer that 30.30% of variants in the consequent variable has been explained by the combination of these 3 causal variables.

Revelation

Cropping intensity is the provider of food and nutrition as well as integrated to the food security. So, the higher is the crop the higher has been the cropping intensity and for the growers, it offers better security for them. The livestock count is an indicator for directing and determining the spill over amount for the livestock raisers' own family.

So these three variables in terms of managing hunger can be traced as marker variables which has contributed collectively 30.30 per cent variance in hunger.

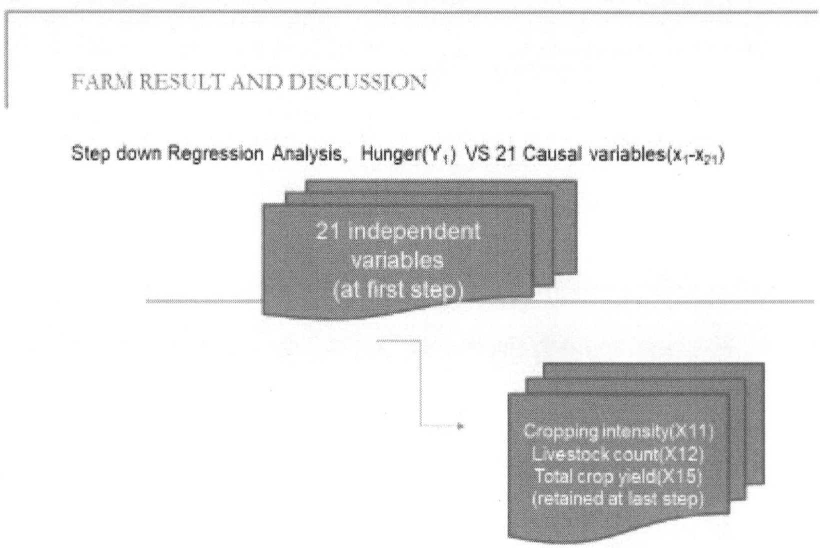

Model: 6.1.5

Table 5.1.8-Step down Regression Analysis, Poverty (y_2) VS 21 Causal variables (x_1-x_{21})

Notation	Variables	Beta value	B value	Std. error	T value
x_1	age	0.165	89.429	87.971	1.017
x_2	education	-0.056	-74.454	213.638	-0.349
x_3	Family size	0.178	840.511	678.739	1.238
x_4	Economic motivation	0.037	391.392	1728.638	0.226
x_5	Risk orientation	0.054	1365.982	3416.982	0.400

x_6	Management orientation	-0.039	-77.472	275.235	-0.281
x_7	Stress perception on hunger	0.034	87.900	340.206	0.258
x_8	Stress perception on poverty	0.002	4.599	272.606	0.017
x_9	Stress perception on voice	-0.307	-1312.316	552.998	-2.373
x_{10}	Size of holding	-0.611	-988.995	564.719	-1.751
x_{11}	Cropping intensity	0.080	5.278	10.629	0.497
x_{12}	Livestock count	0.182	986.282	755.781	1.305
x_{13}	Livestock yield	0.102	265.207	322.371	0.823
x_{14}	Pond and fish	0.246	7173.616	4315.766	1.662
x_{15}	Total crop yield	-0.606	-0.207	0.172	-1.201
x_{16}	Cost of cultivation	1.327	0.064	0.025	2.560
x_{17}	No. of fragments	0.103	137.591	283.418	0.485
x_{18}	Communication variables	0.350	1331.124	453.612	2.934
x_{19}	Marketed surplus	0.153	23.774	28.642	0.830
x_{20}	Energy consumption	-0.053	-111.592	300.456	-0.371
x_{21}	BMI	-0.093	-101.389	149.647	-0.678

R square value 44.5%

SE 0.20

Table 5.1.9 -Regression Analysis, Poverty (y_2) vs 3 causal variables (x_{12}, x_{16}, x_{18})

S. no.	Variables	Beta value	B value	Std. error	t value
1.	Cropping intensity (x_{11})	0.223	1211.547	570.409	2.124
2.	Cost of cultivation (x_{16})	0.375	0.018	0.005	3.600
3.	Communication variables (x_{18})	0.206	785.422	399.821	1.964

R square value 28.3%

SE 0.25

Table 5.1.8 presents the multiple regression analysis between exogenous variable poverty (y_2)vs. 21 causal variables. It has been found that the variable cropping intensity (x_{11}), cost of cultivation (x_{16}) and communication variables (x_{18}) has contributed to the substantive variance embedded with the consequent variable y_2.

The R^2 value being 0.445, it is to infer that 44.50% of variants in the consequent variable has been explained by the combination of these 21 causal variables.

Table 5.1.9 presents the step wise regression and it has been depicted that the 3 causal variables that are cropping intensity (x_{11}), cost of cultivation (x_{16}) and communication variables (x_{18}) have been retained at the last step.

The R^2 value being 0.283, it is to infer that 28.30% of variants in the consequent variable has been explained by the combination of these 3 causal variables.

Revelation

The farmers are suffering from poverty because in most cases for the area under study, are vulnerable to low yield, poor communication access to critical inputs and rising cost of cultivation. These three menaces collectively has contributed a deleterious impact on their economy, social well being and livelihood. Thus three variables can be considered as marker variables and together has contributed 28.30 per cent variance in poverty.

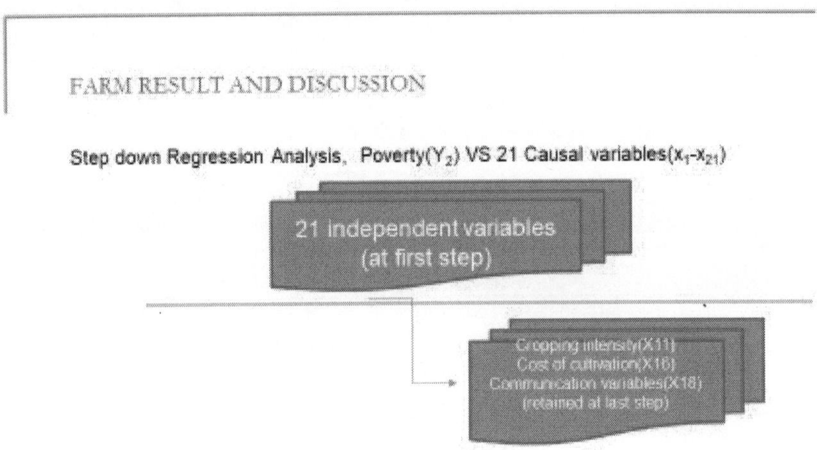

Model 5.1.6

Table 5.1.10-Step down Regression Analysis, Voice (y_3) VS 21 Causal variables (x_1-x_{21})

Notation	Variables	Beta value	B value	Std. error	t value
x_1	age	0.118	0.012	0.018	0.684
x_2	education	-0.081	-0.020	0.043	-0.474
x_3	Family size	0.179	0.151	0.137	1.177
x_4	Economic motivation	0.358	0.713	0.348	2.047
x_5	Risk orientation	-0.221	-1.065	0.688	-1.547
x_6	Management orientation	-0.051	-0.019	0.055	-0.344

x_7	Stress perception on hunger	-0.011	-0.005	0.069	-0.079
x_8	Stress perception on poverty	0.565	0.219	0.055	3.983
x_9	Stress perception on voice	-0.020	-0.016	0.111	-0.148
x_{10}	Size of holding	-0.292	-0.090	0.114	-0.790
x_{11}	Cropping intensity	-0.039	0.001	0.002	-0.226
x_{12}	Livestock count	-0.077	-0.079	0.152	-0.522
x_{13}	Livestock yield	0.276	0.136	0.065	2.096
x_{14}	Pond and fish	0.001	0.002	0.869	0.002
x_{15}	Total crop yield	-0.063	-4.060	0.001	-0.117
x_{16}	Cost of cultivation	-0.070	-6.458	0.001	-0.128
x_{17}	No. of fragments	0.227	0.057	0.057	1.004
x_{18}	Communication variables	0.028	0.020	0.091	0.221
x_{19}	Marketed surplus	-0.030	-0.001	0.006	-0.154
x_{20}	Energy consumption	0.168	0.067	0.061	1.107
x_{21}	BMI	0.201	0.042	0.030	1.387

R square value 37.7%

SE 0.67

Table 5.1.11-Regression Analysis, Voice (y_3) vs 3 causal variables (x_4, x_8, x_{13})

S.no.	Variables	Beta value	B value	Std. error	t value
1.	Economic motivation (x_4)	0.282	0.582	0.224	2.504
2.	Stress perception on poverty (x_8)	0.514	0.199	0.224	4.782
3.	Livestock yield (x_{13})	0.243	0.120	0.056	2.133

R square value 27.7%

SE 0.23

Table 5.1.10 presents the multiple regression analysis between exogenous variable voice (y_3) vs. 21 causal variables. It has been found that the variable economic motivation (x_4), stress perception on poverty (x_8) and livestock yield (x_{13}) has contributed to the substantive variance embedded with the consequent variable y_3.

The R^2 value being 0.377, it is to infer that 37.70% of variants in the consequent variable has been explained by the combination of these 21 causal variables.

Table 5.1.11 presents the step wise regression and it has been depicted that the 3 causal variables that are variable economic motivation (x_4), stress perception on poverty (x_8) and livestock yield (x_{13}) have been retained at the last stage.

Revelation

The on going punitive and its long drawn impact on the bio-physical and psychological status of the farmers has conspicuously being reflected through poor economic motivation, fragile mental ability and feeling of being dejected subsequently enrooted into a variables character. May be livestock in addition to agriculture can help to get over the stressing pursuits of agriculture.

So these variables can be traced as strong determinants for characterizing voice of oppressed.

The rural people in a response to increasing chaos and stress being inflicted into their life and livelihood, they have become a relentless hunter for options, choices, experiments and moving from an initial combination of crop-crop to crop-horticulture-livestock-fishery in a cascading manner to ultimately move from integration to co integration. In their endless hunching for a better livelihood and earning a decent status in the community, the voices of the subalterns have come up as critical as inputs and machineries, money and motivation.

The R2 value being 0.277, it is to infer that 27.70% of variants in the consequent variable has been explained by the combination of these 3 causal variables.

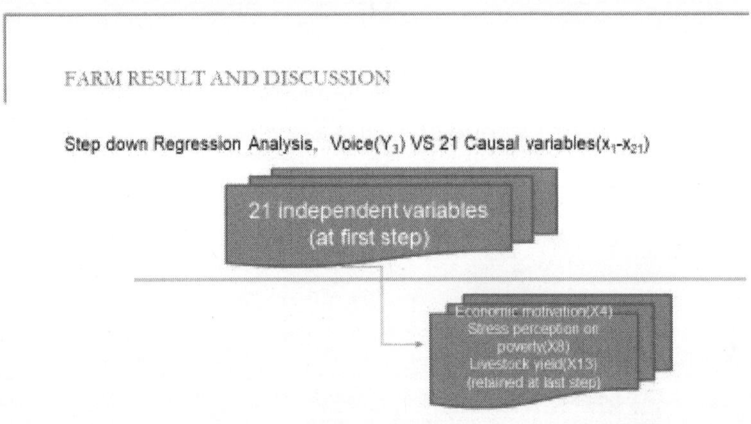

Model 5.1.7

Table 5.1.12-Step down Regression Analysis, Cognitive differential (y_4) VS 21 Causal variables (x_1-x_{21})

Notation	Variables	Beta value	B value	Std.error	t value
x_1	age	0.304	0.021	0.012	1.667
x_2	education	0.273	0.046	0.030	1.512

x_3	Family size	-0.129	-0.077	0.096	-0.804
x_4	Economic motivation	0.203	0.269	0.245	1.094
x_5	Risk orientation	0.038	0.123	0.485	0.254
x_6	Management orientation	-0.139	-0.035	0.039	-0.884
x_7	Stress perception on hunger	0.339	0.111	0.048	2.289
x_8	Stress perception on poverty	0.057	0.015	0.039	0.382
x_9	Stress perception on voice	0.245	0.133	0.079	1.691
x_{10}	Size of holding	0.125	0.026	0.080	0.319
x_{11}	Cropping intensity	0.208	0.002	0.002	1.152
x_{12}	Livestock count	0.047	0.033	0.107	0.303
x_{13}	Livestock yield	0.238	0.078	0.046	1.710
x_{14}	Pond and fish	-0.114	-0.422	0.613	-0.689
x_{15}	Total crop yield	-0.174	-7.503	0.001	-0.307
x_{16}	Cost of cultivation	0.110	6.716	0.001	0.189
x_{17}	No. of fragments	-0.002	0.001	0.040	-0.008
x_{18}	Communication variables	-0.036	-0.017	0.064	-0.265
x_{19}	Marketed surplus	-0.015	0.001	0.004	-0.075
x_{20}	Energy consumption	-0.238	-0.063	0.043	-1.478
x_{21}	BMI	0.055	0.008	0.021	0.358

R square value 58.70%

SE 0.32

Table 5.1.13-Step down Regression Analysis, Cognitive differential (y_4) VS 4 Causal variables (x_7, x_9, x_{11}, x_{13})

S. No.	Variables	Beta value	B value	Std.error	t value
1	Stress perception on hunger (x_7)	0.278	0.091	0.035	2.611
2	Stress perception on voice (x_9)	0.213	0.115	0.057	2.022
3	Cropping intensity (x_{11})	0.301	0.003	0.001	2.802
4	Livestock yield (13)	0.213	0.070	0.035	1.984

R square value 47.70%

SE 0.27

Table 5.1.12 presents the multiple regression analysis between exogenous variable cognitive development (y_4) vs. 21 causal variables. It has been found that the variables stress perception on hunger (x_7), stress perception on poverty (x_8), cropping intensity (x_{11}) and livestock yield (x_{13}) have contributed to the substantive variance embedded with the consequent variable y_4.

The R^2 value being 0.587, it is to infer that 58.70% of variants in the consequent variable has been explained by the combination of these 21 causal variables.

Table 5.1.13 presents the step wise regression and it has been depicted that the 4 causal variables that are stress perception on hunger (x_7), stress perception on poverty (x_8), cropping intensity (x_{11}) and livestock yield (x_{13})variables have been retained at the last step.

The R^2 value being 0.477, it is to infer that 47.70% of variants in the consequent variable has been explained by the combination of these 4 causal variables.

Revelation

The cognitive differential or knowledge variance among the respondents can well be predicted by two clusters of independent variable which again are attuned to each other and these are motivational factors and agro managerial factors. When the respondents are exposed to agro managerial factors, their cognitive changes are discernable in terms of package of practices, ecological functions and techno managerial pursuits. Given these are properly done, they can generate a higher level of confidence to manage both stress and stressors. So these are the marker variables and collectively have contributed to 47.70 per cent of variance in cognitive differential.

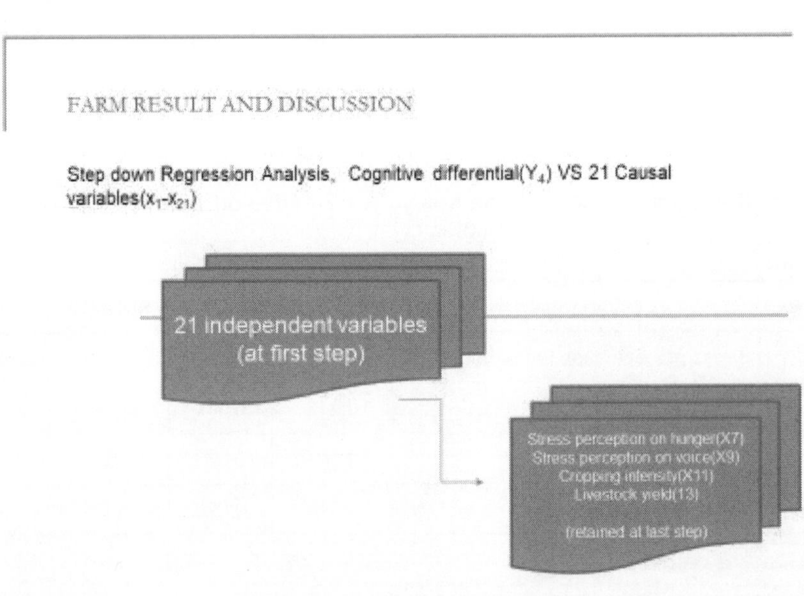

Model 5.1.8

Table 5.1.14- Factor Analysis –Conglomeration of 21 variables in 4 Factors.

Factors	Variables	Factor Loading	% of variance	Cumulative %	Factors Renamed
Factor 1	education (x_2) stress perception on poverty (x_8) size of holding (x_{10}) cropping intensity (x_{11}) total crop yield (x_{15}) cost of cultivation (x_{16}) no. of fragments (x_{17}) marketed surplus (x_{19})	0.490 -0.487 0.907 0.550 0.850 0.888 0.805 0.781	24.162	24.162	farm efficiency.
Factor 2	Age (x_1) Family size (x_3) Economic motivation (x_4) Management orientation (x_6) Pond and fish (x_{14}) Communication variables (x_{18}) Energy consumption (x_{20})	0.570 -0.506 -0.535 -0.556 0.674 -0.447 0.518	12.694	36.856	family efficiency
Factor 3	Livestock count (x_{12}) Livestock yield (x13)	-0.706 -0.466	10.030	46.886	livestock economy.
Factor 4	Risk orientation (x_5) Stress perception on hunger (x_7)	0.363 0.717	6.632	61.960	stress psychol ogy.
Factor 5	Stress perception on voice (x_9) BMI (x_{21})	0.798 0.379	6.081	68.042	psychosomatic communication.

Table 5.1.14 presents the factor analysis, wherein 21 numbers of independent variables have been conglomerated into 5 dominant factors.

Factor one is consisting of eight variables those are education (x_2), stress perception on poverty (x_8), size of holding (x_{10}), cropping intensity (x11), total crop yield (x_{15}), cost of cultivation (x_{16}), no. of fragments (x_{17}) and marketed surplus (x_{19}) and is renamed as **farm efficiency.**

Factor two is consisting of seven variables those are Age (x_1), Family size (x_3), Economic motivation (x_4), Management orientation (x_6), Pond and fish (x_{14}), Communication variables (x_{18}) and Energy consumption (x_{20}) and is renamed as **family efficiency.**

Factor three is consisting of two variables those are Livestock count (x_{12}) and Livestock yield (x_{13}) and is renamed as **livestock economy.**

Factor four is consisting of two variables those are Risk orientation (x_5) and Stress perception on hunger (x_7) and is renamed as **stress psychology.**

Factor five is consisting of two variables those are stress perception on voice (x_9) and

BMI (x_{21}) and is renamed as **psychosomatic communication**.

Henceforth and onwards these 21 variables can be operationally conceived in five factors which inputs offers a strategic conglomeration of apparently different variables into strategic constellation of five factors.

Factor analysis offers a data reduction process wherein the sprawling variables are brought under small homogeneous groups of variables called factors. These conglomeration is based on varimax rotation for reducing the standard error and Eigen values for enhancing the integrations amongst variables. Factor analysis offers immense strategic importance for an efficient management of different variables, dimensions, characters and variance, putting them all into a commonality of interactions.

Model 5.1.9

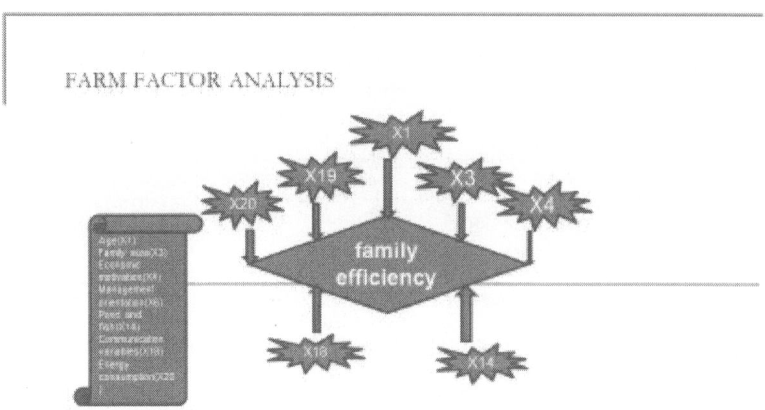

Model 5.1.10

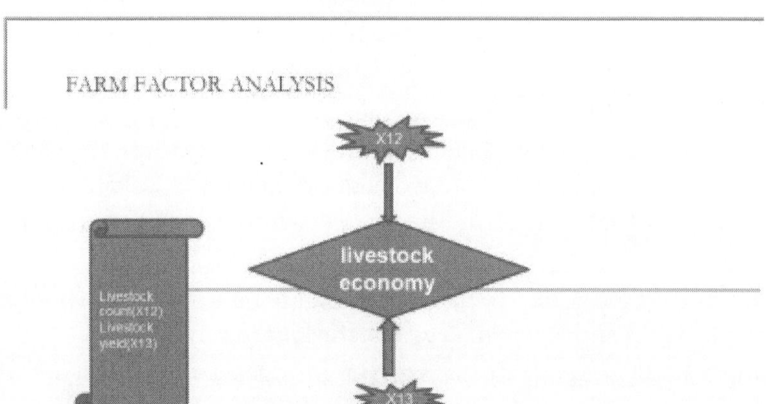

Model 5.1.11

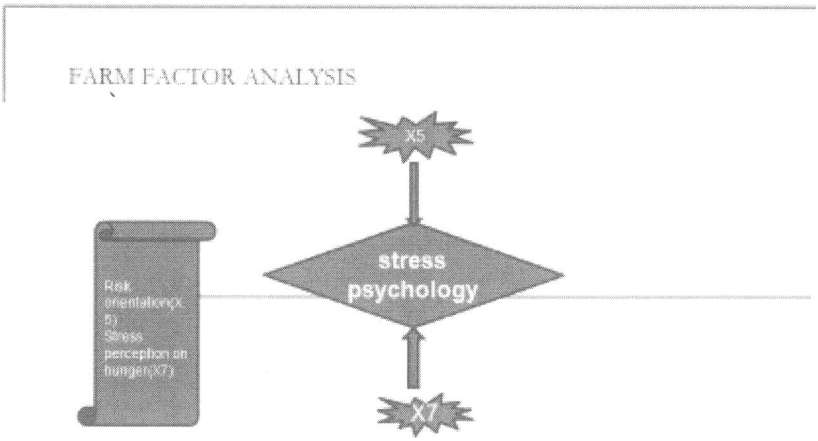

Model 5.1.12

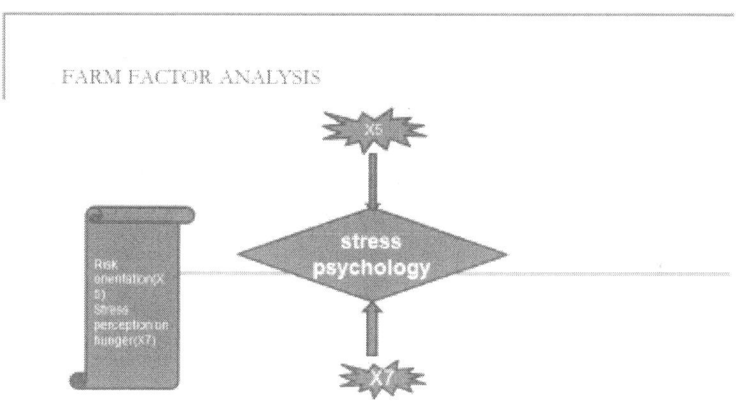

Model 5.1.13

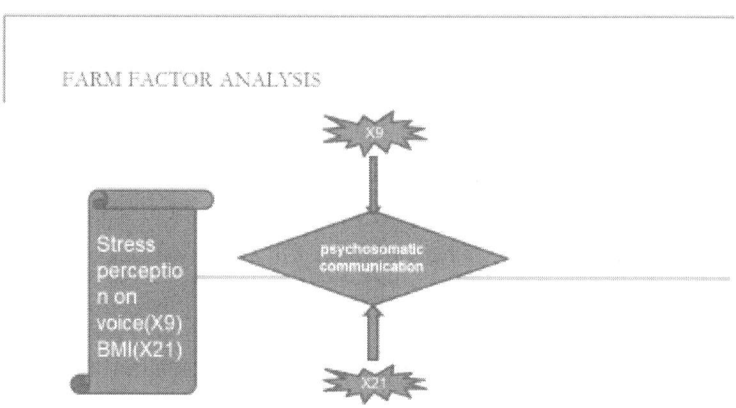

Table 5.1.15: Path analysis: Decomposition of total effect (r) into Direct, Indirect and Residual effect [Hunger VS 21 consequent variables (x_1-x_{21})]

Notation	Variables	Total Effect	Direct Effect	Indirect Effect	Highest Indirect Effect
x_1	age	-0.116	-0.189	0.073	-0.094 (x_5)
x_2	education	0.043	-0.056	0.099	-0.205 (x_{15})
x_3	Family size	-0.046	-0.074	0.028	-0.081 (x_{15})
x_4	Economic motivation	0.171	-0.069	0.24	-0.304 (x_{15})
x_5	Risk orientation	-0.276	**-0.457**	0.181	0.198 (x_{15})
x_6	Management orientation	-0.009	0.085	-0.094	-0.149 (x_{15})
x_7	Stress perception on hunger	-0.150	-0.147	-0.003	0.156 (x_{15})
x_8	Stress perception on poverty	0.127	0.22	-0.093	0.244 (x_{15})
x_9	Stress perception on voice	0.103	0.09	0.013	-0.062 (x_{12})
x_{10}	Size of holding	-0.137	0.337	**-0.474**	-0.717 (x_{15})
x_{11}	Cropping intensity	0.130	**0.486**	-0.356	-0.304 (x_{15})
x_{12}	Livestock count	-0.228	-0.49	0.262	0.134 (x_{11})
x_{13}	Livestock yield	-0.084	0.038	-0.122	-0.157 (x_{12})
x_{14}	Pond and fish	0.010	0.033	-0.023	0.108 (x_{11})
x_{15}	Total crop yield	-0.103	-0.829	0.726	0.343 (x_{16})
x_{16}	Cost of cultivation	-0.099	0.356	**-0.455**	0.302 (x_{10})
x_{17}	No. of fragments	-0.012	-0.08	0.068	0.285 (x_{11})
x_{18}	Communication variables	-0.014	0.013	-0.027	-0.065 (x_{11})
x_{19}	Marketed surplus	-0.103	-0.166	0.063	-0.449 (x_{15})
x_{20}	Energy consumption	-0.183	-0.129	-0.054	-0.246 (x_{15})
x_{21}	BMI	-0.198	-0.053	-0.145	-0.181 (x_{15})

Residual effect: 0.575

The above table presents the path analysis wherein the total effect has been decomposed into direct, indirect and residual effect. Here it has been found that the variables risk orientation (x_5), cropping intensity (x_{11}) and total crop yield (x_{15}) have exerted substantive direct effect on the dependent variable hunger (y_1). So, these three variables have got strategic impact in dealing with the issues of hunger.

The best way in redressing hunger can be possible by augmenting the productivity of the operating farm of the respondents. Whenever farm productivity goes low, the problem of hunger and poverty would turn more complex and untreatable. That is why total crop yield, cropping intensity and risk orientation have recorded the direct and substantive impact on hunger status. Whenever the owner of marginal and fragmented holding keep loosing their confidence to confront against risk, they turn more vulnerable to hunger stress. So, productivity, cropping intensity and risk orientation are moving together interovenly to generate impact on hunger.

Cost of cultivation, total crop yield and size of holding, the other set of independent variable,have exerted substantive indirect effect; it is good enough to conclude that these variables are to be treated isochronously whenever we would be considering the direct effect of such three substantive variables (x_5, xx_{11}, x_{15}).

The residual effect being 0.575, it is to infer that 57.50 per cent variable in the dependent variable that is hunger could not be explained even though there has been a combination of 21 exogenous variables. This would rather suggest that some collateral or peripheral variables or even some few core variables have not been considered or selected either. That is why the spectrum of residual effect could not be rationalized. If these comes through more core variables and some other natural or peripheral variables would have been considered, then this level of residual effect could have been lesser.

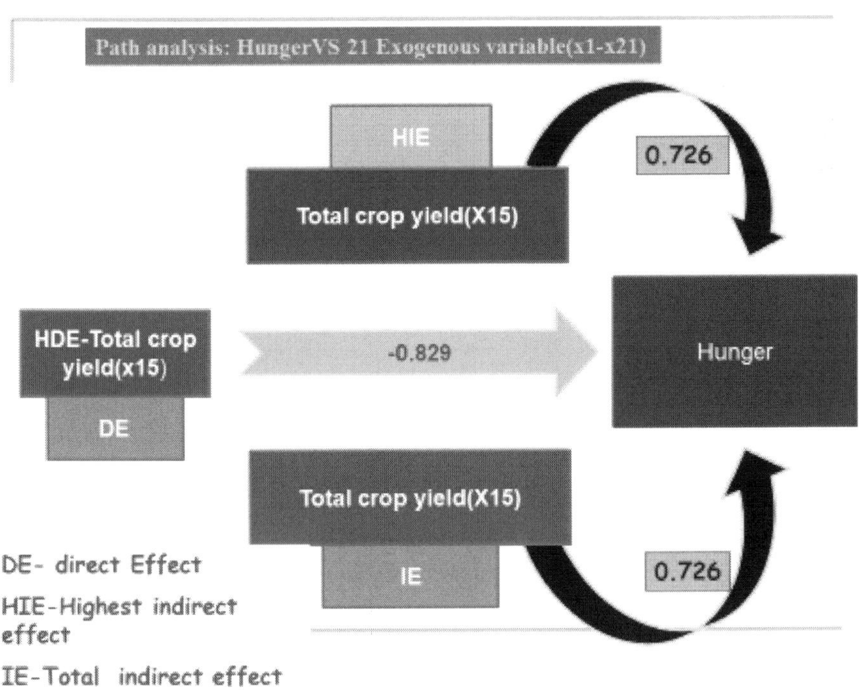

Model 5.1.14

Table 5.1.16:Path analysis: Decomposition of total effect (r) into Direct, Indirect and Residual effect[poverty VS 21 consequent variables (x_1-x_{21})]

Notation	Variables	Total Effect	Direct Effect	Indirect Effect	Highest Indirect Effect
x_1	age	0.107	0.165	-0.058	0.044 (x_{14})
x_2	education	0.196	-0.052	0.248	0.411 (x_{16})
x_3	Family size	0.104	0.178	-0.074	0.179 (x_{16})
x_4	Economic motivation	-0.023	0.037	-0.06	0.409 (x_{16})
x_5	Risk orientation	-0.018	0.058	-0.076	-0.276 (x_{16})
x_6	Management orientation	-0.229	-0.042	-0.187	0.177 (x_{16})
x_7	Stress perception on hunger	-0.07	0.033	-0.103	-0.220 (x_{16})
x_8	Stress perception on poverty	-0.115	0.001	-0.116	-0.399 (x_{16})
x_9	Stress perception on voice	-0.29	-0.303	0.013	-0.058 (x_{18})
x_{10}	Size of holding	0.302	**-0.617**	**0.919**	-0.493 (x_{15})
x_{11}	Cropping intensity	0.354	0.075	0.279	0.496 (x_{16})
x_{12}	Livestock count	0.145	0.184	-0.039	-0.135 (x_{16})
x_{13}	Livestock yield	0.115	0.1	0.015	-0.069 (x_{10})
x_{14}	Pond and fish	0.185	0.249	-0.064	0.100 (x_{16})
x_{15}	Total crop yield	0.288	**-0.571**	**0.859**	-0.533 (x_{10})
x_{16}	Cost of cultivation	0.354	**1.295**	**-0.941**	-0.553 (x_{10})
x_{17}	No. of fragments	0.362	0.108	0.254	0.823 (x_{16})
x_{18}	Communication variables	0.262	0.351	-0.089	-0.058 (x_{14})
x_{19}	Marketed surplus	0.324	0.157	0.167	0.795 (x_{16})
x_{20}	Energy consumption	0.171	-0.052	0.223	-0.169 (x_{15})
x_{21}	BMI	0.052	-0.095	0.147	0.310 (x_{16})

Residual effect: 0.494

The above table presents the path analysis wherein the total effect has been decomposed into direct, indirect and residual effect. Here it has been found that the variables size of holding (x_{10}), total crop yield (x_{15}), and cost of cultivation (x_{16}) have exerted substantive direct effect on the dependent variable poverty (y_2). So, these three variables have got strategic impact in dealing with the issues of poverty.

The path analysis shows that size of holding still comes as a strong socio economic determinant to decide on poverty and quite related to it two other variables, total crop yield and cost of cultivation are also exerting substantive direct effect on poverty and interestingly the set of these three variables have also exerted the substantive indirect effect on the consequent variable poverty. So, in the social ecology of poverty the system variables like size of holding, crop yield and cost of cultivation stand so relevant as well as virulent in terms of dealing with the poverty issues of the respondents.

The residual effect being 0.494, it is to infer that 49.40 per cent variable in the dependent variable that is poverty could not be explained even though there has been a combination of 21 exogenous variables were operated in this interaction. This would rather suggest that some collateral or peripheral variables or even some few core variables should be considered or selected in some future study. That is why the spectrum of residual effect could not be rationalized. If these comes through more core variables and some other natural or peripheral variables would have been considered, then this level of residual effect could have been lesser.

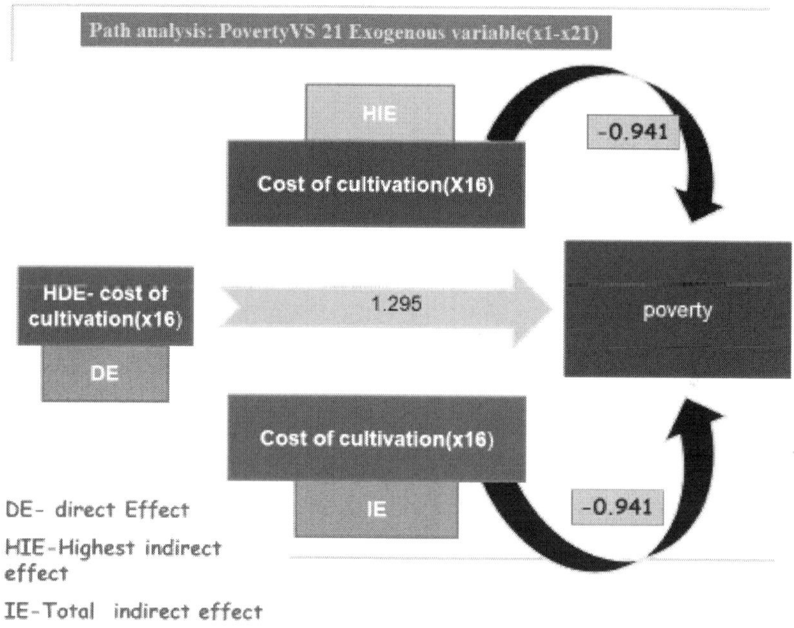

Model 5.1.15

Table 5.1.17:Path analysis: Decomposition of total effect (r) into Direct, Indirect and Residual effect. [voice VS 21 consequent variables (x_1-x_{21})]

Notation	Variables	Total Effect	Direct Effect	Indirect Effect	Highest Indirect Effect
x_1	age	0.035	0.118	-0.083	-0.138 (x_4)
x_2	education	-0.168	-0.081	-0.087	-0.201 (x_8)
x_3	Family size	0.066	0.179	-0.113	-0.091 (x_8)

Notation	Variables	Total Effect	Direct Effect	Indirect Effect	Highest Indirect Effect
x_4	Economic motivation	0.150	0.359	-0.209	-0.094 (x_{13})
x_5	Risk orientation	-0.081	-0.22	0.139	0.169 (x_8)
x_6	Management orientation	-0.090	-0.051	-0.039	0.128 (x_4)
x_7	Stress perception on hunger	0.042	-0.01	0.052	0.211 (x_8)
x_8	Stress perception on poverty	0.443	**0.564**	-0.121	0.088 (x_{10})
x_9	Stress perception on voice	0.051	-0.02	0.071	0.037 (x_{21})
x_{10}	Size of holding	-0.177	**-0.293**	0.116	0.168 (x_{17})
x_{11}	Cropping intensity	-0.041	-0.039	-0.002	-0.114 (x_{10})
x_{12}	Livestock count	-0.038	-0.075	0.037	-0.119 (x_4)
x_{13}	Livestock yield	0.053	**0.276**	**-0.223**	-0.123 (x_4)
x_{14}	Pond and fish	-0.061	0.003	-0.064	-0.087 (x_3)
x_{15}	Total crop yield	-0.159	-0.064	-0.095	-0.253 (x_{10})
x_{16}	Cost of cultivation	-0.179	-0.067	-0.112	-0.262 (x_{10})
x_{17}	No. of fragments	0.001	0.227	**-0.226**	-0.217 (x_{10})
x_{18}	Communication variables	0.030	0.028	0.002	0.071 (x_4)
x_{19}	Marketed surplus	-0.130	-0.031	-0.099	-0.156 (x_8)
x_{20}	Energy consumption	0.040	0.166	-0.126	-0.047 (x_3)
x_{21}	BMI	-0.068	0.200	**-0.268**	-0.238 (x_8)

Residual effect: 0.555

The above table presents the path analysis wherein the total effect has been decomposed into direct, indirect and residual effect. Here it has been found that the variables stress perception on poverty (x_8), size of holding (x_{10}) and livestock yield (x_{13}) have exerted substantive direct effect on the dependent variable voice (y_3). So, these three variables have got strategic impact in dealing with the issues of voice.

A person can raise their voice when he/she is free from stress and the size of holding is bigger, associated with an ancillary income from the size of holding. The variables livestock yield , no. of fragments and BMI have exerted substantial indirect effect to infer that these variables do retain in associating properties in order to characterize the performance of the consequent variable voice.

Interestingly no. of fragment and BMI are impacting conjointly on the voice character of the respondents. The poor and suffering people remain silent when economy goes fragile, getting more and more uncertain and thereby inviting more and more stress on the psychic domain. So poverty the alleviation programmes , as these are being done usually in a developing economy, should not deal with the economy of a target group in isolation. For example providing critical credit support to the marginal farmers cannot go amply successful unless it being associated with motivational programmes. Any change intervention in any socio ecological transformation process miss technology input credit ,but above all a psychic energy and motivational back up to translate this digitized entity into an euphoria of success.

The residual effect being 0.555, it is to infer that 55.50 per cent variable in the dependent variable that is voice could not be explained even though there has been a combination of 21 exogenous variables were operated in this interaction. This would rather suggest that some collateral or peripheral variables or even some few core variables should be considered or selected in some future study. That is why the spectrum of residual effect could not be rationalized. If these comes through more core variables and some other natural or peripheral variables would have been considered, then this level of residual effect could have been lesser.

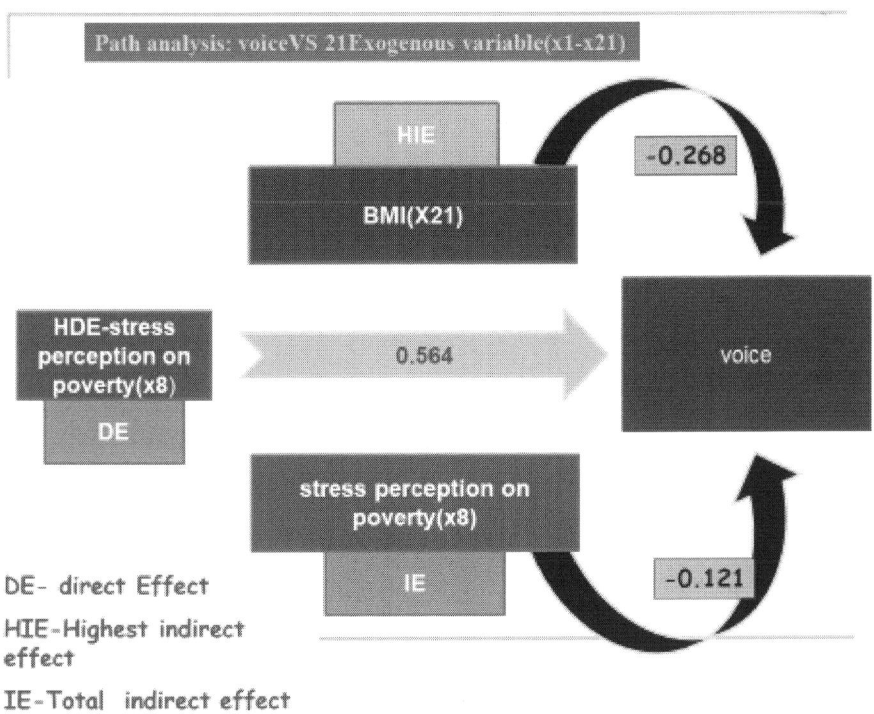

Model 5.1.16

Table 5.1.18:Path analysis: Decomposition of total effect (r) into Direct, Indirect and Residual effect[cognitive differential VS 21 consequent variables (x_1-x_{21})]

Notation	Variables	Total Effect	Direct Effect	Indirect Effect	Highest Indirect Effect
x_1	age	0.148	**0.303**	-0.155	-0.145 (x_2)
x_2	education	0.099	**0.271**	-0.172	-0.162 (x_1)
x_3	Family size	-0.018	-0.13	0.112	-0.085 (x_7)
x_4	Economic motivation	-0.135	0.204	**-0.339**	-0.116 (x_1)
x_5	Risk orientation	0.026	0.039	-0.013	0.106 (x_7)
x_6	Management orientation	-0.057	-0.139	0.082	-0.077 (x_1)
x_7	Stress perception on hunger	0.275	0.34	-0.065	-0.056 (x_4)
x_8	Stress perception on poverty	0.059	0.057	0.002	0.127 (x_7)
x_9	Stress perception on voice	0.206	0.245	-0.039	-0.044 (x_2)
x_{10}	Size of holding	0.109	0.124	-0.015	-0.154 (x_{15})
x_{11}	Cropping intensity	0.315	0.207	0.108	-0.065 (x_{15})
x_{12}	Livestock count	0.171	0.05	0.121	0.076 (x_{13})
x_{13}	Livestock yield	0.289	0.238	0.051	-0.070 (x_4)
x_{14}	Pond and fish	0.029	-0.113	0.142	0.053 (x_1)
x_{15}	Total crop yield	0.055	-0.178	**0.233**	0.113 (x_{16})
x_{16}	Cost of cultivation	0.086	0.117	-0.031	-0.171 (x_{15})
x_{17}	No. of fragments	0.223	-0.001	**0.224**	0.121 (x_{11})
x_{18}	Communication variables	-0.176	-0.036	-0.140	-0.101 (x_1)
x_{19}	Marketed surplus	0.193	-0.017	0.210	0.105 (x_{11})
x_{20}	Energy consumption	-0.068	**-0.239**	0.171	0.073 (x_7)
x_{21}	BMI	0.040	0.054	-0.014	0.066 (x_2)

Residual effect: 0.623

The above table presents the path analysis wherein the total effect has been decomposed into direct, indirect and residual effect. Here it has been found that the variables age (x_1), education (x_2) and energy consumption (x_{20}) have exerted substantive direct effect on the dependent variable cognitive differential (y_4). So, these three variables have got strategic impact in dealing with the issues of cognitive differential.

The table of path analysis rightly suggests that the differential cognitive can well be predicted by the variables age, education and energy consumption. The cognitive differential results from lack of experience, educational level and also the mobility in a given social space which has got an energy equivalence of course. So to deal with these consequent variable that is cognitive differential, these three variables to be considered with a well directed operating strategy.

The substantial indirect effect are coming from economic motivation, total crop yield and no. of fragments. So to acquire knowledge and information about different poverty alleviation and other developmental projects, the respondents need both experiential and cognitive leaning. Education is the prime driver for acquiring knowledge and information in a given social ecology. Here energy indicates the consumption of fuel, one of which mobility fuels like diesel, petrol etc. which indicates the mobility and interaction of a respondent in a social ecology.

The residual effect being 0.623, it is to infer that 62.30 per cent variable in the dependent variable that is cognitive differential could not be explained even though there has been a combination of 21 exogenous variables were operated in this interaction. This would rather suggest that some collateral or peripheral variables or even some few core variables should be considered or selected in some future study. That is why the spectrum of residual effect could not be rationalized. If these comes through more core variables and some other natural or peripheral variables would have been considered, then this level of residual effect could have been lesser.

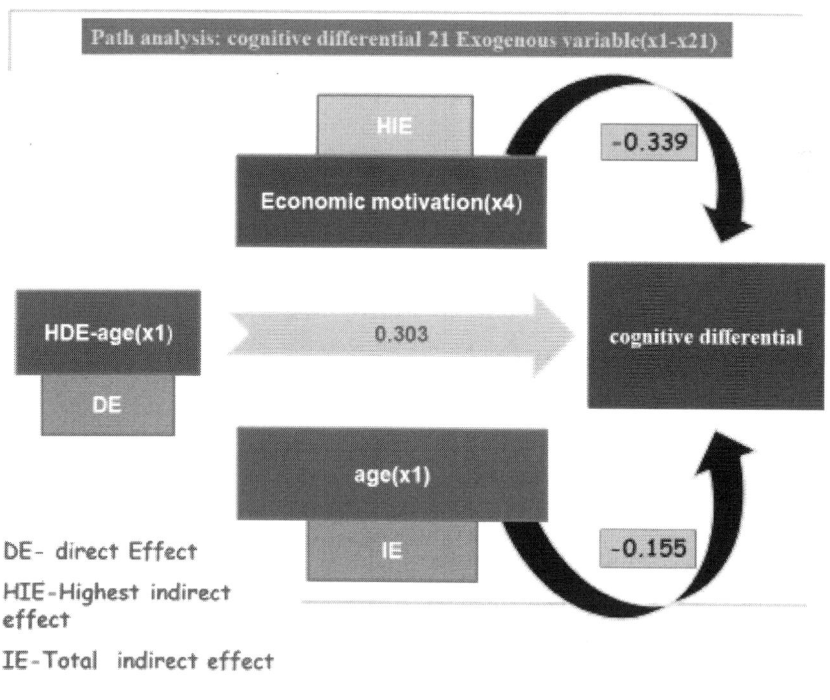

Model 5.1.17

Note: Basically whenever qualitative variables like economic motivation, stress perception, risk orientation etc. are dominating it is very difficult to quantify them and hence residual effect will be more just to infer that more cognate and relevant variables with ability to be quantified, should have been there in the array of variables, exogenous or consequent.

Canonical Covariate Analysis

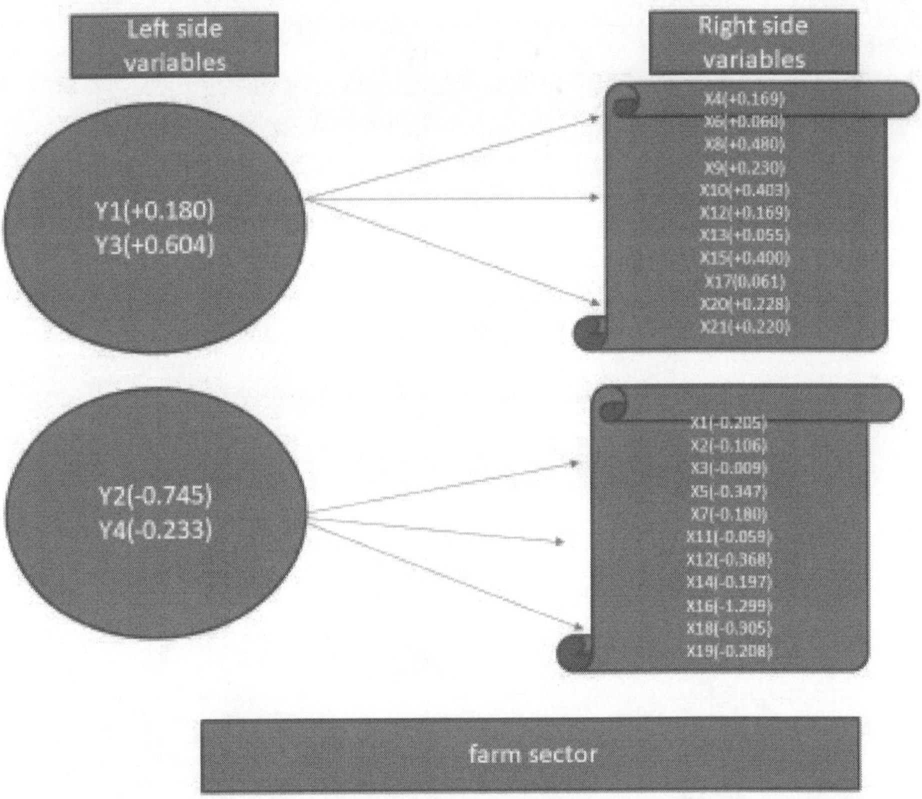

Model 5.1.18

The above diagram presents the CCA wherein a parallel interaction between x and y set of variables have delineated through cross locality. It has been found that two dependent variables y_1 and y_3 are forming a diode to move together. These two dependent variables have shown a clear choice and affinity for these selected right hand side variables. These interaction between x and y sets of variables hold good while the other sets of interaction are in place.

Hunger offers both physiological and psychological stress. Combination of which is lethal and deleterious to humanity. That is why the hunger and associating stress are moving together to offer them a propensity of forming a psychological diode and these two are substantially characterized by the contribution made from the x set of variables (x_4, x_6, x_8, x_9, x_{10}, x_{12}, x_{13}, x_{15}, x_{17}, x_{20}, x_{21}).

Similarly the two dependent variables y_2 and y_4 are moving together to generate an apparent inference that the poorer people are assessing or can assess less information regarding sources of income or livelihoods and even different projects meant for their welfare and well being. So, it is more important to access or build up knowledge about the welfare projects rather than just drive the means of welfare to an unclear destination by the government or NGO or any other institution and these diode of independent variables have got a distinct affinity towards the set of these x variables (x_1, x_2, x_3, x_5, x_7, x_{11}, x_{12}, x_{14}, x_{16}, x_{18}, x_{19}).

7 CHAPTER

EMPIRICAL STUDY: OFF FARM REALITY

The result of the study is revealed, interpreted and discussed here under;

DESCRIPTIVE STATISTICS

Table 6.2.1: Descriptive statistics of independent variable with respect to Mean, Standard Deviation of values and Coefficient of variance:

Notation	Variables	Mean	SD	CV
x_1	age	32.1	10.72739	33.41866
x_2	education	8.8857	3.27283	36.83255
x_3	Family size	4.9286	1.12058	22.73627
x_4	Economic motivation	1.3051	0.58956	45.17355
x_5	Risk orientation	0.4321	0.16983	39.3034
x_6	Management orientation	5.55	1.94889	35.11514
x_7	Stress perception on hunger	3.5571	1.1438	32.15541
x_8	Stress perception on poverty	4.4643	1.63149	36.54526
x_9	Stress perception on voice	5.3571	1.06711	19.91955
x_{10}	Homestead land	0.1393	0.10838	77.8033
x_{11}	occupation	14.15	8.68255	61.36078
x_{12}	Ancillary income	83.4851	99.08871	91.6903
x_{13}	expenditure	1203.212	631.1021	52.45145
x_{14}	savings	143.0747	178.7072	94.9048

x_{15}	Bank account	0.376	0.14128	37.57447
x_{16}	Communication variables	1.4427	1.46329	98.4272
x_{17}	Management variables	118.5414	101.0068	85.20802
x_{18}	Energy consumption	5.1753	3.20383	61.90617
x_{19}	BMI	25.3731	5.70089	22.46824

It has been found from the study that the mean Age group(x_1) is 32.1 years with standard deviation 10.723 for the total distribution taken for the study. Coefficient of variation of Age (x_1) is 33.42% which shows a very high level of consistency in its distribution

The independent variable Education (x_2) of farmer, has been found from the study that the mean 8.86 years of schooling with standard deviation, 3.27 for the total distribution taken for the study. Coefficient of variation of Education (x_2) is 36.83% which shows a high level of consistency in its distribution.

The independent variable Family size (x_3) of farmer, has been found from the study that the mean 4.93with standard deviation, 1.12 for the total distribution taken for the study. Coefficient of variation of Family size (x_3) is 22.74% which shows a high level of consistency in its distribution.

The independent variable economic motivation (x_4) of farmer, has been found from the study that the mean 1.31 bigha with standard deviation, 0.59 for the total distribution taken for the study. Coefficient of variation of size of holding (x_4) is 45.17% which shows a high level of consistency in its distribution.

The independent variable Risk orientation (x_5) of farmer, has been found from the study that the mean 0.43 with standard deviation, 0.17 for the total distribution taken for the study. Coefficient of variation of Risk orientation (x_5) is 39.30% which shows a high level of consistency in its distribution.

The independent variable Management orientation (x_6) of farmer, has been found from the study that the mean 5.55 with standard deviation, 1.95 for the total distribution taken for the study. Coefficient of variation of Management orientation (x_6) is 35.11% which shows a high level of consistency in its distribution.

The independent variable Stress Perception on Hunger (x_7) of farmer, has been found from the study that the mean 3.56 with standard deviation, 1.14 for the total distribution taken for the study. Coefficient of variation of Stress Perception on Hunger (x_7) is 32.15% which shows a high level of consistency in its distribution.

The independent variable Stress Perception on Poverty (x_8) of farmer, has been found from the study that the mean 4.46 with standard deviation, 1.63 for the total distribution taken for the study. Coefficient of variation of Stress Perception on Poverty (x_8) is 36.54% which shows a high level of consistency in its distribution.

The independent variable Stress Perception on Voice (x_9) of farmer, has been found from the study that the mean 5.36 with standard deviation, 1.07 for the total distribution taken for the study. Coefficient of variation of Stress Perception on Voice (x_9) is 19.92% which shows a high level of consistency in its distribution.

The independent variable Homestead Land (x_{10}) of farmer, has been found from the study that the mean 0.14 with standard deviation, 0.11 for the total distribution taken for the study. Coefficient of variation of Homestead Land (x_{10}) is 77.80% which shows a low level of consistency in its distribution.

The independent variable Occupation (x_{11}) of farmer, has been found from the study that the mean 14.15 with standard deviation, 8.68 for the total distribution taken for the study. Coefficient of variation of Occupation (x_{11}) is 61.36% which shows a moderately High level of consistency in its distribution.

The independent variable Ancillary Income (x_{12}) of farmer, has been found from the study that the mean 83.48 with standard deviation, 99.09 for the total distribution taken for the study. Coefficient of variation of Ancillary Income (x_{12}) is 91.69% which shows a Low level of consistency in its distribution.

The independent variable Expenditure (x_{13}) of farmer, has been found from the study that the mean 1203.21 with standard deviation, 631.10 for the total distribution taken for the study. Coefficient of variation of Expenditure (x_{13}) is 52.45% which shows a Moderately High level of consistency in its distribution.

The independent variable Savings (x_{14}) of farmer, has been found from the study that the mean 143.07 with standard deviation, 178.71 for the total distribution taken for the study. Coefficient of variation of Savings (x_{14}) is 94.90% which shows a Low level of consistency in its distribution.

The independent variable Bank Account (x_{15}) of farmer, has been found from the study that the mean 0.38 with standard deviation, 0.14 for the total distribution taken for the study. Coefficient of variation of Bank Account (x_{15}) is 37.57% which shows a High level of consistency in its distribution.

The independent variable Communication Variables (x_{16}) of farmer, has been found from the study that the mean 1.44 with standard deviation, 1.46 for the total distribution taken for the study. Coefficient of variation of Communication Variables (x_{16}) is 98.43% which shows a Low level of consistency in its distribution.

The independent variable Management Variables (x_{17}) of farmer, has been found from the study that the mean 118.54 with standard deviation, 101 for the total distribution taken for the study. Coefficient of variation of Management Variables (x_{17}) is 85.20% which shows a Low level of consistency in its distribution.

The independent variable Energy Consumption (x_{18}) of farmer, has been found from the study that the mean 5.17 with standard deviation, 3.20 for the total distribution taken for the study. Coefficient of variation of Energy Consumption (x_{18}) is 61.90% which shows a Moderately High level of consistency in its distribution.

The independent variable BMI (x_{19}) of farmer, has been found from the study that the mean 25.37 with standard deviation, 5.70 for the total distribution taken for the study. Coefficient of variation of BMI (x_{19}) is 22.46% which shows a High level of consistency in its distribution.

Table 6.2.2: Coefficient of Correlation (r): Hunger.(y_1)vs.19 independent variables(x_{1}-x_{19})

Notation	Variables	Coefficient of correlation (r)	Remarks At .05& .01 levels
x_1	age	-0.077	
x_2	education	-0.154	
x_3	Family size	-0.339	**
x_4	Economic motivation	0.016	
x_5	Risk orientation	0.132	
x_6	Management orientation	0.008	
x_7	Stress perception on hunger	-0.232	*
x_8	Stress perception on poverty	-0.208	*
x_9	Stress perception on voice	0.142	
x_{10}	Homestead land	0.201	*
x_{11}	occupation	-0.301	**
x_{12}	Ancillary income	0.358	**
x_{13}	expenditure	0.851	**
x_{14}	savings	0.594	**
x_{15}	Bank account	0.400	**
x_{16}	Communication variables	-0.045	
x_{17}	Management variables	0.134	
x_{18}	Energy consumption	-0.025	
x_{19}	BMI	0.330	**

The above table mentioned the correlation in between the 19 independent variables and the dependent variable hunger (y_1). Here, family size (x_3), stress perception on hunger (x_7), stress perception on poverty (x_8), homestead land (x_{10}), occupation (x_{11}), ancillary income (x_{12}), expenditure (x_{13}), savings (x_{14}), bank account (x_{15}) and BMI (x_{19}) are significantly correlated with hunger.

So, we can say that in the off farm social ecology, the family size is very much important because higher the family size higher would be the income requirement. But as an old saying goes, that population does not only means one stomach, but it also means a pair of hand, it is also true for family size. Higher the number of members, higher would be the working hands. But if the number of eating members are much higher than the number of earning members, then there would be a gap in between what they need and what they earn. Quite logically the stress per perception is organically linked with hunger. A hungry stomach suffers from stress and a stressed mind is suffered from esurience vis a vis hunger. Similarly, nature of occupation is related to the total volume of income as well as the nature of income. If the family has some ancillary income rather than the main source of income, they are surely getting a chance of having better livelihood and better economic security. Less expenditure and higher savings ignite a person to open a bank account and to save more for future. Homestead is certainly another predictor for the livelihood status of a respondent. Higher the possession lesser would be the economic and social vulnerability. Lastly BMI is directly linked with the hunger and nutritional status but it does not mean that higher BMI will always mean higher nutrition. In our overall study may it be in case of farm sector or off farm sector, the mean BMI is higher than that of the normal range, because they are having more carbohydrate in their diet, that is why they are bulky but not strong as their protein requirement is hardly fulfilled.

Off FARM RESULT AND DISCUSSION

Coefficient of correlation: Hunger(Y1) Vs. 21 independent variables

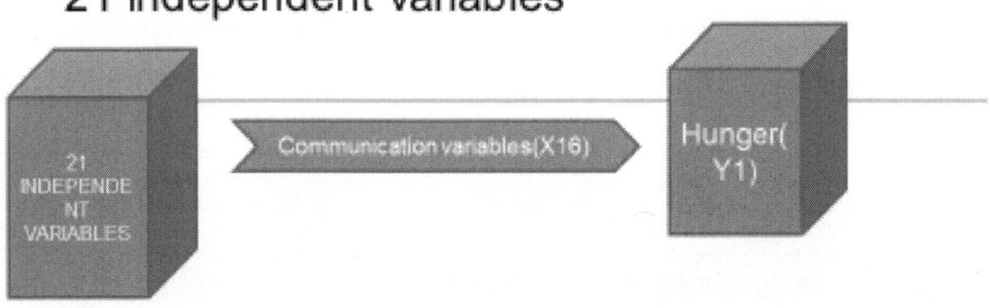

Model 6.2.1

Table 6.2.3: Coefficient of Correlation (r): Poverty.(y_2)vs.19 independent variables($x_{1-}x_{19}$)

S. no.	Variables	Coefficient of correlation (r)	Remarks At .05& .01 levels
x_1	age	0.081	
x_2	education	-0.063	

x_3	Family size	0.051	
x_4	Economic motivation	-0.258	*
x_5	Risk orientation	0.176	
x_6	Management orientation	0.347	**
x_7	Stress perception on hunger	0.362	**
x_8	Stress perception on poverty	.099	
x_9	Stress perception on voice	0.265	*
x_{10}	Homestead land	0.288	*
x_{11}	occupation	0.027	
x_{12}	Ancillary income	0.410	**
x_{13}	expenditure	0.068	
x_{14}	savings	0.264	**
x_{15}	Bank account	0.066	
x_{16}	Communication variables	0.180	
x_{17}	Management variables	0.247	*
x_{18}	Energy consumption	0.225	*
x_{19}	BMI	-0.108	

The above table mentioned the correlation in between the 19 independent variables and the dependent variable poverty (y_2). Here, economic motivation (x_4), management orientation (x_6), stress perception on hunger (x_7), stress perception on voice (x_8), homestead land (x_{10}), ancillary income

(x_{12}), savings (x_{14}), management variables (x_{17}) and energy consumption (x_{18}) are significantly correlated with poverty.

Due to poverty and lack of economic motivation, the selected respondents could not exert their entrepreneurial skills and this affects their livelihood improvement negatively. Similarly the management orientation is strongly related to the poverty status. Lack of competency in managing resources would impact on the economic status. The respondents who are managing their resources and skill well, (for example there were some people who are doing some side business like bidi binding, tailoring, livestock management etc.) they are getting some ancillary income as well as can save more and this ultimately improves their economic status by eradicating the poverty to some extent. Stress perception is also an indicator of bad economic status and poverty and status are both way linked. In a poverty ridden economy, the people would always suffer from stress and stress ridden society would suffer from lack of energy and positive vibes that ultimately leads to lower economic development. Stress perception on voice here is an important marker variable.

Poor people are hesitant to speak about their rights and needs, they suffer from dejection which culminates into poorer economic status. Homestead land is another marker variable of poverty status because there is a common perception that higher the homestead land higher would be the economic status. Interestingly energy consumption is correlated with poverty. Energy consumption is related to the status of mobility, the power consumption status in households and in business sectors. Energy consumption may be considered as a neo normal variable with the pace of modernization where the energy has got a tremendous role to play. The more would be the fuel and power consumption, the more incensement in social and economic status vis a vis income and security would be seen.

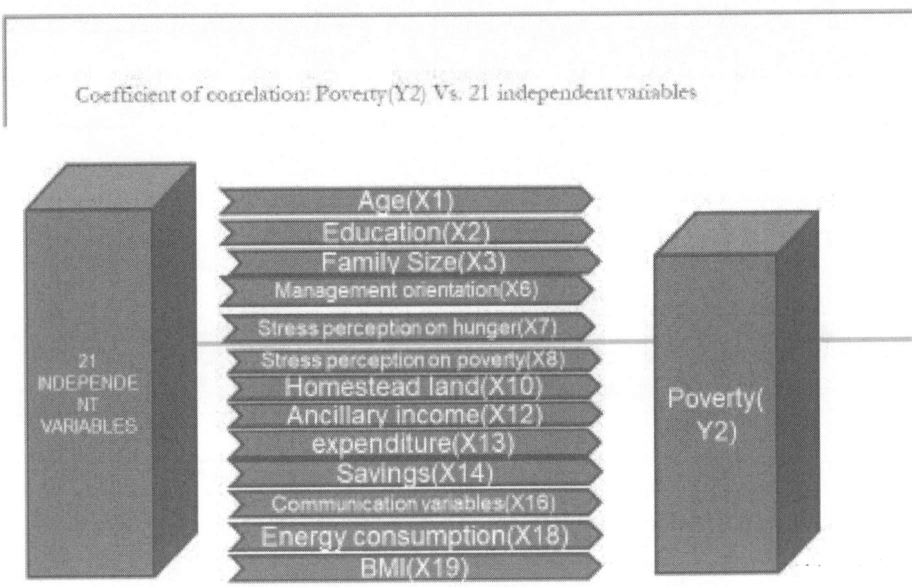

Model 6.2.2

Table 6.2.4: Coefficient of Correlation (r): voice (y_3) vs.19 independent variables(x_1-x_{19})

S.no.	Variables	Coefficient of correlation (r)	Remarks At .05& .01 levels
x_1	age	-0.056	
x_2	education	0.062	
x_3	Family size	0.239	*
x_4	Economic motivation	0.451	**
x_5	Risk orientation	0.021	

x_6	Management orientation	0.905	**
x_7	Stress perception on hunger	0.385	**
x_8	Stress perception on poverty	0.218	*
x_9	Stress perception on voice	0.051	
x_{10}	Homestead land	0.209	*
x_{11}	occupation	0.412	**
x_{12}	Ancillary income	0.031	
x_{13}	expenditure	-0.019	
x_{14}	savings	-0.055	
x_{15}	Bank account	0.358	**
x_{16}	Communication variables	0.410	**
x_{17}	Management variables	0.120	
x_{18}	Energy consumption	-0.042	
x_{19}	BMI	0.064	

The above table mentioned the correlation in between the 19 independent variables and the dependent variable voice (y_3). Here, family size (x_3),economic motivation (x_4), management orientation (x_6), stress perception on hunger (x_7), stress perception on poverty(x8), homestead land (x_{10}), occupation (x_{11}), bank account (x_{15}) and communication variables (x_{16}) are significantly correlated with voice.

Family size has also been figured up in the other variable hunger. There hunger in one hand can be decreased if the number of working hands are more, or the hunger status can be increased if the number of stomachs to feed is more. Same way, if the family size is more in one hand the voices can be loud, but in another hand if the hunger status is increased then stress perception due to hunger and poverty would increase and this would ultimately ends up with lowering of legitimate voices. Ancillary income, homestead land, economic motivation, access to banks and management orientation; these are the variables lack of which economic status is downtrodden and with the poor economy voices are also lower. Now in a non farm economy, those who are having higher entrepreneurial skills, they would have higher voices that is why the transformation from farm economy to off farm economy, may be in some cases signifies the transformation from lower voiced population to a louder one. Last but not the least, communication variables are quite obviously directly impactful on voice because more the communication, more would be the gathering of information, more would be the participation and more would be the voice.

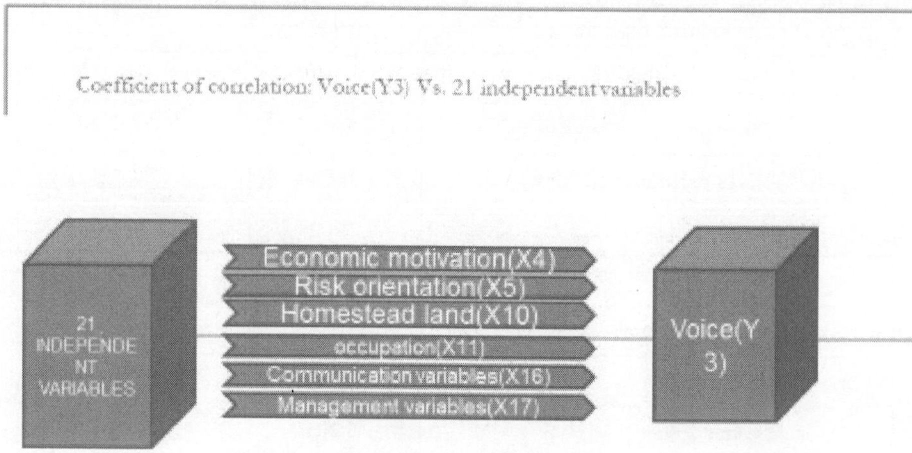

Model 6.2.3

Table 6.2.5: Coefficient of Correlation (r): cognitive differential (y_4) vs.19 independent variables(x_1-x_{19})

S. no.	Variables	Coefficient of correlation (r)	Remarks At .05& .01 levels
x_1	age	-0.014	
x_2	education	0.142	
x_3	Family size	0.039	
x_4	Economic motivation	0.152	
x_5	Risk orientation	-0.062	
x_6	Management orientation	0.160	
x_7	Stress perception on hunger	0.106	
x_8	Stress perception on poverty	0.073	
x_9	Stress perception on voice	0.074	
x_{10}	Homestead land	0.126	
x_{11}	occupation	-0.052	
x_{12}	Ancillary income	0.120	
x_{13}	expenditure	0.150	
x_{14}	savings	0.056	
x_{15}	Bank account	0.159	
x_{16}	Communication variables	0.209	*

x_{17}	Management variables	0.145	
x_{18}	Energy consumption	0.207	*
x_{19}	BMI	0.215	*

The above table mentioned the correlation in between the 19 independent variables and the dependent variable cognitive differential (y_4). Here, communication variables (x_{16}), energy consumption (x_{18}) and BMI (x_{19}) are significantly correlated with cognitive differential.

Cognitive differential actually means the gap between what are the information available and what are the information accessed. Here, communication variables are playing a pivotal role, as better the communication, narrower would be the cognitive differential. Energy consumption and BMI are the indirect indicator of cognitive differentials. Poorer BMI apparently indicates a bad health. So there would be lesser energy consumption, lesser mobility, lesser communication and lesser gathering of information. The culminating point is broader cognitive differential.

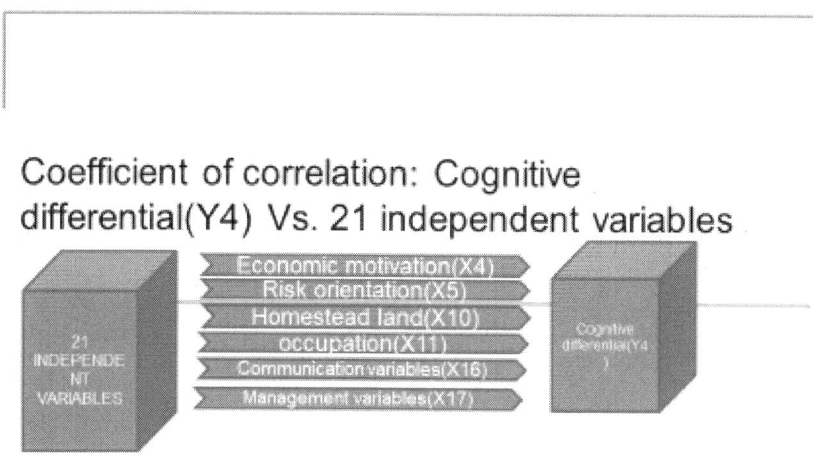

Model 6.2.4

Table 6.2.6-Step down Regression Analysis, Hunger(y_1) VS 19 Causal variables(x_1-x_{19})

S. no.	Variables	Beta value	B value	Std.error	t value
x_1	age	0.575	15.614	6.021	2.593
x_2	education	0.373	33.200	12.647	2.625
x_3	Family size	0.297	77.111	41.345	1.865
x_4	Economic motivation	0.110	54.123	71.593	0.756
x_5	Risk orientation	0.001	1.128	23.799	.005

x_6	Management orientation	-0.088	-13.162	20.002	-0.658
x_7	Stress perception on hunger	0.059	15.146	43.976	0.344
x_8	Stress perception on poverty	0.056	10.077	24.862	0.405
x_9	Stress perception on voice	-0.174	-47.557	37.629	-1.264
x_{10}	Homestead land	0.016	41.681	35.848	0.116
x_{11}	occupation	-0.117	-3.937	7.177	-0.549
x_{12}	Ancillary income	-0.156	-0.458	0.421	-1.087
x_{13}	expenditure	0.270	0.124	0.071	1.764
x_{14}	savings	-0.095	-0.155	0.224	-0.690
x_{15}	Bank account	0.180	70.338	34.465	1.078
x_{16}	Communication variables	0.089	17.750	31.191	0.569
x_{17}	Management variables	-0.006	-0.017	0.428	-0.041
x_{18}	Energy consumption	0.049	4.420	13.807	0.320
x_{19}	BMI	0.101	5.139	7.362	0.698

R square value 47.90%

SE 0.24

Table 6.2.7-Step down Regression Analysis, Hunger (y_1) vs 3 causal variables (x_1, x_2 and x_9)

S. no.	Variables	Beta value	B value	Std. error	t value
1.	Age (x_1)	0.502	13.631	2.780	4.904
2.	Education (x_2)	0.330	29.381	9.043	3.249
3.	Stress perception on voice (x_9)	-0.199	-54.232	27.240	-1.991

R square value 35.70%

SE 0 .23

Result

The above full table presents the multiple regression analysis between exogenous variable Hunger (y_1) vs. 19 causal variables. The R2 value being 0.479, it is to infer that 47.90% of variants in the consequent variable has been explained by the combination of these 21 causal variables.

Table 6.2.7 presents the step wise regression and it has been depicted that the 3 causal variables that are age (x_1), education (x_2) and stress perception on voice (x_9) have been retained at the last step.

The R2 value being 0.357, it is to infer that 35.70% of variants in the consequent variable has been explained by the combination of these 3 causal variables.

Revelation

It has been found that three variables age (x_1), education (x_2) and stress perception on voice (x_9) have been retained in the last step of step wise regression analysis to infer that, these three variables are important determinant in estimating hunger of the respondents dwelling in an off farm social ecology. Now age is a very important determinant of Hunger. Hunger is basically a physiological demand. With the age maturity grows and with the maturity subsequently the consciousness about the hunger status increases. Coming to education, it is quite obvious that those who have poorer education, they have a lesser accessibility of economic opportunities. So, In order to redress the problem of hunger; the basic education, non formal education , informal education or the education as a whole should be improved among the respondents. Hunger is both physiological as well as psychological stress. With a prolong hungry status, there comes a mind set of dejection and silence.

So these three variables in terms of managing hunger can be traced as marker variables which has contributed collectively 35.70 per cent variance in hunger.

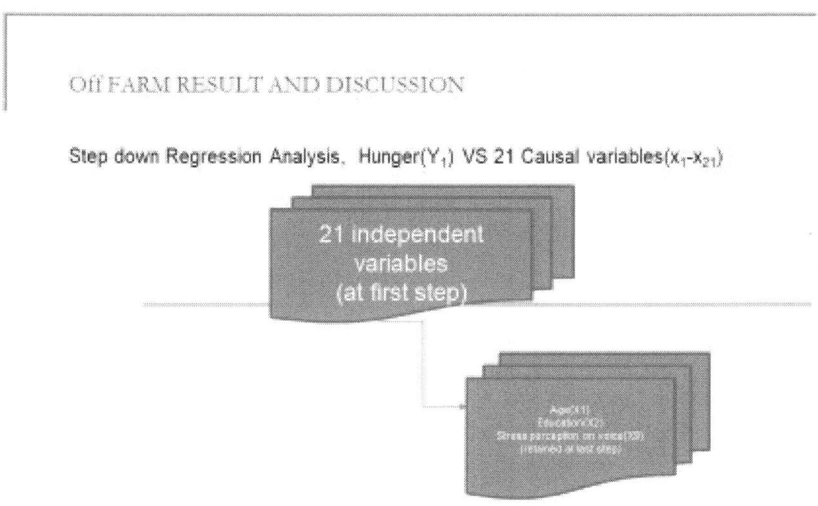

Model 6.2.5

Table 6.2.8-Step down Regression Analysis, Poverty(y_2) VS 19 Causal variables (x_1- x_{19})

S. no.	Variables	Beta value	B value	Std. error	t value
x_1	age	-0.077	-6.090	8.768	-0.695
x_2	education	-0.008	-2.131	18.416	-0.116

x_3	Family size	-0.119	-90.274	60.203	-1.499
x_4	Economic motivation	0.029	41.323	41.246	0.396
x_5	Risk orientation	-0.018	-91.754	32.873	-0.282
x_6	Management orientation	0.013	5.792	29.125	0.199
x_7	Stress perception on hunger	-0.046	-34.078	64.033	-0.532
x_8	Stress perception on poverty	-0.056	-29.275	36.201	-0.809
x_9	Stress perception on voice	0.034	26.743	54.791	0.488
x_{10}	Homestead land	-0.093	-72.204	51.060	-1.394
x_{11}	occupation	0.109	10.681	10.450	1.022
x_{12}	Ancillary income	0.158	1.354	0.613	2.206
x_{13}	expenditure	0.602	0.809	0.103	7.880
x_{14}	savings	0.320	1.520	0.327	4.649
x_{15}	Bank account	0.075	42.532	50.118	0.905
x_{16}	Communication variables	-0.155	-89.681	45.417	-1.975
x_{17}	Management variables	-0.046	-0.389	0.624	-0.624
x_{18}	Energy consumption	0.068	18.122	20.105	0.901
x_{19}	BMI	0.095	14.208	10.719	1.326

R square value 60.30%

SE 0.67

Table6.2. 9-Step down regression Analysis, Poverty (y_2) vs 2 causal variables (x_{14}, and x_{16})

S. no.	Variables	Beta value	B value	Std. error	t value
1.	Savings (x_{14})	0.300	1.424	0.261	5.467
2.	Communication variables (x_{16})	-0.091	-52.994	31.326	-1.692

R square value 58.50%

SE 0.63

Result

The above full table presents the multiple regression analysis between exogenous variable poverty (y_2) vs. 19 causal variables . The R2 value being 0.603, it is to infer that 60.30% of variants in the consequent variable has been explained by the combination of these 19 causal variables.

Table 6.2.9 presents the step wise regression and it has been depicted that the 2 causal variables that are savings (x_{14}) and communication variables (x_{16}) and have been retained at the last step.

The R2 value being 0.585, it is to infer that 58.50% of variants in the consequent variable has been explained by the combination of these 2 causal variables.

Revelation

Savings and communication variables have retained in the last step to imply that both these variables have a strong and determinant impact on predicting the consequent variable poverty. It is quite obvious that the poor people will have a meagre savings and due to their lower income, their communication modes would be constrained. For example, their mobility would be lesser, accessibility of mass media would be tough and also due to their more struggle to earn their bread and butter, they get less chance to interact with their peen groups. Therefore, this constraint of communication creates a barrier to know about the government schemes and probation which they could get benefited from.

So these two variables in terms of managing poverty can be traced as marker variables which has contributed collectively 58.50 per cent variance whereas the 19 variables together explained 60.30 per cent variance in the concerned consequent variable poverty. That means these two (x_{14} and x_{16}) together has explained 97.01 per cent{ (58.50/60.30)*100) }of the total variance explained by all those 19 variables.

We can site here the famous Nobel laureate economist Abhijit V. Banerjee when he said " A lot of us have been saying that we need a stimulation package.....Giving money in the hands of everybody, so that they can buy in stores or they buy consumer goods...I think spending is the easiest way to revive the economy" (Source an edited video posted by the national president of congress Rahul Gandhi on twitter).

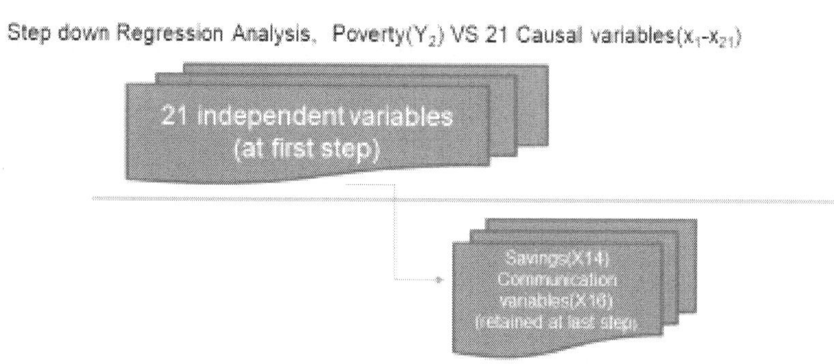

Model 6.2.6

Table 6.2.10-Step down Regression Analysis, Voice(y_3) VS 19 Causal variables(x_1-x_{19})

S. no.	Variables	Beta value	B value	Stnd. error	t value
x_1	age	-0.298	-0.025	0.016	-1.539
x_2	education	-0.196	-0.055	0.035	-1.580
x_3	Family size	-0.031	-0.026	0.113	-0.226
x_4	Economic motivation	-0.102	-0.159	0.196	-0.809
x_5	Risk orientation	0.078	0.421	0.613	0.688
x_6	Management orientation	0.429	0.201	0.055	3.674
x_7	Stress perception on hunger	0.486	0.389	0.120	3.227
x_8	Stress perception on poverty	-0.050	-0.028	0.068	-0.415
x_9	Stress perception on voice	-0.031	-0.027	0.103	-0.261
x_{10}	Homestead land	0.150	1.264	0.980	1.290
x_{11}	occupation	0.071	0.007	0.020	0.379
x_{12}	Ancillary income	0.262	0.002	0.001	2.091
x_{13}	expenditure	0.053	7.677	0.001	0.398
x_{14}	savings	0.052	0.001	0.001	0.436
x_{15}	Bank account	-0.099	-0.643	0.941	-0.684
x_{16}	Communication variables	-0.105	-0.066	0.085	-0.771
x_{17}	Management variables	0.314	0.003	0.001	0.436
x_{18}	Energy consumption	0.092	0.026	0.038	0.695
x_{19}	BMI	-0.203	-0.033	0.020	-1.618

R square 32.5%

SE 0.44

Table 6.2.11-Step wise regression Analysis, voice (y_3) vs 4 causal variables (x_1, x_2, x_4 and x_5)

S.no	Variables	Beta value	B value	Stnd. error	t value
1	Age (x_1)	-0.228	-0.019	0.010	-1.938
2	Education (x_2)	-0.217	-0.061	0.028	-2.153
3	Economic motivation (x_4)	-0.157	-0.061	0.157	-1.602
4	Risk orientation (x_5)	0.054	0.293	0.543	0.540

R square 18.9%

SE 0.42

Result

The above full table presents the multiple regression analysis between exogenous variable voice (y_3) vs. 19 causal variables. The R2 value being 0.325, it is to infer that 32.50% of variants in the consequent variable has been explained by the combination of these 19 causal variables.

Table 6.2.11 presents the step wise regression and it has been depicted that the 4 causal variables that are age (x_1), education (x_2), economic motivation (x_4) and risk orientation (x_5) have been retained at the last step.

The R2 value being 0.189, it is to infer that 18.90% of variants in the consequent variable has been explained by the combination of these 4 causal variables.

Revelation

Whenever the age is changing with the level of maturity, the voices will automatically encrypted with the personality. At the same time with the education, self confidence and empowerment level would be up. As far as the economic motivation is concerned, if someone wants to have a change in their life and livelihood altogether to make it a better economic proposition, their voices would be up quite logically. Now coming on to the risk orientation, higher the risk, better would be the gain. But the problem with the respondents concerned is that they don't know how to comply with the risk, therefore scared to take risks in their life and occupations. Due to such timid attitude they are not keen to take credit from banks to do a better exposure. Here, the risk orientation and taking credit from banks can go in compliance with each other. Hence, bank credits can be considered as an important dummy variable.

So, with the increasing age, better education and scaling up of both economic motivation and risk orientation, voices would automatically higher.

These four variables in terms of managing voice can be traced as marker variables which has contributed collectively 18.90 per cent variance in voice.

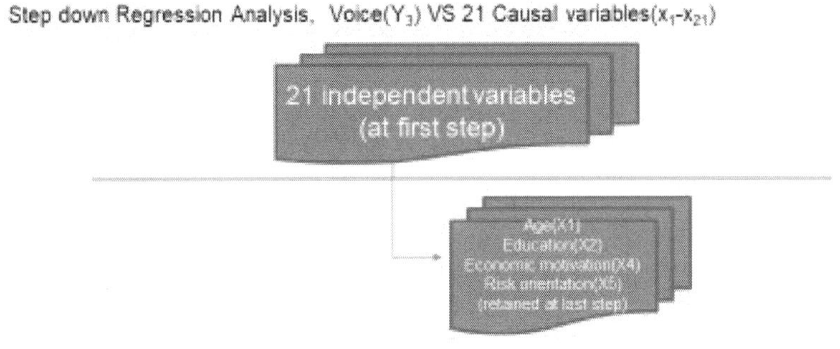

Model 6.1.7

Table 6.2.12-Step down Regression Analysis, Cognitive differential (y_4) VS 19 Causal variables(x_1-x_{19})

S. no.	Variables	Beta value	B value	Stnd. error	t value
x_1	age	0.074	0.003	0.011	0.293
x_2	education	0.158	0.022	0.023	0.977
x_3	Family size	0.241	0.099	0.074	1.333
x_4	Economic motivation	0.035	0.028	0.129	0.215
x_5	Risk orientation	-0.150	-0.407	0.403	-1.009
x_6	Management orientation	0.164	0.039	0.036	1.074
x_7	Stress perception on hunger	0.209	0.084	0.079	1.064
x_8	Stress perception on poverty	0.163	0.046	0.045	1.027
x_9	Stress perception on voice	-0.004	-0.002	0.068	-0.028
x_{10}	Homestead land	0.084	0.356	0.644	0.552
x_{11}	occupation	-0.081	-0.004	0.013	-0.334
x_{12}	Ancillary income	-0.122	-0.001	0.001	-0.748
x_{13}	expenditure	0.332	0.001	0.001	1.910
x_{14}	savings	-0.219	-0.001	0.001	-1.398
x_{15}	Bank account	0.243	0.792	0.618	1.281
x_{16}	Communication variables	0.337	0.106	0.056	1.890
x_{17}	Management variables	0.072	0.001	0.001	0.428
x_{18}	Energy consumption	-0.031	-0.005	0.025	-0.182
x_{19}	BMI	0.321	0.026	0.013	1.957

R square value 56.7%

SE 0.45

Table 6.2.13-Regression Analysis, Cognitive Differential (y_4) vs 4 causal variables (x_3, x_{15}, x_{16} and x_{19})

S. no.	Variables	Beta value	B value	Stnd. error	t value
1	Family size (x_3)	0.278	0.059	0.114	1.936
2	Bank account (x^{15})	0.314	1.024	0.465	2.203
3	Communication variables (x_{16})	0.315	0.099	0.037	2.692
4	BMI (x_{19})	0.291	0.023	0.009	2.506

R square value 46.60%

SE 0.23

Result

The above full table presents the multiple regression analysis between exogenous variable cognitive differential (y_4) vs. 19 causal variables . The R2 value being 0.567, it is to infer that 56.70% of variants in the consequent variable has been explained by the combination of these 19 causal variables.

Table 6.2.13 presents the step wise regression and it has been depicted that the 4 causal variables that are family size (x_3), bank account (x_{15}), communication variables (x_{16}) and BMI (x_{19}) have been retained at the last step.

The R2 value being 0.466, it is to infer that 46.60% of variants in the consequent variable has been explained by the combination of these 4 causal variables.

Higher the family size, higher would be the no. of ears to gather information and higher would be the no. of brains to process the information, so the cognitive differential would be more. Quite logically Bank account is a very good estimator to elucidate the cognitive differential. Because going to bank gives a chance to interact with people with cognitive differentials and this subsequently gives them a chance to gather information and come up with a different level of perception. Here, we can clearly see that two causal variables one is bank account and the other is communication variables are also linked together. Lastly, whenever the respondents are having poorer body mass index, they would be suffering from psycho-physiological stresses and their cognitive persuasion and cognitive differential, both will be affected impacted on.

So these four variables in terms of managing cognitive differentials can be traced as marker variables which has contributed collectively 46.60 per cent variance in cognitive differentials.

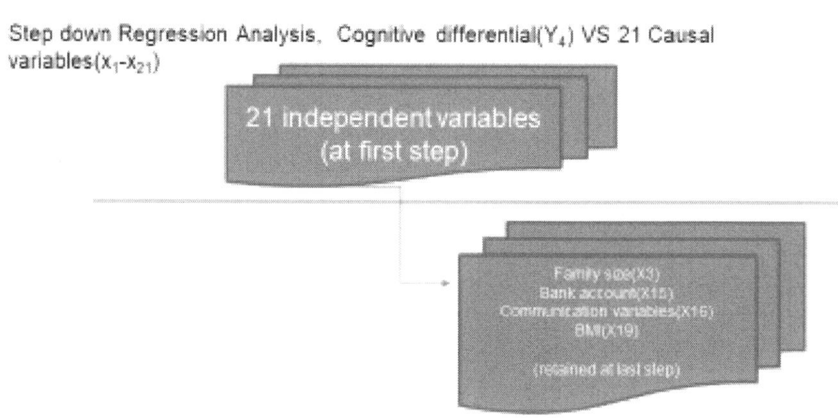

Model 6.2.8

Table 6.2.14- Factor Analysis –Conglomeration of 19 variables in 7 Factors.

Factors	Variables	Factor Loading	% of variance	Cumula-tive %	Factors Renamed
Factor 1	Age (x_1) occupation (x_{11})	0.838 0.805	**15.146**	**15.146**	**Experimental learning**
Factor 2	Homestead land (x_{10}) Ancillary income(x_{12}) expenditure (x_{13}) savings (x_{14}) management variables (x_{17})	**0.458** **0.593** **0.762** **0.590** **0.417**	**13.205**	**28.351**	**Off farm economy**
Factor 3	Education (x_2) Family size (x_3) Bank account (x_{15})	0.425 **0.737** **-0.628**	**10.901**	**39.252**	**Family capacity**
Factor 4	Management orientation(x_6) Stress perception on hunger (x_7)	0.544 0.717	9.175	48.427	Stress management
Factor 5	Energy consumption (x_{18}) BMI (x_{19})	0.668 0.697	8.143	56.570	Energy metabolism
Factor 6	Risk orientation (x_5) Stress perception on voice (x_9) Communication variables (x_{16})	0.412 0.463 -0.443	7.647	64.217	Communication stress
Factor 7	Economic motivation (x_4) Stress perception on poverty (x_8)	0.445 0.715	6.502	70.719	Motivational stress

Table 15 presents the factor analysis, wherein 19 numbers of independent variables have been conglomerated into 7 dominant factors.

Factor 1 is consisting of two variables those are Age(x_1) and occupation (x_{11}) and is renamed as **Experimental learning.** According to the age and the occupation (that is skill) better training, better capacity development , entrepreneurial skill development can help them in getting a better status and livelihood.

Factor 2 is consisting of five variables those are Homestead land (x_{10}), Ancillary income (x_{12}), expenditure (x_{13}), savings (x_{14}) and management variables (x_{17}) and is renamed as **off farm economy.** This is the era where the transformation from farm economy to off farm economy is quite frequent. The benefit of off farm economy is that the income or wage is accessed in a regular interval (like daily, weekly or monthly) which is not possible for farm economy (seasonal income). So off farm economy helps people in rural ecology to get a regular and fixed income which gives them social and economic security. By this way off farm economy is playing a dominant and decisive role as far as the income of the respondents are concerned.

Factor 3 is consisting of three variables those are Education (x_2), Family size (x_3) and Bank account (x_{15}) and is renamed as **family capacity.** It means better access to the surrounding world, compatibility to the rest of the world and capacity to earn. Highly competent family would be able to deal with the hunger, poverty and voice issues better as well as would narrow down the cognitive differential.

Factor 4 is consisting of two variables those are Management orientation (x_6) and Stress perception on hunger (x_7) and is renamed as **stress management.** Whenever stress is managed through better resource management, well directed motivation and better value addition to the capacity, it creates a strident strategic impact on hunger and poverty status.

Factor 5 is consisting of two variables those are Energy consumption (x_{18}) and BMI (x_{19}) and is renamed as **energy metabolism.** In a human physiology or in a farm metabolism; the mechanism is always same. If a human body is consuming less food and exerting more work energy, then the whole mechanism would be ruptured. Same way, if a farm is consuming less energy and exerts higher energy, it would definitely create a negative energy balance. So, optimum calorie consumption and a normal Body mass index would be helpful in dealing with the consequent variables.

Factor 6 is consisting of three variables those are Risk orientation(x_5), Stress perception on voice (x_9) and Communication variables (x_{16}) and is renamed as **communication stress.** Higher the communication stress, lower would be the voice and higher would be the cognitive differential.

Factor 7 is consisting of two variables those are Economic motivation (x_4) and Stress perception on poverty (x_8) and is renamed as **motivational stress.** Lower economic motivation and higher stress perception means lesser strive to improve the economic status.

So, we can say that, out of these 19 variables, only seven dominating factors are extracted by combining the related variables together. That means seven conspicuous variables are actually there. These seven variables, out of this empirical research, are offering tremendous strategic importance at the micro level situation and reality. It is well discernible that these all strategic implications are restricted to the research locale selected for study. To get higher customized and standardized result this research needs to be replicated in different geo spatial economic and ecologic status. Higher the variance across the ecologies, higher would be the reliability. So, it is also one of the recommendations that we need to replicate this study to a higher core of ecological sets to get and extrapolate the micro level variance into a more standardize protocol.

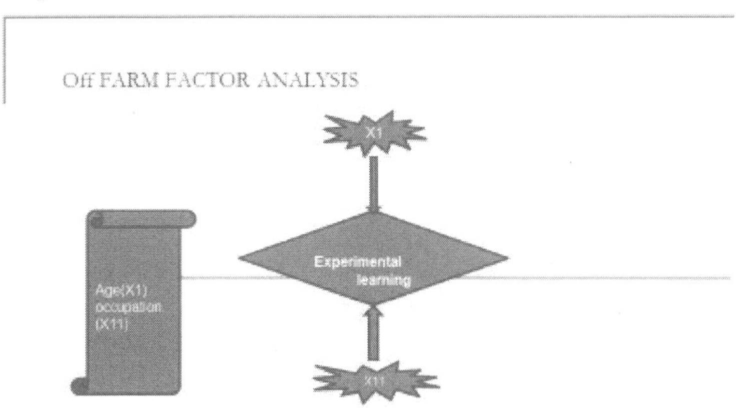

Model 6.2.9

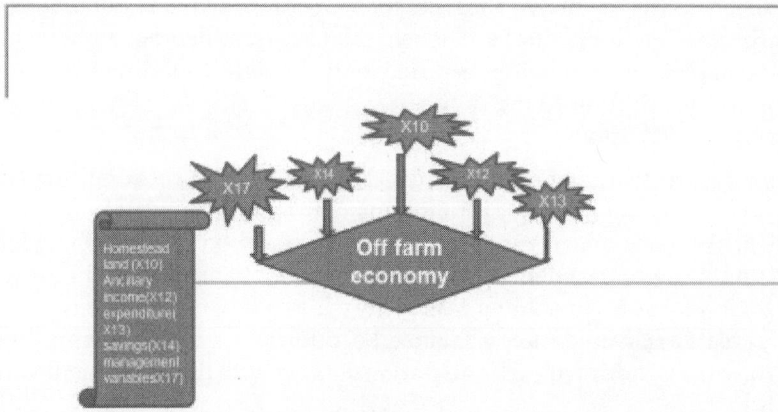

Model 6.2.10

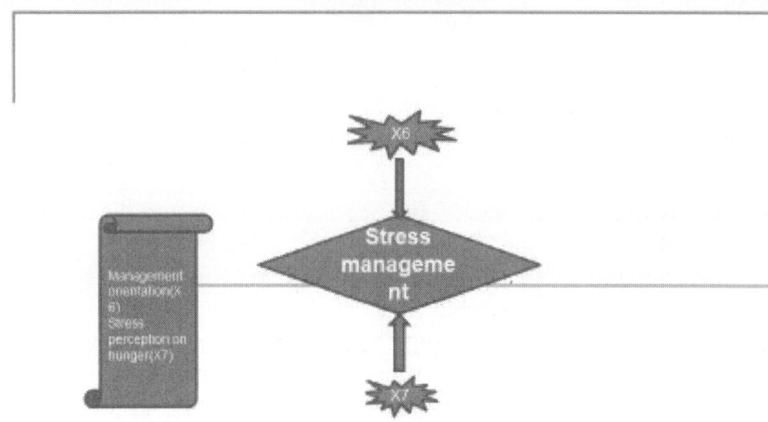

Model6.2.11

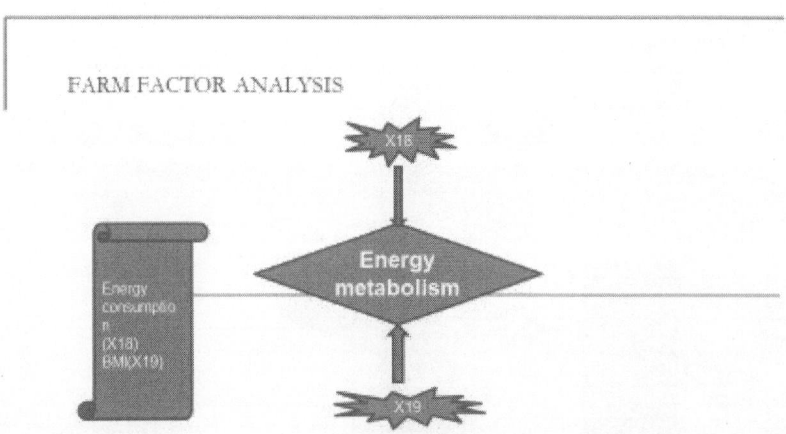

Model 6.2.12

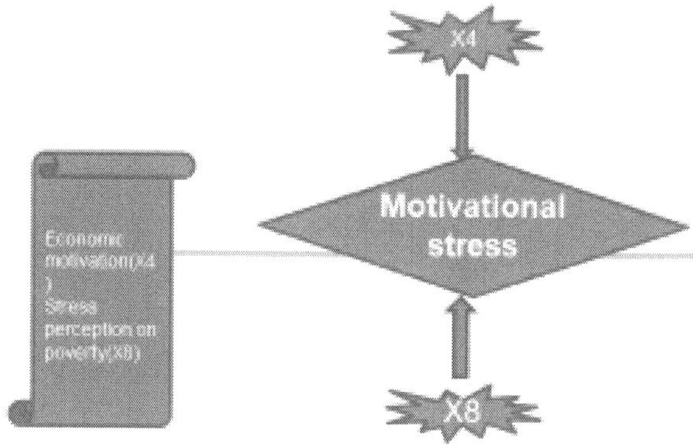

Model 6.2.13

Table 6.2.15 :Path analysis: Decomposition of total effect (r) into Direct, Indirect and Residual effect[Hunger VS 19 consequent variables(x_1-x_{19})]

S. no.	Variables	Total effect	Direct effect	Indirect effect	Highest indirect effect
x_1	age	0.451	**0.575**	-0.123	x_{11}(-0.093)
x_2	education	0.220	**0.373**	-0.153	x_{13}(-0.057)
x_3	Family size	0.188	0.297	-0.108	x_{15}(-0.111)
x_4	Economic motivation	-0.106	0.110	**-0.216**	x_1(-0.233)
x_5	Risk orientation	0.020	0.001	0.019	x_1(0.148)
x_6	Management orientation	-0.453	-0.088	-0.542	x_1(-0.117)
x_7	Stress perception on hunger	0.207	0.059	0.267	x_1(0.249)
x_8	Stress perception on poverty	0.012	0.056	0.068	x_{13}(-0.053)
x_9	Stress perception on voice	-0.239	-0.174	**0.413**	x_1(-0.074)
x_{10}	Homestead land	0.045	0.016	0.061	x_1(0.104)
x_{11}	occupation	0.300	-0.117	0.183	x_1(0.458)

x_{12}	Ancillary income	0.138	-0.156	-0.018	x_1(0.103)
x_{13}	expenditure	0.119	0.270	-0.150	x_2(-0.079)
x_{14}	savings	0.014	-0.095	-0.080	x_{14}(-0.095)
x_{15}	Bank account	0.126	0.180	-0.053	x_3(-0.183)
x_{16}	Communication variables	-0.059	0.089	-0.148	x_1(-0.141)
x_{17}	Management variables	0.167	-0.006	0.173	x_{14}(-0.095)
x_{18}	Energy consumption	0.187	0.049	0.138	x_{18}(0.046)
x_{19}	BMI	0.136	0.101	0.035	x_{13}(0.060)

Residual 52.13%

The above table presents the path analysis wherein the total effect has been decomposed into direct, indirect and residual effect. Here it has been found that the variables age (x_1), education (x_2), and family size (x_3) have exerted substantive direct effect on the dependent variable hunger (y_1). So, these three variables have got strategic impact in dealing with the issues of hunger.

Higher family size, poorer education at any stage of growth (age) are impacting on hunger. The other variables like economic motivation (x_4), management orientation (x_6) and stress perception on voice (x_9) have clearly recorded associating properties i.e. these variables will ultimately characterize the behaviour of hunger. So, poor motivation, less management orientation and higher stress perception are offering a reticulate interactive behaviour with other variables to ultimately deciding on the severity of hunger and enduring ability of the person being inflicted with the hit of hunger.

The residual effect being 52.13 per cent is inferring that even with the combination of nineteen variables, a substantive portion of variance in hunger could not be explained with the participation with these nineteen variables in the entire wave of interaction.

So, the CCA can be an option to derive more strategic interaction beyond a decomposition approach as followed in path analysis.

The variable has enrooted the highest indirect effect of thirteen variables in characterizing hunger to justify and testify its imbibing properties with other variables.

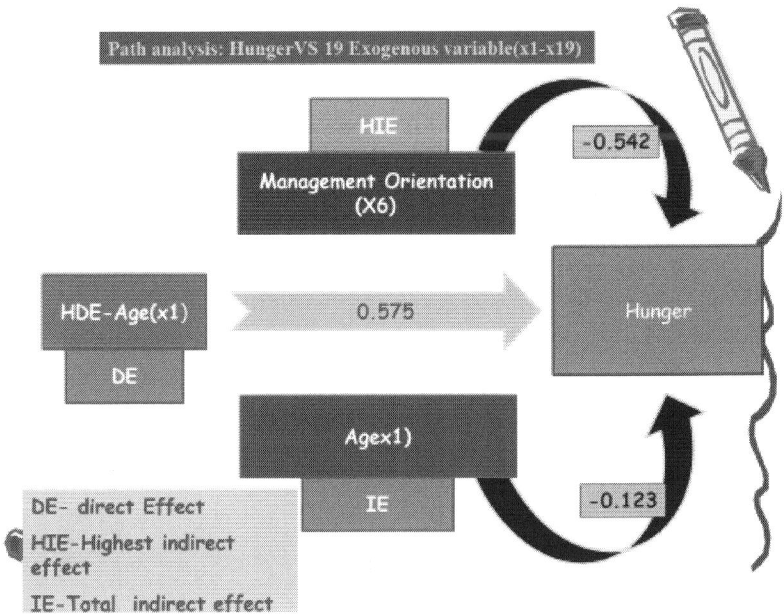

Model 6.1.14

Table 6.2.16:Path analysis: Decomposition of total effect (r) into Direct, Indirect and Residual effect [Poverty VS 19 consequent variables (x_1-x_{19})]

S. no.	Variables	Total effect	Direct effect	Indirect effect	Highest indirect effect
x_1	age	0.0022	-0.077	0.079	x_{11}(0.087)
x_2	education	-0.145	-0.008	-0.153	x_{13}(-0.127)
x_3	Family size	-0.238	-0.119	-0.119	x_{15}(-0.046)
x_4	Economic motivation	-0.0113	0.029	-0.017	x_{14}(-0.052)
x_5	Risk orientation	0.150	-0.018	0.168	x_{13}(0.072)
x_6	Management orientation	0.007	0.013	0.005	x_{14}(0.089)
x_7	Stress perception on hunger	-0.232	-0.046	-0.186	x_{13}(-0.192)
x_8	Stress perception on poverty	-0.202	-0.056	-0.146	x_{13}(-0.118)
x_9	Stress perception on voice	0.142	0.034	0.108	x_{13}(0.093)
x_{10}	Homestead land	0.200	-0.093	**0.293**	x_{13}(0.185)
x_{11}	occupation	0.090	0.109	-0.018	x_1(-0.061)
x_{12}	Ancillary income	0.359	**0.158**	0.201	x_{13}(0.231)
x_{13}	expenditure	0.347	**0.602**	-0.254	x_{14}(0.122)

x_{14}	savings	0.593	**0.320**	**0.273**	x_{13}(0.229)
x_{15}	Bank account	0.404	0.075	**0.329**	x_{13}(0.180)
x_{16}	Communication variables	-0.048	-0.155	0.106	x_{14}(0.040)
x_{17}	Management variables	0.133	-0.046	0.179	x_{13}(0.096)
x_{18}	Energy consumption	-0.023	0.068	0.045	x_{13}(-0.061)
x_{19}	BMI	0.328	0.095	0.233	x_{13}(0.134)

Residual 13%

The above table presents the path analysis wherein the total effect has been decomposed into direct, indirect and residual effect. Here it has been found that the variables ancillary income (x_{12}), expenditure (x_{13}), and savings (x_{14}) have exerted substantive direct effect on the dependent variable poverty (y_2). So, these three variables have got strategic impact in dealing with the issues of poverty.

Expenditure is a very important determinant in shaping of the level of poverty. people are poor because their expenditure is more over their income. If they are compelled to spend more in spite of their low income, then there comes the poverty situation. On the other hand, to fight poverty everyone needs an ancillary income apart from their main source of income. Here in an off farm social ecology, ancillary income plays a very important and determinant role as far as the poverty status of the people is concerned. In today's circumstances, the off farm economy is more stronger than that of the on farm one and sometimes these two become inter changeable. Now coming on to savings part. It is quite obvious that if one needs to get over the brunt of poverty, he/she needs savings. The daily wage earners are the most vulnerable category under any undesirable economic situation. For instance we can take the example of COVID 19 pandemic situation, where we have seen that the most affected class is the migrant labourer category, because they are the daily wage earners with a meagre savings. To fight poverty and to increase the savings of the poor and deprived class, Jandhan accounts were opened by the government of India in 2015 which really a commendable step towards the eradication of poverty by increasing people's savings. Savings also means the possibility of investment. If a person has savings, then he/she can invest and take moderate risk and that is where the opportunity of becoming an entrepreneur lies. Very interestingly we may notice that here bank account is exerting the highest indirect effect on poverty which rightly proves the importance of opening bank accounts of poor people. Another important variable which is having the highest indirect effect is homestead land. Bigger the homestead land higher would be the opportunity to complete in the realm of economy as in today's era of modernization and urbanization, the cost of homestead land is getting higher.

The residual effect being 13 per cent is inferring that even with the combination of nineteen variables, a substantive portion of variance in poverty could not be explained with the participation with these nineteen variables in the entire wave of interaction. Though 87 per cent of the variance is explained in this case which shows the less error occurrence.

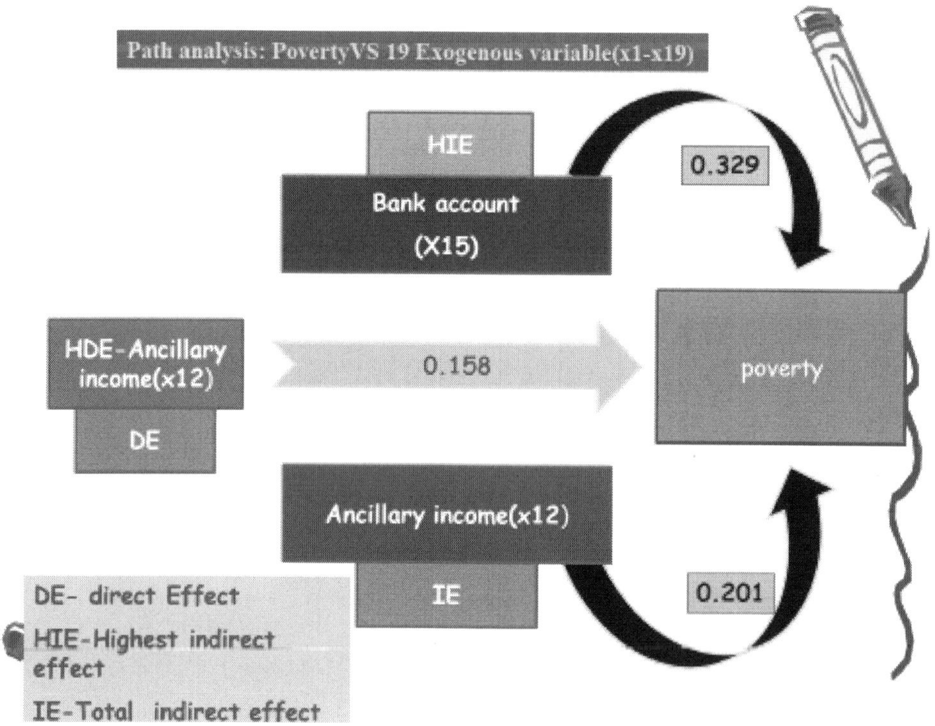

Model 6.2.15

Table 6.2.17:Path analysis: Decomposition of total effect (r) into Direct, Indirect and Residual effect [Voice VS 19 consequent variables (x_1-x_{19})]

S. no.	Variables	Total effect	Direct effect	Indirect effect	Highest indirect effect
x_1	age	0.082	-0.298	**0.380**	x_6(-0.087)
x_2	education	-0.065	-0.196	0.131	x_6(0.070)
x_3	Family size	0.050	-0.031	0.081	x_{15}(0.061)
x_4	Economic motivation	-0.253	-0.102	-0.151	x_1(0.120)
x_5	Risk orientation	0.175	0.078	0.097	x_1(-0.076)
x_6	Management orientation	0.348	**0.429**	-0.081	x_{17}(-0.061)
x_7	Stress perception on hunger	0.364	**0.486**	-0.121	x_1(0.486)
x_8	Stress perception on poverty	0.100	-0.050	0.150	x_1(0.138)

x_9	Stress perception on voice	0.263	-0.031	**0.294**	x_{12}(0.062)
x_{10}	Homestead land	0.285	0.150	0.135	x_{12}(0.118)
x_{11}	occupation	0.022	0.071	-0.048	x_1(-0.237)
x_{12}	Ancillary income	0.408	0.262	0.146	x_{17}(0.129)
x_{13}	expenditure	0.065	0.053	0.013	x_{17}(0.050)
x_{14}	savings	0.262	0.052	0.210	x_6(0.119)
x_{15}	Bank account	0.064	-0.099	0.163	x_7(0.096)
x_{16}	Communication variables	0.180	-0.105	**0.285**	x_{17}(0.059)
x_{17}	Management variables	0.379	**0.314**	-0.065	x_{17}(0.313)
x_{18}	Energy consumption	0.223	0.092	0.131	x_7(0.134)
x_{19}	BMI	-0.113	-0.203	0.090	x_7(-0.083)

Residual 39.68%

The above table presents the path analysis wherein the total effect has been decomposed into direct, indirect and residual effect. Here it has been found that the variables management orientation (x_6), stress perception on hunger (x_7), and management variables (x_{17}) have exerted substantive direct effect on the dependent variable voice (y_3). So, these three variables have got strategic impact in dealing with the issues of voice.

Management orientation basically is a psychic, corporeal and operational engagement which is also associated with mobility, communication and interaction. Management orientation has a sizable contribution in raising the voice. When someone is very good in management process, the confidence would go up and when one's confidence is boosted, his/ her voice would automatically start rippling up. Good in management orientation means he/she is good in dealing with the risk, harvesting the opportunity and setting a plan for the future goals. Now, as far as the stress perception is concerned stress perception and voice are reciprocally related. A person with higher stress be there morose and exhausted, so couldn't raise voice. Now as far as the management variables are concerned, better handling of the variables of management gives lesser stress and lesser stress helps in better communication, better communication leads to better information gathering ability, which ultimately leads to a confident raising of voice. Age is also a variable which is exerting a good impact on voice and it is effecting the voice indirectly. With the age maturity to handle stress is growing up and entrepreneual vision becomes keen. So, together with other variables, age exerts a swashbuckling impact on voice.

The residual effect being 39.68 per cent is inferring that even with the combination of nineteen variables, a substantive portion of variance in voice could not be explained with the participation with these nineteen variables in the entire wave of interaction.

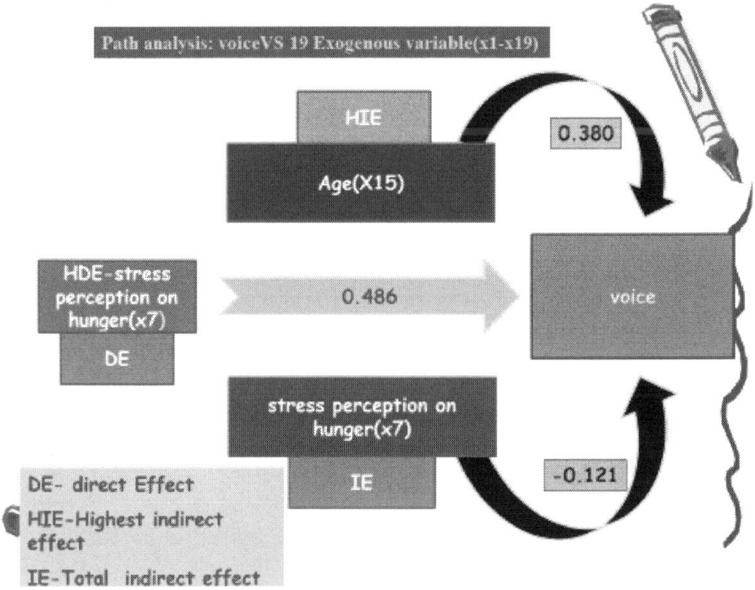

Model 6.2.16

Table 6.2.18:Path analysis: Decomposition of total effect (r) into Direct, Indirect and Residual effect [Cognitive differential VS 19 consequent variables (x_1-x_{19})]

S. no.	Variables	Total effect	Direct effect	Indirect effect	Highest indirect effect
x_1	age	0.112	0.074	**0.410**	x_7(0.090)
x_2	education	-0.065	0.158	0.131	x_5(0.053)
x_3	Family size	0.051	**0.**241	0.082	x_{15}(-0.111)
x_4	Economic motivation	-0.253	0.035	-0.151	x_4(-0.062)
x_5	Risk orientation	0.175	-0.150	0.097	x_2(-0.056)
x_6	Management orientation	0.162	0.164	-0.267	x_{13}(-0.030)
x_7	Stress perception on hunger	0.001	0.209	**-0.484**	x_{13}(-0.086)
x_8	Stress perception on poverty	-0.029	0.163	0.020	x_{13}(-0.053)
x_9	Stress perception on voice	-0.226	-0.004	-0.195	x_6(0.061)
x_{10}	Homestead land	-0.006	0.084	-0.156	x_{12}(-0.070)
x_{11}	occupation	-0.106	-0.081	-0.177	
x_{12}	Ancillary income	-0.060	-0.122	-0.322	x_9(-0.041)
x_{13}	expenditure	0.054	**0.332**	-0.001	x_3(-0.060)
x_{14}	savings	0.023	-0.219	0.028	x_{13}(0.102)

x_{15}	Bank account	0.058	0.243	0.157	x_3(-0.149)
x_{16}	Communication variables	0.017	**0.337**	0.122	x_9(0.043)
x_{17}	Management variables	-0.017	0.072	0.297	x_{12}(-0.064)
x_{18}	Energy consumption	0.156	-0.031	0.064	x_9(0.043)
x_{19}	BMI	0.121	**0.321**	**0.324**	x_{18}(0.060)

Residual 52.13%

The above table presents the path analysis wherein the total effect has been decomposed into direct, indirect and residual effect. Here it has been found that the variables expenditure (x_{13}), communication variables (x_{16}), and BMI (x_{19}) have exerted substantive direct effect on the dependent variable cognitive differential (y_4). So, these three variables have got strategic impact in dealing with the issues of cognitive differential.

Cognitive differential basically means difference in knowledge among the different respondents. Our basic perception in this study was, there would be a difference in cognation in between the stakeholders and the mangers and this would lead to an endemic hunger, poverty and voice situation. Expenditure in this case is a very important indicator. Higher the expenditure, higher would be the numbers of point of interaction. Communication plays an important role in reducing the cognitive differential. Higher the level of communication, lesser would be the cognitive differential and that is why the people should be getting the opportunity to speak. As far as the BMI is concerned, people with better health and nutritional food habit, backed up with better economy, income and livelihood would be more confident and wishful to get information and reduce the ignorance, in other words they would then eager to decrease the cognitive differentials. The impactful indirect effects are age, stress perception on hunger and BMI. Age is directed related to maturity and maturity is closely related to stress. These variables when linked with other variables like income, savings, homestead land etc. then they are exerting a significant effect on the cognitive differential. Here BMI is a variable which is having both significant direct effect and significant indirect effect. BMI solo can effect on cognitive differential and also it is highly impactful in combination with stress, income, expenditure, saving etc.

The residual effect being 52.13 per cent is inferring that even with the combination of nineteen variables, a substantive portion of variance in voice could not be explained with the participation with these nineteen variables in the entire wave of interaction.

Note : The residual effect can be reduced by-

1. Inclusion of more new relevant variables.
2. Exclusion of errors by more standardization and normalizations of data.
3. Increasing the number of respondents.

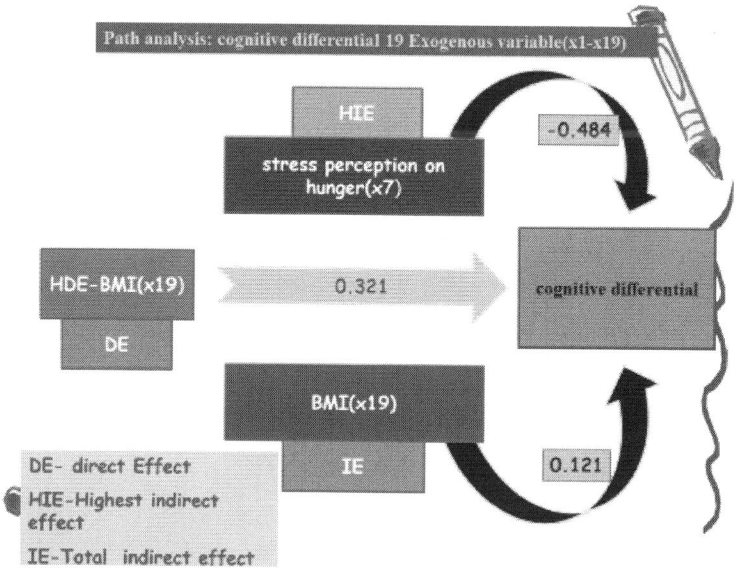

Model 6.2.17

Canonical covariate analysis

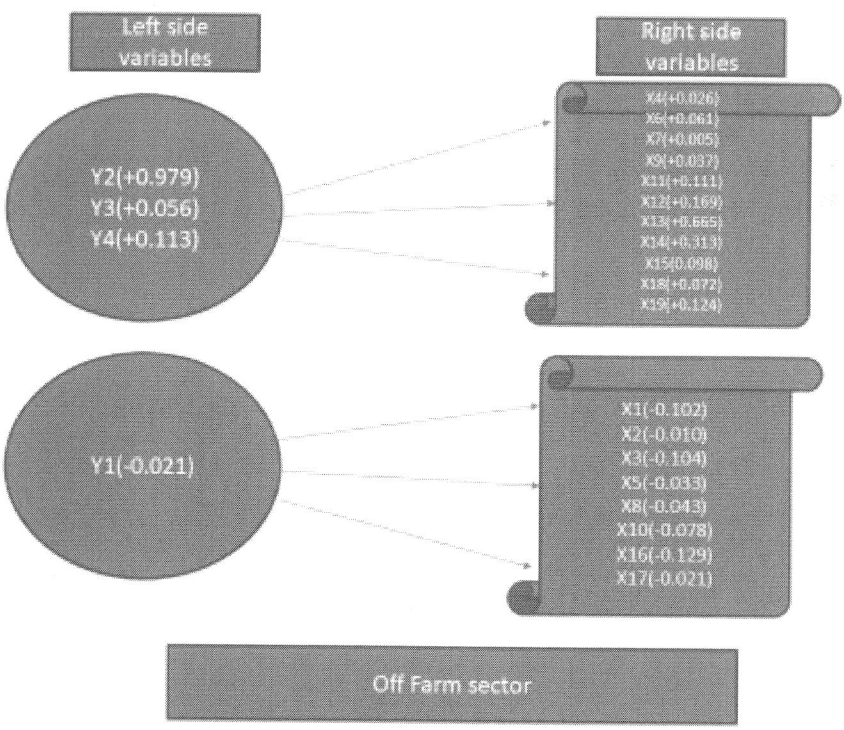

Model 6.2.18

The paradigm for canonical covariate suggests that the three predicted characters viz. poverty (y_2), voice (y_3) and cognitive differential (y_4) are moving together to create an echelon of predicted characters for the respondents hailing from off farm social ecology.

So, the off farm economy and ecology as have been found in the semi urban or peri urban ecosystems suggest that there is income variabilities, expenditure variabilities, multi layered poverty syndromes and differential exposure to sources of information. So, variability in information access, poverty status and voice have formed a complex cob web and they have together presented a selective choices from the right side echelon. Similar things have gone for the other predicted variable hunger (y_1) to suggest that hunger has got a unique character for off farm economy. The off farm occupations are characterized with, while compared with farm based ecology, it can be observed that income is here daily or weekly, less seasonal and less speculative. That is why hunger has moved solitarily here and at the same time has got a clear choice from right side echelons.

Crux of the overall discussion

Whatever may it be the hunger, poverty, voice or cognitive differential they are interrelated to each other and mutually interactive. Of course the result evinces that whenever a set of variables have got the highest direct effect, the others have the indirect or substantive effect on the dependent variables. A very few variables which are coming again and again during the discussion, (like age, economic motivation, stress perception, risk orientation, management orientation, commutation variables, BMI etc.) those variables can be considered as the marker variables or the variables with high strategic impact.

8 CHAPTER

POOLED RESULTS AND DISCUSSION

Result and discussion of Pooled data

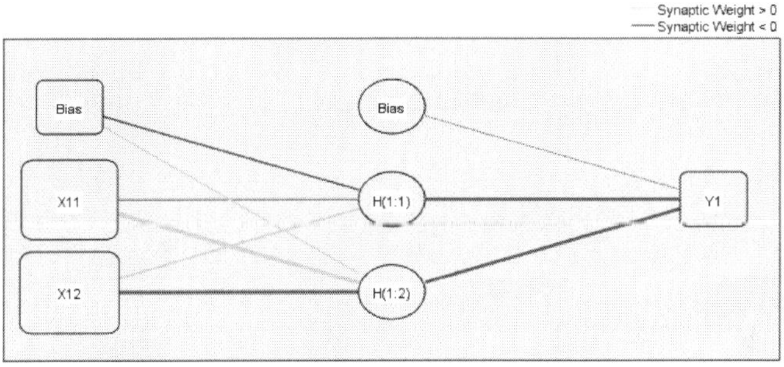

Hidden layer activation function: Hyperbolic tangent

Output layer activation function: Identity

Model 7.3.1

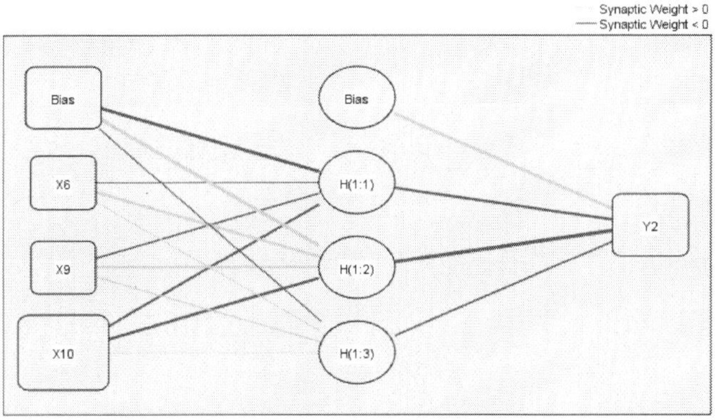

Hidden layer activation function: Hyperbolic tangent

Output layer activation function: Identity

Model 7.3.2

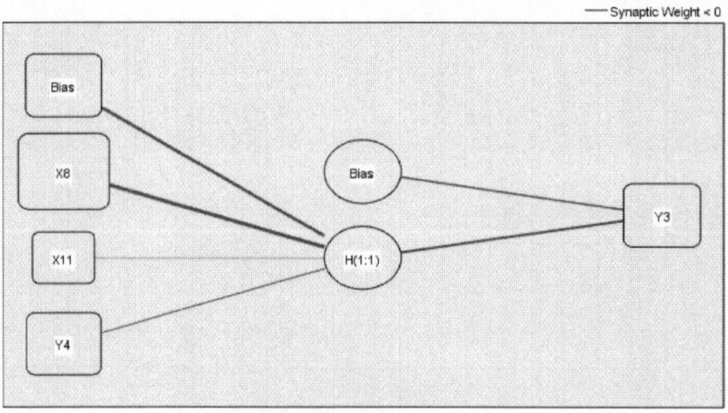

Hidden layer activation function: Hyperbolic tangent

Output layer activation function: Identity

Model 7.3.3

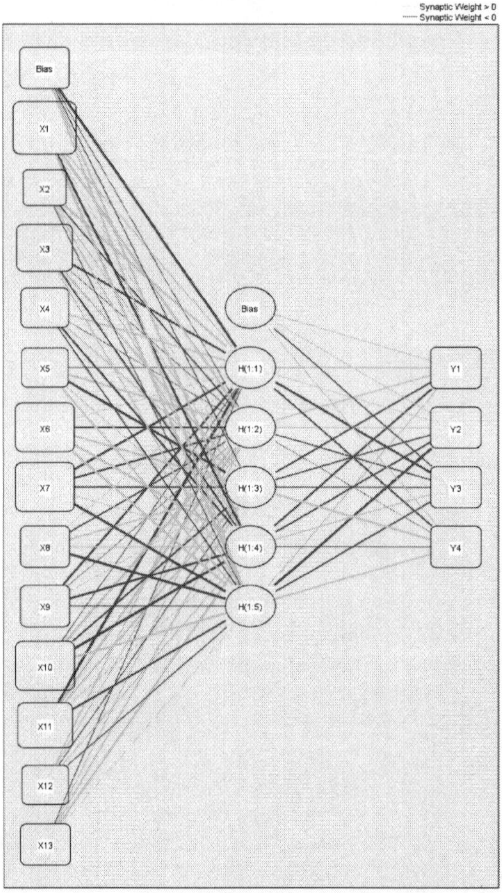

Hidden layer activation function: Hyperbolic tangent

Output layer activation function: Identity

Model 7.3.4

Q based distribution- Y1 vs. X11

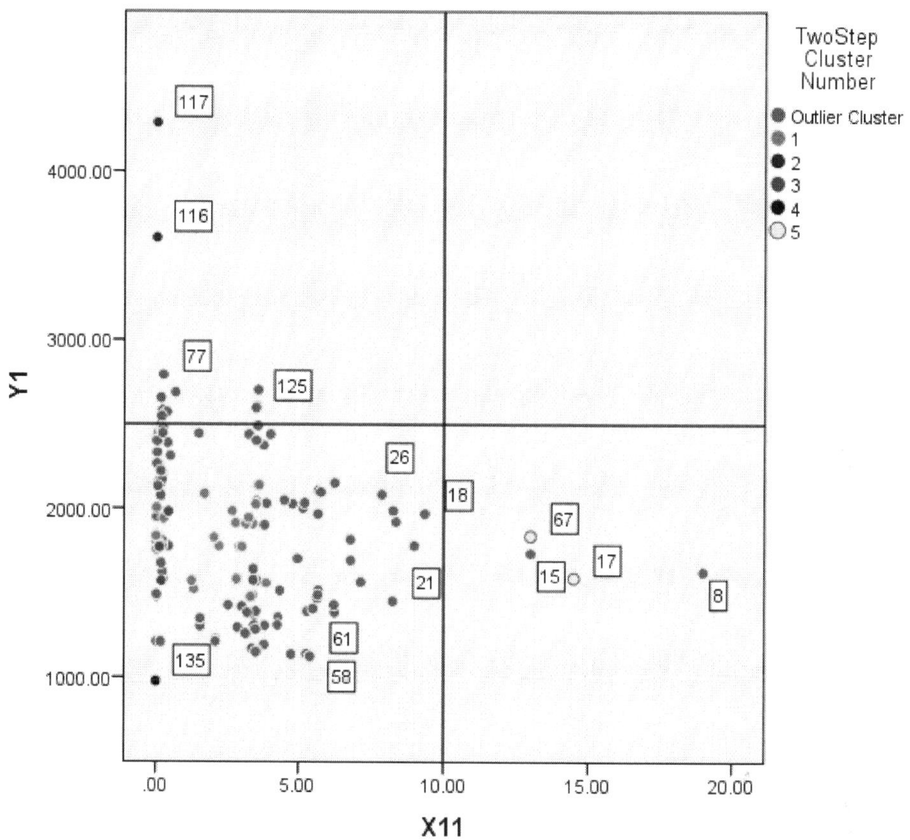

Model 7.3.5

Q based distribution Y2 Vs. X10

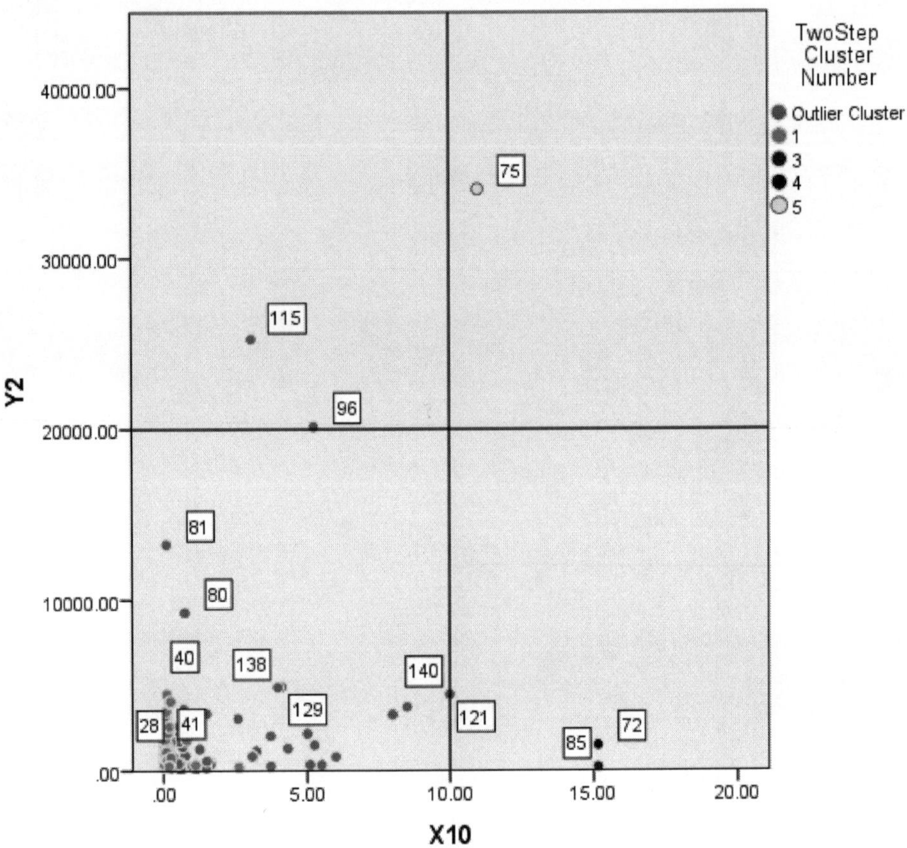

Model 7.3.6

Q based distribution Y3 Vs. X11

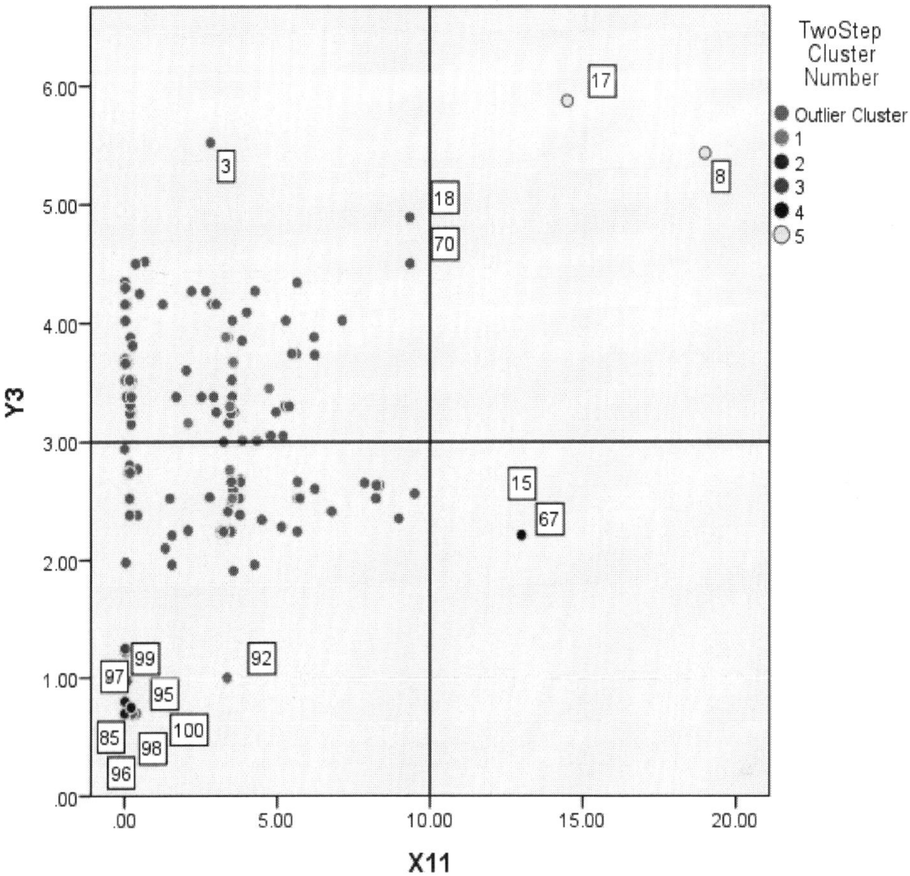

Model 7.3.7

Q based distribution Y4 Vs. Y3

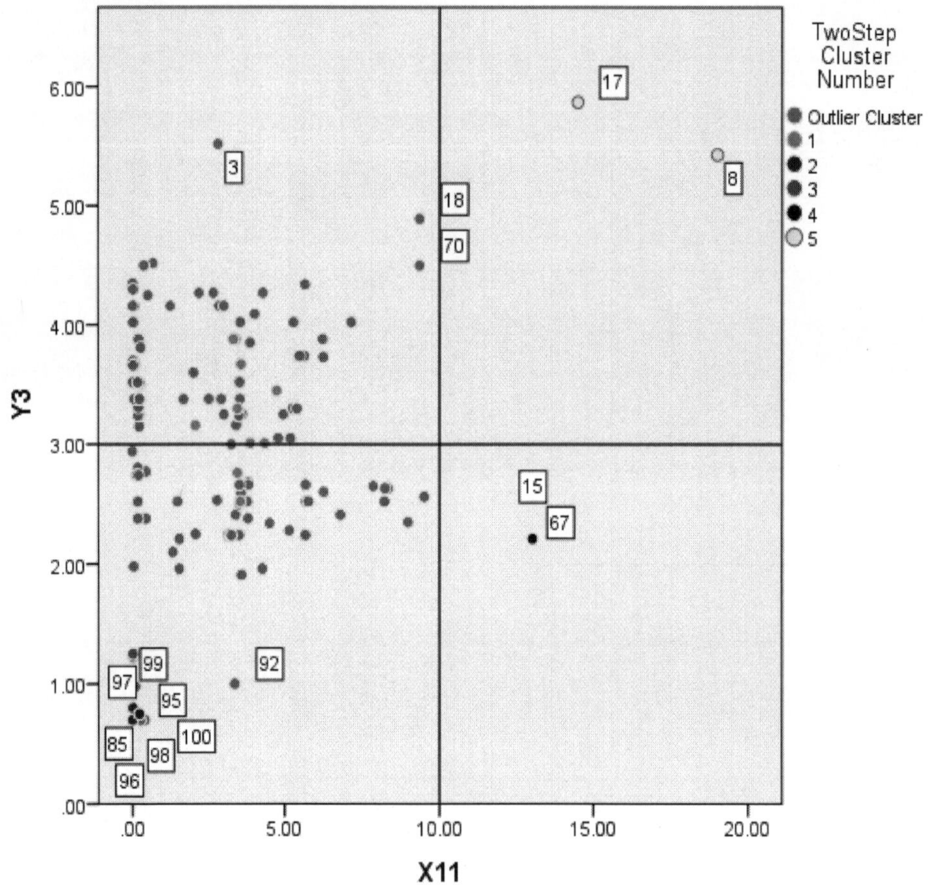

Model 7.3.8

Artificial neural network helps us to interpret and estimate the non linear pathway of selected input variables on the output variables. Three distinct layers are over here, one is input layer the second is hidden layer wherein the errors are minimized through activation and the output layers or the dependent variables. For the graphical presentation of the first neural network we can find that X11 that is communication variables,X12 that is energy consumption variable have rooted the distinct and dominant impact on the output variable Y1 that is hunger. It has passed through the two hidden layers for the rationalization vis a vis the reduction of error. So we can predict hunger with least of errors from these two input variables one is X11 that is communication variables and X12 that is energy consumption. So hunger has been contributed characteristically and dominantly by the communication variables and at the same time it has been picked up energy consumption pattern as one of the important predictor for the level of hunger of the respondents.

Now we are switching over to the second output variable poverty. Through the same way we can found that the management orientation X6 variable, stress perception on voice X9 variable and homestead land X10 variable have contributed in a decisive manner on the output variable poverty. Interestingly all these variables have been passed through three hidden layers for the management of their errors vis a vis estimation of their effect on the output variables.

Now we can switch over to the neural third neural network graph, here we can find that the predicted characters of the output variable that is cognitive differential, different level of knowledge and consciousness about different projects which are meant for poverty eradication and redressal of hunger. We can find that two variables that are X8 (stress perception on poverty) and X11(communication variables) have in a swashbuckling manner characterized the output variable Y3 that is voice. Here only one hidden layer would be considered through which these two variables have passed.

Now we are moving to words wherein all these output variables have been considered together and at the same time all the input variables have been taken care of. Here we can find that five hidden layers are there and through which all these input variables have generated their impact on the consequent variables those are hunger, poverty, voice and cognitive differential. We take one case for example that is X8. Here we can find that the X8 variable that is stress perception on poverty has passed through two hidden layers, one is hidden layer 1: 2 and the other is hidden layer 1:5 and after rooted through these two hidden layers as well as being reduced in terms of error these variable X8 has impacted on the output variable Y2 that is poverty. So we can see that whenever the respondents are under serious stress they are morose and don't show the interest to interact with others. Then they undergo a kind of withdrawal syndrome and then they lack the information about the projects simply because of there lethargy to access the same. So these variables if we see that to be taken care of to characterize the variable poverty. In another case we can find that the variable X11 that is communication variable after being passed through the hidden layer 1:1 and 1:5, they have been characterized the output variable Y4 that is cognitive differential. So, it is interesting to note that the communication variables are the main reason why the respondents have developed their different levels of cognitive differentials on the awareness of different government programmes. These four neural network paradigms have been successful in isolating the dominant variables, determining the number of hidden layers and ultimately how they have characterize the respective and corresponding output variables.

Now based on these four categories of responses we have got four quadrants and whenever we are taking care of the communication variables(X11) and it is considered as an integral part of the respondents, we can find that for the quadrant third most of the respondents are conglomerated. So the skewness of these respondents which are basically dominant in the quadrant 3, we can find that the respondents 135, 61, 58, 21 and 26 are so closely, intensely and integrally links with these variable Y1 that is hunger. Hence we can go strategically to identify these respondents to deal with the issues of communication vis a vis hunger. It is also true for the second picture where there is as usual four quadrants are there and for the third quadrant we can find that respondents 28, 41,129,140,148,40, 80 and 81, they are organically linked with the variable homestead land(X10). So for these respondents and in order to deal with the issues of poverty(Y2), homestead land can be considered as an important predictor.So the social ecology wherein the respondents are haling from, they are quite conspicuously having the size of their homestead land in estimating the poverty. Accordingly for the third graphics we can find that the respondents are almost equally distributed for the first two quadrants and for the third quadrant we will find that the communication variables have been integrated for the responds 85,92,95,96,97,98,99 and 100 are characterized for the output variable voice. So in order to deal with the problem of voices why the respondents are so shaky or they are being dominant by withdrawal syndrome, we can take care of these respondents because the communication variables are organically linked with them. For the last graphics we can find that almost all the respondents are equally distributed among the four quadrants (++,+-,-+ and --). So, cognitive differential is a kind of dependent character which is almost universal for all the respondents. Therefore we can deal with this dependent variable for all the respondents in a uniform manner as well.

So these are the analysis of artificial neural network graphs and in a concluding manner we can say that we need not bother about all the variables while selecting. We have to let the input variables passed through the hidden layers and select few ones. So, for all the dependent variables a few selected variables are there. Therefore for any kind of neural network it is interesting to isolate the dominant variables out of the trivial variables and all these interactions and integrated characters again can be distributed based on their beta regression co efficient and the way they are characterizing the Y1, Y2,Y3 and Y4 all these can be divided into four quadrants. For each quadrants, unique respondents are retaining and for them we can go for unique treatments without bothering about all at a time. Therefore ANN is a very useful and effective technique in delineating and describing the micro level strategical interventions.

Partial Correlation Analysis

Partial correlation is a primary tool to study the non linear path way of the instances and impact of several independent variables under study.

Hunger (Y1)Vs. 13 independent variables

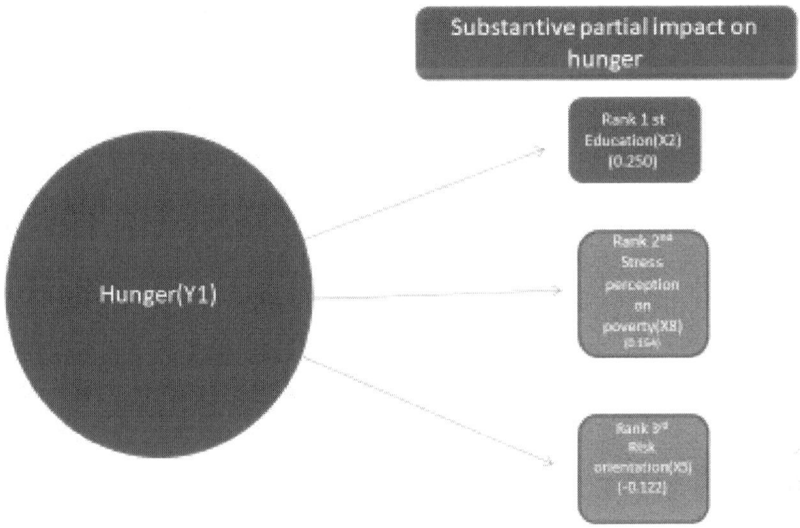

Model 7.3.9

It has been seen that the risk orientation , stress perception and education are creating highest partial impact as far as the hunger status is concerned. Things may so happened that due to poorer education the respondents are having less risk bearing ability and more psycho somatic stresses and thus collectively they effect on hunger in a substantive manner.

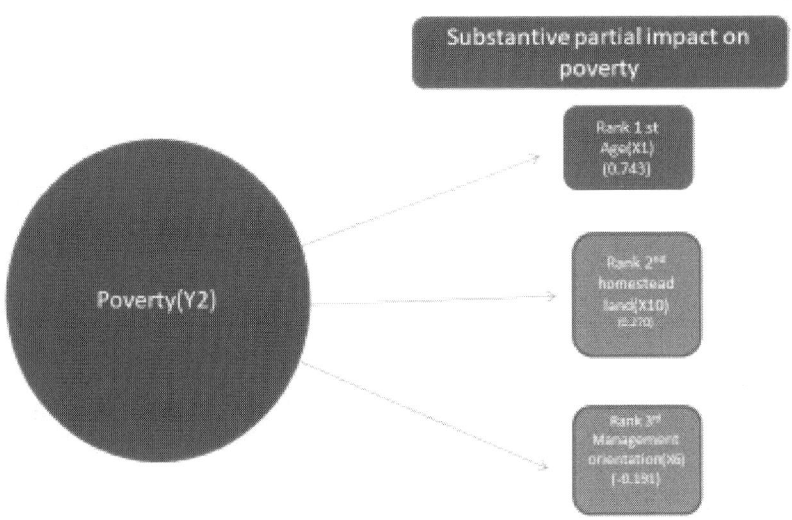

Model 7.3.10

Poverty (Y2) Vs. 13 independent variables

It has been seen that the age , homestead land and management orientation are creating highest partial impact as far as the poverty status is concerned. These three variables have the highest substantive effect than all the remaining independent variables on the said dependent variable. with the age maturity comes and with the maturity the sense of status comes. People starts understanding their actual condition. On the other hand homestead land is an indicator of the status of a person whether he/she is from farm or off farm back ground. The third substantially impactful variable is management orientation. Be it in a farm situation or in an off farm situation the well managed resources are always the key factor. The entrepreneurs or the agripreneurs now a days are more successful than a simple shop owner or a farmer. Indian farmers are poor because they are poor in agripreneurship. They are innovatively poor. They are not managing their resources like the Vietnam or the Israel farmers are doing. So, why the indian farmers are poor ? is it only because they are monetarily poor or resource- poor? Or because they a poor in ideas or innovation?

Hence, from this above discussion one thing can be clearly concluded that if we want to change the status of our poorer population, we have to train them, groom them, motivate them and encapacited them to generate ideas and to utilize their resources the fullest.

Voice (Y3) Vs. 13 independent variables

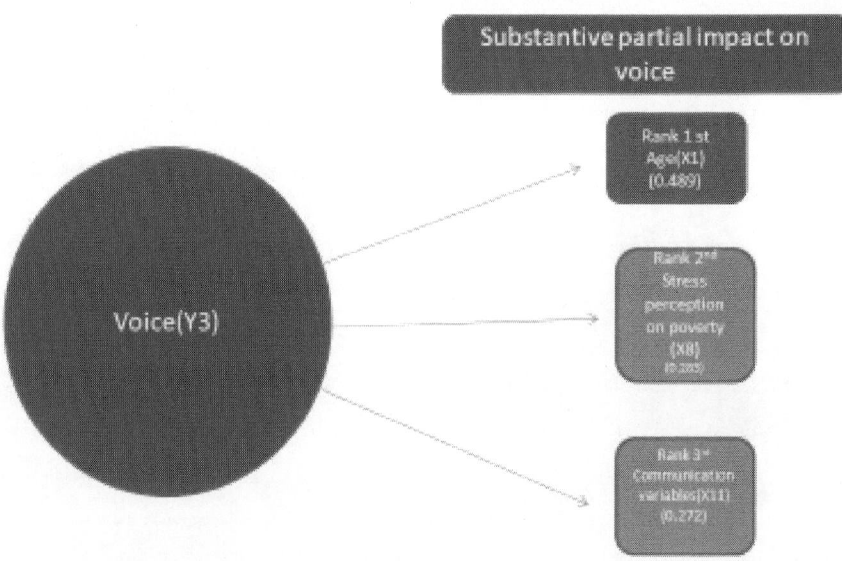

Model 7.3.11

Here in the above model it has been seen that the age, strees perception on poverty and communication variables are creating highest partial impact as far as their capacity to raise voice is concerned. As with the age the sense of status is grown, in the same manner with the age the desire and capacity to break silence is also elevated. Stress perception is also a stimulator in this issue. If the stress is high, people are frustrated, depressed, morose and finally suffer from withdrawal syndrome where they are totally indifferent about their surrounding ambience. Lastly, the communication variables. They are basically the precursors of one's voice. With communication, information come. And with these information the urge and need for voice in generated.

Cognitive differential (Y4) Vs. 13 independent variables

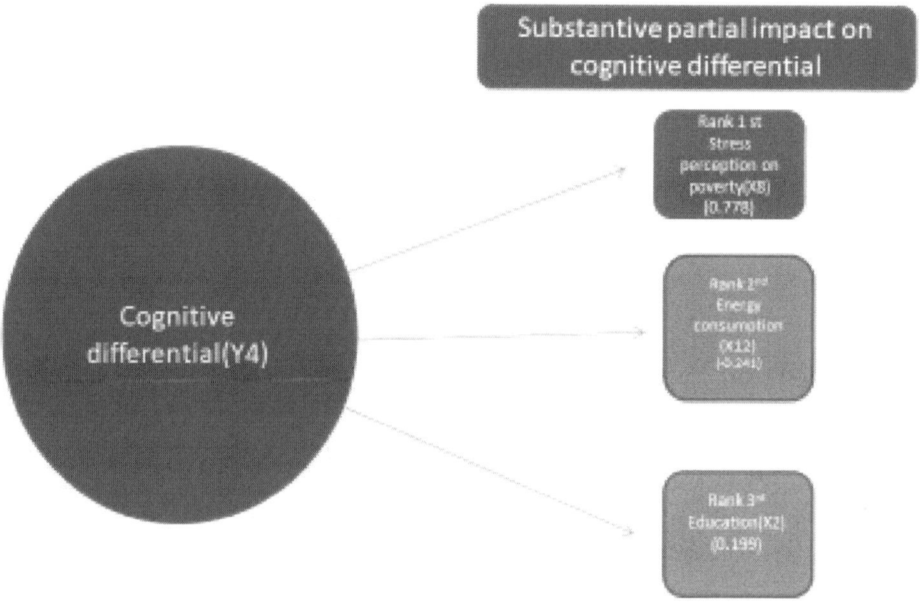

Model 7.3.12

Here in the model it has been seen that the stress perception on voice, energy consumption and education are creating highest partial impact as far as the cognitive differential of the respondents are concerned. Like we have discussed earlier, stress causes depression, inactive nature and lastly withdrawal syndrome. So, it is very different for a person in stress to increase their informative behaviour and awareness. Energy consumption is basically the indicator of a person's overall mobility. Energy consumption is becoming a very important and predominant factor in the entire dynamics of the ecological changes in the sector of agriculture. Now, as far as education is concerned, lesser the education more would be cognitive differential. You are less educated(non formal and informal education is also considered) means you are less aware. Sometimes you can be ignored rather. Why you are not allowed to speak, is can simply because you are a drop out.

Canonical co variate analysis

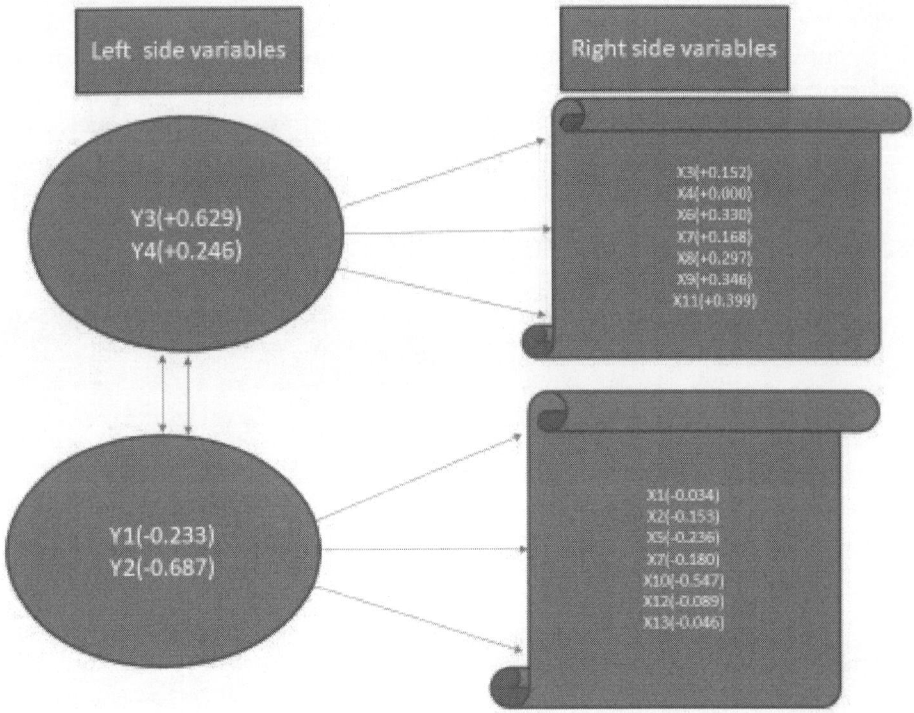

Model 7.3.13

In statistics, CCA is a way of inferring information from cross-covariance matrices. If we have two vectors, $X=(X_{1,......,}X_n)$ and $Y=(Y_{1,......,}Y_n)$ of random variables and there are correlations among the variables, then canonical correlation analysis will find linear combinations of the$X_i$and Y_j which have maximum correlation with each other. T.R Knapp notes that" virtually all of the commonly encountered parametric tests of significance can be treated as special cases of canonical correlation analysis, which is the general procedure for investing the relationships between two sets of variables". The method was first introduced by Harold Hotelling in 1936.

CCA for this study has been applied to extract the canonical covariates between two sets of variables. The left side variables and the right side variables. It has been observed that the LS variable again has formed two conglomerations further. Here, voice(Y3) and cognitive differential (Y4) both have picked together seven independent variables those are family size(X3), economic motivation(X4) , management orientation(X6), stress perception on hunger(X7), stress perception on poverty(X8), stress perception on voice(X9) and Communication variables(X11). On the other hand hunger(Y1) and poverty(Y2) have picked together the remaining six independent variables those are age(X1), education(X2) , risk orientation(X5), stress perception on hunger(X7), Homesteadland(X10), Energy consumption(X12), BMI(X13).

So, from CCA we have come to know that the groups of Y variables have got precise selectivity to ultimately form a splendid strategy as to cater component related interaction to characterize the perception hunger, poverty and voice.

Multiple stepwise logistic forward regressions

The result of stepwise logistic regression analysis is presented in Table 1 that shows the factors influencing perceived hunger score. The similar procedure was used for the perceived poverty score, voice score and development cognitive score. The summary table for rural sector is presented in table 1 whereas the same for semi-urban region is given in Table 2.

Table 7.3.1: Summary of stepwise logistic selection of various identified variables affecting different Social stigma in rural sector:

Step	Effect Entered	df	Score Chi-square	Pr.>Chi square
		Hunger		
1	Risk orientation	1	5.4861	0.0192
2	Size of holding	1	5.7822	0.0162
		Poverty		
1	Marketed surplus	1	10.6494	0.0011
2	Pond and fish	1	5.4263	0.0198
3	Age	1	4.4597	0.0347
		Voice		
1	Stress perception on poverty	1	5.6322	0.0176
2	Livestock yield	1	4.5708	0.0325

3	No. of fragments	1	6.3001	0.0121
4	Size of holding	1	5.7547	0.0164
	Cognitive differential			
1	Livestock yield	1	6.8684	0.0088
2	Cropping intensity	1	3.9131	0.0479

(The Pr.>Chi square is >0.05 for all cases as per Hosmer and Lemeshow Goodness-of- Fit test)

Table 7.3.2: Profile of Odds Ratio Estimates found under stepwise logistic selection of different identified variables in rural sector:

Effect	Odds ratio estimate	95% Wald Confidence Limits	
		Lower	Upper
Hunger			
Risk orientation	0.002	<0.001	0.102
Size of holding	0.879	0.624	1.237
Poverty			
Marketed surplus	1.023	0.995	1.051
Pond and fish	>999.999	1.489	>999.999
Age	1.325	1.025	1.713
Voice			
Stress perception on poverty	1.636	1.188	2.254
Livestock yield	1.785	0.968	3.290
No. of fragments	2.363	1.378	4.052
Size of holding	0.490	0.281	0.853
Development cognitive			
Livestock yield	1.393	1.026	1.891
Cropping intensity	1.007	1.000	1.013

Table 7.3.3: Summary of stepwise logistic selection of various identified variables affecting different Social stigma in Semi Urban sector:

Step	Effect Entered	df	Score Chi-square	Pr.>Chi square
Hunger				
1	Energy consumption	1	8.0688	0.0045
2	Age	1	4.1767	0.0410
3	Marketed surplus	1	3.9325	0.0474
4	Communication variables	1	3.5188	0.0607
Poverty				
1	Livestock yield	1	16.7263	<.0001
2	Pond and fish	1	7.3699	0.0066
Voice				
1	Livestock yield	1	8.8001	0.0030
2	Education	1	8.3770	0.0038
3	Economic motivation	1	5.8549	0.0155
4	Stress perception on hunger	1	5.8594	0.0155
Development cognitive				
1	Marketed surplus	1	8.6266	0.0033
2	Stress perception on poverty	1	3.7734	0.0521
3	Management orientation	1	5.2000	0.0226
4	Economic motivation	1	5.6234	0.0177
5	Energy consumption	1	4.8590	0.0275

(The Pr.>Chi square is >0.05 for all cases as per Hosmer and Lemeshow Goodness-of- Fit test)

Table 7.3.4: Profile of Odds Ratio Estimates found under stepwise logistic selection of different identified variables in Semi urban sector:

Effect	Odds ratio estimate	95% Wald Confidence Limits	
		Lower	Upper
Hunger			
Energy consumption	28.928	0.384	>999.999

Age	1.291	0.943	1.767
Marketed surplus	2.347	1.084	5.082
Communication variables	0.446	0.186	1.068
Poverty			
Livestock yield	1.011	1.000	1.022
Pond and fish	1.026	0.975	1.079
Voice			
Livestock yield	1.307	0.878	1.946
Education	0.025	<0.001	4.855
Economic motivation	<0.001	<0.001	>999.999
Stress perception on hunger	>999.999	<0.001	>999.999
Development cognitive			
Marketed surplus	4.963	0.792	31.101
Stress perception on poverty	<0.001	<0.001	2.530
Management orientation	48.744	0.353	>999.999
Economic motivation	60.539	0.389	>999.999
Energy consumption	0.005	<0.001	4.572

Table 7.3.2 and 6.3.4 lists the odds ratio estimates along with 95% Wald confidence limits found under stepwise logistic selection of different identified variables in rural and Semi urban sector. The factors where the Odds Ratio Estimates is >1 indicated the likelihood of average hunger >1. The odd ratio (<1) indicated an inverse association of such factor with the likelihood of average hunger. It means Risk orientation, Size of holding in rural sector are inversely associated with the likelihood of average hunger as the odds ratio <1. On the other hand , in case of semi-urban sector, energy consumption, age , marketed surplus indicated the likelihood of average hunger >1 but for Communication variables it is inversely associated with the likelihood of average hunger as the odds ratio <1. Similarly, for all other variables it was followed.

Multidimensional scaling (MDS) technique was employed to draw conclusions about the proximity/closeness among the social constraints. By MDS a map was created displaying the relative positions of different variables like poverty, hunger, voice, and development cognitive causing social stigma in both rural and semi-urban areas using a distance matrix. The perceptual map consisted of two dimensions to interpret the dissimilarity matrix (lower triangular) of those variables on a two-dimensional scatter plot (Shepard diagram). The stress values in both the cases are found satisfactorily (0.00000) to do further analysis. Final Co-ordinates for perceptual mapping for both rural and semi-urban sector are given in table 5 and Table 7. In this example, (in both Fig. 1 and Fig.2) it is evident that the respondents considered voice and cognitive development to be closer. **They development also considered voice and hunger to be closer. Poverty appears quite different from Hunger, cognitive and voice.**

It also appears from MDS analysis the underlying factors that respondents used when they created these dissimilarities. In the Perceptual mapping for rural sector (Fig.1) identifies that voice and cognitive development are in the negative side of dimension 1 (implies the efforts to reduce the social stigma) where as voice and cognitive development are in positive side of dimension 1 in case of Perceptual mapping (Fig.2) for semi-urban areas. Dimension 2 implies the impact of social stigma that causes social entropy. It is clear that (From dimension 2) in case of rural sector, the combined effects of poverty, voice and cognitive development results hunger which is the driving force of social entropy in the region. But in case of semi-urban sector, the effect of poverty and cognitive development in mass scale cause the hunger.

Table 7.3.5: Final Co-ordinates for perceptual mapping for rural sector

Final Coordinates		
	Dimension	
	1	2
Hunger	-0.142	-0.496
Poverty	0.891	0.111
Voice	-0.302	0.013
Cognity	-0.446	0.371

Table 7.3.6: Final Co-ordinates for perceptual mapping for Semi-urban sector

Final Coordinates		
	Dimension	
	1	2
Hunger	-0.355	-0.419
Poverty	-0.683	0.334
Voice	0.413	-0.124
Cognity	0.625	0.209

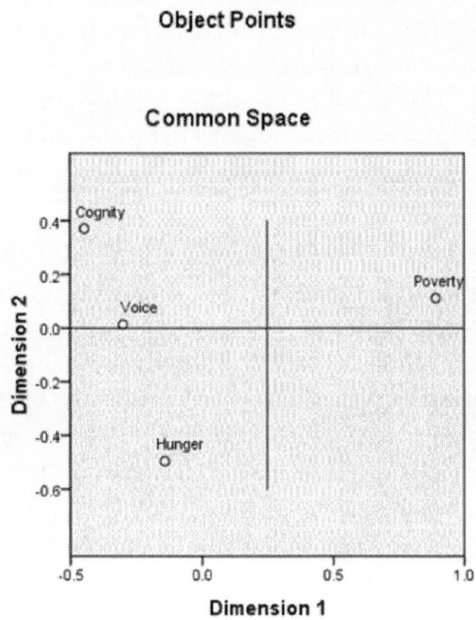

Model 7.3.14: Perceptual mapping for rural sector

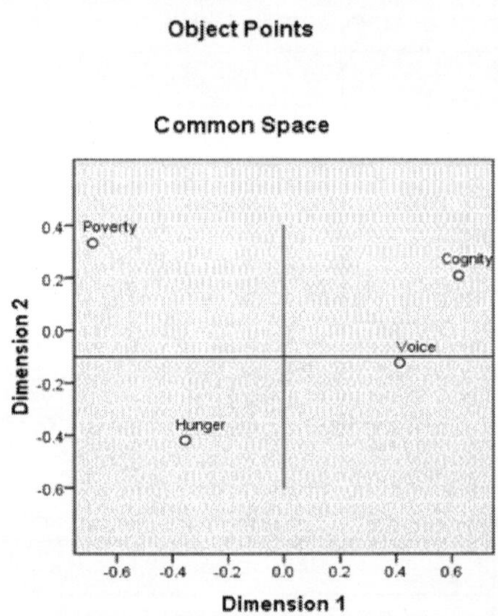

Model 7.3.15: Perceptual mapping for Semi Urban sector

CHAPTER 8

PARICIPATORY ANALYSIS AND EMPOWERMENT

PRA IN FARM AND NON-FARM SECTOR

The Concept

The philosophical roots of participatory rural appraisal techniques can be traced to activist adult education methods such as those of Paulo Freire and the study clubs of the Antagonist Movement. In this view, an actively involved and empowered local population is essential to successful rural community development. Robert Chambers, a key exponent of PRA, argued that the approach owes much to "the Freirian theme that poor and exploited people can and should be enabled to analyze their own reality."

By the early 1980s, there was growing dissatisfaction among development experts with both the reductionism of formal surveys, and the biases of typical field visits. In 1983, Robert Chambers, a Fellow at the Institute of Development Studies (UK), used the term *rapid rural appraisal* (RRA) to describe techniques that could bring about a «reversal of learning», to learn from rural people directly. Two years later, the first international conference to share experiences relating to RRA was held in Thailand.[8] This was followed by a rapid acceptance of usage of methods that involved rural people in examining their own problems, setting their own goals, and monitoring their own achievements. By the mid 1990s, the term RRA had been replaced by a number of other terms including *participatory rural appraisal* (PRA) and *participatory learning and action* (PLA).

Here in this empirical study, different PRA(Participatory Rural Appraisal) tools have been executed to elicit the root cause of poverty, migration, perception on project impact, Kissan Credit Card(KCC) and their overall views on empowerment and entitlement in the mainstream social activism.

PRA OF FARM SECTOR UNDER STUDY (BERABERI VILLAGE, HABRA II BLOCK)

MIGRATION MAPPING : OCCUPATIONAL MIGRATION

Here the length of the arrows are showing the distance of the migration. As the arrow become longer it shows that the respondents are moving away from the village .The black circle represents the state. The arrows within the black circle shows that the respondents migrate within the state, the white circle represents the country and the arrows show that the respondents migrate elsewhere in the country. Lastly the orange line is the border between inside and outside the country and the arrow shows that the respondent migrates outside India.

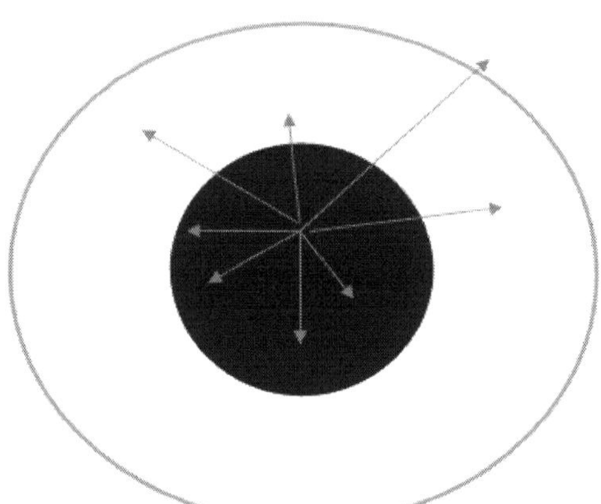

Model 6.4.1

PROJECT ACCESS AND EMPOWERMENT RANKING

Access to project and empowerment have gone co genital by nature. So in ranking project access, the empowerment level also elucidated isochronously. The access to project had been calculated by following 10 point scale subsequently the mean score was calculated.

Table 6.4.1: Project Access and Empowerment ranking(farm sector)

Projects	Scoring by the key informants	Mean score and rank	Remarks
	A B C D E F		
Mid day meal	9 10 9 10 10 10	9.66(rank 1)	Implementation and popularity of the project within this locality is excellent.
KrishiSinchayYojana	7 8 8 6 8 7	7.33(rank 4)	Implementation and popularity of the project within this locality is not so good.
Indira AwasYojana (IAY)	5 4 7 6 5 5	5.33(rank 6)	Implementation and popularity of the project within this locality is very poor.
Kishan Credit Card	8 9 8 8 9 9	8.5 (rank 2)	Implementation and popularity of the project within this locality is very good.
Farm Implement Distribution at Subsidy(FSSM)	6 6 7 8 6 6	6.5 (rank 5)	The popularly of this scheme is gradually increasing but implementation needs betterment.
SashyaBimaYojana	9 7 8 7 9 8	8 (rank 3)	Popularity of this scheme is increased pretty much within last few years. Implementation is also good.

WHY MID-DAY MEAL IS GOING GREAT?

Success criteria:

1. Workers can represent the target community with a scope of improvement, especially by improving their communication skill and knowledge base.
2. Service delivery mechanism is not so smooth. So an enabling environment needs to be created.
3. Material support stands adequate but it needs quality management and timely delivery.

WHY KCC IS GOING GREAT?

Success criteria:

1. Farmers are getting low interest loan, so they are now out of the mercy of the local money lenders.
2. No middle man involvement in the total procedure
3. The local Agriculture Development Officers and the bank officials are doing their job quite sincerely.

WHY SASHYABIMA YOJANA IS GOING GOOD?

Success criteria:

1. The local level campaigning is good to popularize the scheme.
2. The penalty amount is reaching to the farers promptly and effectively.

WHY KRISHISINCHAYYOJANA IS GOING MERELY GOOD?

Moderate Success criteria:

1. Local level campaigning is not that effective but gradually it is getting pace.
2. Local level implementation is not so good.

WHY FSSM IS GOING BAD?

Failure criteria:

1. Lack of custom hiring centres.
2. Lack of campaigning.

WHY IAY IS GOING TOO BAD?

Failure criteria:

1. Lack of communication by the actual needy people.
2. More political interventions.
3. Lack of proper documentation of BPL card holders.

ENTERPRISE PREFERENCE RANKING:

For organizing the enterprise preference ranking we have selected five key informants. Those key informants were asked to score the following attributes like income, risk orientation, dependence of livelihood and ease of doing business for the four enterprises like farming, bidi binding,toto pulling, shop keeping,. The total score given by the key informants to the each enterprise and each attributes was summed up then mean score was brought out and this mean score is used in the following ranking matrix.

Table 6.4.2 Enterprise preference ranking (farm sector)

Attributes	E1	E2	E3	E4
income	9.64	7.60	7.50	5.55
Risk orientation (higher the score lower the risk)	6.40	6.55	6.56	6.65
Dependence of livelihood	8.45	4.15	7.45	6.00
Ease of doing business	7.33	8.38	7.00	5.65
Total score	31.81	26.68	28.51	23.85
Mean score	**7.95**	**6.67**	**7.13**	**5.96**
Remarks	Highest ranking enterprise in terms of preference of the respondents. (E1)	Third highest ranking enterprise in terms of preference of the respondents. (E2)	Second highest ranking enterprise in terms of preference of the respondents. (E3)	Lowest ranking enterprise in terms of preference of the respondents.(E4)

- E1- farming
- E2- bidi binding
- E3- toto pulling
- E4- shop keeping

PRA OF NON-FARM SECTOR UNDER STUDY (BILKANDA VILLAGE, BARRACKPORE II BLOCK)

Migration Mapping of Occupational Migration

Here the length of the arrows are showing the distance of the migration. As the arrow become longer it shows that the respondents are moving away from the village .The black circle represents the state. The arrows within the black circle shows that the respondents migrate within the state, the white circle represents the country and the arrows show that the respondents migrate elsewhere in the country. Lastly the orange line is the border between inside and outside the country and the arrow shows that the respondent migrates outside India.

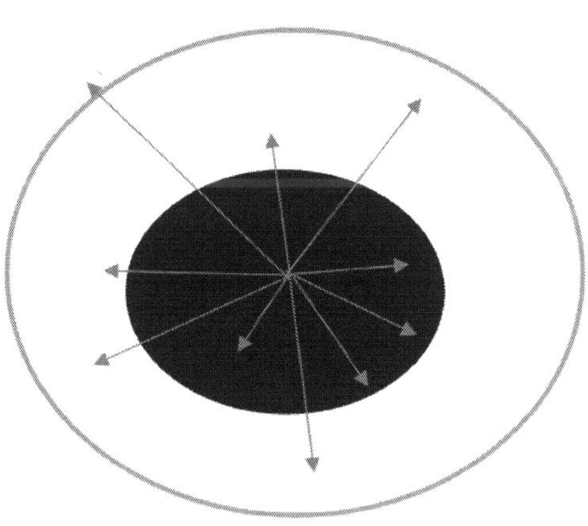

Model 6.4.2

PROJECT ACCESS AND EMPOWERMENT RANKING

Access to project and empowerment have gone co genital by nature. So in ranking project access, the empowerment level also elucidated isochronously. The access to project had been calculated by following 10 point scale subsequently the mean score was calculated.

Table 6.4.3 :Project Access and Empowerment ranking(off farm sector)

Projects	Scoring by the key informants	Mean score and rank	Remarks
	A B C D E F		
Mid day meal	9 7 8 7 9 8	8(rank 2)	Implementation and popularity of the project within this locality is good.
ICDS	9 10 9 10 10 10	9.66(rank 1)	Implementation and popularity of the project within this locality is excellent.
IAY	5 4 7 6 5 5	5.33(rank 3)	Implementation and popularity of the project within this locality is not so effective.

WHY ICDS GOES EXCELLENT?

Success criteria:

1. ICDS workers are very much representative to the community and community can identify them with their own needs.
2. Service delivery mechanism is working smoothly.
3. Material support stands adequate and timely.

WHY DOES MID-DAY MEAL GO MERELY GOOD?

Success criteria:

1. Workers can represent the target community with a scope of improvement, especially by improving their communication skill and knowledge base.
2. Service delivery mechanism is not so smooth. So an enabling environment needs to be created.
3. Material support stands adequate but it needs quality management and timely delivery.

WHY DOES NOT IAY (INDIRA AWAAS YOJANA) GO EFFECTIVE

Limiting factors:

1. Unabated interference from middlemen.
2. Political interference goes indomitable.
3. Lack of technical support.

ENTERPRISE PREFERENCE RANKING

For organizing the enterprise preference ranking we have selected five key informants. Those key informants were asked to score the following attributes like income, risk orientation, dependence of livelihood and ease of doing business for the four enterprises like shop keeping, toto or van rickshaw pulling, vendor and tailoring. The total score given by the key informants to the each enterprise and each attributes was summed up then mean score was brought out and this mean score is used in the following ranking matrix.

Table 6.4.4 : Enterprise preference ranking (off farm sector)

Attributes	E1	E2	E3	E4
Income	7.33	7.6	8.54	6.55
Risk orientation (higher the score lower the risk)	8.40	5.55	7.56	7.65
Dependence of livelihood	5.45	6.15	4.45	6.33
Ease of doing business	8	8.33	7.86	5.65
Total score	29.18	27.63	28.41	26.18

Mean score	7.30	6.90	7.10	6.55
Remarks	Highest ranking enterprise in terms of preference of the respondents.(E1)	Third highest ranking enterprise in terms of preference of the respondents.(E2)	Second highest ranking enterprise in terms of preference of the respondents.(E3)	Lowest ranking enterprise in terms of preference of the respondents.(E4)

- E1- shop keeping
- E2- toto or van rickshaw pulling
- E3-vendor
- E4-tailoring

10
CHAPTER

COVID REALITY AND HUNGER

HUNGER, POVERTY AND VOICE DURING COVID-19 PANDEMIC: GLIMPSES AND REFLECTION

Coronaviruses are a group of related RNA viruses that cause diseases in mammals and birds. In humans, these viruses cause respiratory tract infections that can range from mild to lethal. Mild illnesses include some cases of the common cold (which is also caused by other viruses, predominantly rhinoviruses), while more lethal varieties can cause SARS, MERS, and COVID-19. Symptoms in other species vary: in chickens, they cause an upper respiratory tract disease, while in cows and pigs they cause diarrhea. There are as yet no vaccines or antiviral drugs to prevent or treat human coronavirus infections.

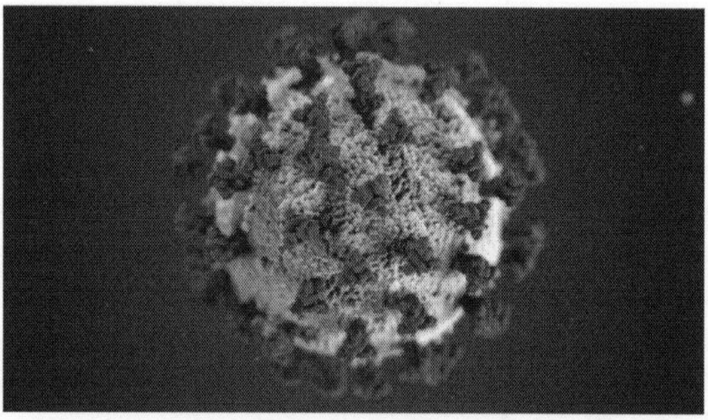

Source: internet

In the December month of last year (2019) this corona virus has started showing its presence in the Wuhan province of China. Within two months it has started spreading in the over all world through human to human transmission. For its rapid spreading and human to human transmitting nature most of the countries has started locking down their economic, social, political and educational sectors.

Having no difference the Indian government has also started the process of lockdown by strictly prohibiting the movement and gathering of its people.

This decision was basically taken in a health emergency situation and definitely for the sake of the safely of the vast population. But this apparently necessary decision has its set back too. Due to the sudden lockdown, all the sectors of economy have been stopped overnight and thousands of the daily wage earners have lost their jobs. The situation is worst for the migrant labours who has one hand lost their way to earn their daily bread and butter and on the other hand they were unable to go back to their native places. As a result the hungry has become more hungry and the poor has become poorer.

THE PEOPLE WITH HUNGRY STOMACH

Stories of hunger deaths of migrants stranded between their villages and big cities have begun to slowly find space in the Indian media. Many of these reports are promptly denied, but in each of these cases the dead never return.

In the north Indian town of Bhadohi, a woman daily wage worker threw her five children into the River Ganges as she reportedly claimed that she was no longer able to provide food for them. Police later claimed starvation was not the reason, but that she was of an unsound mind and acrimonious. In another case, in Delhi, a hungry migrant sold his phone for Rs 2500/-, handed over the money to his wife and then hung himself. The police said he was also mentally unstable.

By conservative estimates, ever since India went into a shutdown to flatten the coronavirus pandemic curve, 189 deaths have taken place due to hunger, exhaustion while walking back home and suicides. These numbers will swell further.

According to government estimates, 1.25 million migrants are stranded between big cities and their villages and are living in government shelters. By another estimate, 33 percent of the total migrant population have no access to food since March 24, the day Indian Prime Minister Narendra Modi announced a comprehensive lockdown that has meant shutting down factories, work sites, small and large businesses and transport. In a matter of a few hours, millions of daily wageworkers lost their jobs and found it difficult to stay in their rented houses.

MASS EXODUS: THE LAST OPTION

Without money and fearing starvation, they wanted to take the first bus or train to their villages, where they hoped to find food and emotional security. Thousands headed to the bus terminal in Delhi. They were stopped by force from police officers who were ordered to keep people away from the roads. Many of these migrants began to walk hundreds of kilometres back home. Some cycled or took lifts to reach their villages – a genuine mass exodus during the coronavirus lockdown. While on his way home, a hungry and tired migrant collapsed in north India's state of Uttar Pradesh. He was 400 kilometers away from his village.

According to *India Spend* – there are 120 million rural urban migrants in most of the big and small cities of the country. Professor Santosh Mehrotra, a labor economist with Delhi's Jawahar Lal Nehru University says: "There exists little basis for the migrants to be invisible to the decision makers. This data was in the public domain when the lockdown was declared."

There have been demonstrations and riots in Mumbai and the textile city of Surat, Gujarat, where an approximate 7,00,000 daily wageworkers, unable to go home, are reportedly in miserable shape. Civil society activists are increasingly coming forward with reports of starvation deaths.

"WE ARE WORKERS WHO WORK TO FEED OURSELVES, AND, NOT BEGGARS."

A survey conducted on April 9 of 11,000 migrant workers in different parts of the country by *Stranded Workers Action Network (SWAN)* revealed: "50% of workers had rations left for less than 1 day. 72 percent of the people said that their rations will finish in 2 days. 96% had not received rations from the government and 70% had not received any cooked food. Around 98% have not received any cash relief from the government."

Long queues for food are visible in Delhi and other cities of the country. Some people are embarrassed that they have to go through this ignominy and beg for food. A news report in an online publication quoted a worker nursing a hurt pride: "We are workers who work to feed ourselves and not beggars."

There have been well-meaning suggestions from Nobel laureates Amartya Sen and Abhijeet Banerjee as well as former Central Bank Governor Raghuram Rajan. In an article in an Indian daily, they advised that the Indian government should open up its considerable food reserves of 77 million tones – three times more than its buffer stock. "At a time of national emergency, it makes perfect sense to dispose of some of the existing stock", they argued. They also recommended to issue temporary ration cards to anyone who wants to stand in the queue and pick up their monthly allocation. What has aggravated the crisis is the understanding that there were about 100 million poor Indians who had been excluded from the public distribution system (PDS) due to out dated census data. Many others were left out from the PDS as their fingerprints could not be read by biometric machines.

The people with empty pockets

It is an understatement to say that our world has changed dramatically over the last few months. The novel coronavirus pandemic has resulted in unimaginable loss to the global economy and the loss of human lives has been unprecedented in an era of global peace. Not only are the numbers alarming but, the speed at which the challenges have emerged is intimidating. Governments and health professionals are constantly calibrating the response to this pandemic and frequent strategic adjustments are being done.

We are now in lockdown, exploring novel medical treatments and are on the hunt for a vaccine that will stop this virus in its tracks. And since the scenario is grim, doctors, scientists, and governments are working on an accelerated mission mode.

Unprecedented losses

First, let us get an understanding of the economic impact of the novel coronavirus. The Asian Development Bank has estimated that the global economic cost of the virus is a whopping $2 to $4 trillion. Going by the damages estimated by some legal firms, the figure goes up to an unimaginable $6.5 trillion. Back home, the Indian economy is being battered as well. Under complete lockdown less than a quarter of India's $2.8 trillion economy is functional. We are estimated to lose over Rs 32,000 crore ($4.5 billion) every day during the lockdown.

Huge human cost

While the impact of novel coronavirus on businesses has been devastating worldwide, the human cost has also been staggering. In India, those with savings and access to shelter and food have managed to weather the storm albeit with difficulty.

However, the impact on a large proportion of the 40 million migrant labourers, those who provide the muscle to power India's construction, agriculture and other sectors, has been very disturbing.

After the extension of lockdown on April 14, in a desperate bid to get home, the upheaval that the vast numbers of migrant workers grappled with, was tragic.

However, helped by local communities and government agencies, they have managed to cope with the lockdown. The saga of their journeys will be told over time and there will be plenty to learn from their experience.

First priority - defeating COVID-19

So, is there any light at the end of the tunnel, and where do we go from here? We may be stating the obvious but it is worthwhile reiterating that the greatest challenge before us is tackling the effects of the COVID-19 pandemic.The top priority is finding a way to defeat or neutralise the effects of the virus. All other battles can be handled once we cross this particular challenge. Thanks to the early response and integrated approach of containment that India initiated, has outperformed most other countries in slowing down the pandemic.

Our frontline - the healthcare sector

As early as in February, the private health sector came forward and has been working closely with the public sector. In an interesting development India ramped up health education and leading organisations introduced online training for COVID-19. Healthcare professionals from around the world have been signing up for these courses and a definitive is that e-training will be the next game-changer in stretching the boundaries of care. In addition to measures of containment and prevention, the big need is testing. Nationwide, testing has ramped up quickly and more than half a million samples have been tested as of April 22. Further across India, a total of 586 hospitals have been marked as dedicated COVID-19 hospitals with the capacity of over 100,000 isolation beds and 11,500 ICU beds reserved for coronavirus patients.

Private sector hospitals have also created dedicated negative pressure beds for COVID-19 patients and in an extraordinary step, the Indian Railways have prepared train compartments to serve as isolation wards. Alongside, hotel chains, corporate houses, and hospitals have partnered to convert hotel rooms into facilities for individuals who need to self-isolate and be in quarantine. Alongside, telemedicine has finally found its place in India's healthcare ecosystem. Further to the recent approval of guidelines for telemedicine by the central government, teleconsultation emerged as a godsend for patients who need to consult with their doctors, without going to the hospital. In the near term, digital health with its inherent potential to scale, offering care at a lower cost, will help millions in need, and be a great tool to manage even the rising prevalence of non-communicable diseases.

It is the people who build a sector and the brave hearts and protagonists of healthcare - India's doctors, nurses, paramedics, technicians, administration, and support staff have shown that they can rise spectacularly to the occasion. I am hopeful that seeing medical professionals make so many sacrifices, the ugly doubts and distrust in doctors and hospitals will be erased, forever.

Get set, go

The list is long and no sector has been spared by the COVID-19 pandemic, yet the captains of industry are cautiously upbeat about the recovery. Next, we must focus on re-booting the Indian economy. Over the past few years there has been a decline in India's GDP growth rates and the problem has been further exacerbated by the COVID-19 pandemic. The Indian economy is expected to languish in the 1% to 2% zone next year. When the crisis ends, and when the economy starts running again, the first few countries to get off the starting block will be at an advantage. Can the industry manage to get up and start sprinting on its own? No, every government in the world will have to lend their industry a hand. We in India will need a high octane financial boost. But we all know that governments have taken a beating as far as tax revenues and foreign remittances are concerned. Therefore, governments will have to focus on those sectors that will deliver the biggest bang for the buck.

Getting India to work again

Luckily for us, our agricultural backbone that accounts for almost 14% of our GDP can recover quickly and in fact, even grow next year.

But they must be supported by logistics and storage. The expected normal monsoon this year will help the sector maintain its momentum. Once the threat of the virus recedes, the service industry, the number one contributor to our GDP will start cruising again. We cannot underplay the impact of the COVID-19 pandemic but unlike geophysical disasters and wars, the physical infrastructure of the industry has survived without damage.

Therefore, most industries can quickly become operational with a new work culture, if they have the labour force back and the working capital to restart their business.

Sectorial boost

The Micro, Small and Medium Enterprises (MSMEs) sector, which contributes to 30% of India's GDP, is one of the key drivers of the Indian economy. Today, almost all MSMEs are out of action due to the lockdown, they are unable to pay their employees and several don't have the financial resources to re-start their businesses. The government is contemplating a Rs 20,000 crore relief package for this sector. The other sectors that need help urgently are tourism, aviation, automobile, and real estate.

These sectors will put people back to work and build some traction in the recovery of our economy. This may all sound simple, but obviously there is no easy fix to this unprecedented crisis.

What is most important for us is to get people safely back to work and that will be possible only when we can provide the best possible healthcare to our citizens, as there cannot be a lockdown on medical care, ever.

To address the need for dedicated COVID-19 hospitals and protect staff across board, the hospitals have been making huge investments in PPEs, consumables, even though they are grappling with a decline in OPDs, elective and international patients. Our economic recovery and rebound hinges upon ensuring that our population is health and productive. Thus, it is paramount that the health sector is supported generously, so that it becomes capable of delivering to its fullest potential in this hour of great need. Civilisation has endured several crises in its history which we have put behind us, we will overcome again. This too shall pass.

The **economic impact of the 2020 coronavirus pandemic in India** has been largely disruptive. The World Bank and rating agencies had initially revised India›s growth for fiscal year 2021 with the lowest figures India has seen in three decades since India›s economic liberalization in the 1990s. However after the announcement of the economic package in mid-May, India's GDP estimates were downgraded even more to negative figures, signalling a deep recession. On 26 May, CRISIL announced that this will *perhaps* be India's worst recession since independence. State Bank of India research estimates a contraction of over 40% in the GDP in Q1 FY21.

Within a month, unemployment rose from 6.7% on 15 March to 26% on 19 April. During the lockdown, an estimated 14 crore (140 million) people lost employment while salaries were cut for many others. More than 45% of households across the nation have reported an income drop as compared to the previous year. The Indian economy was expected to lose over ₹32,000 crore (US$4.5 billion) every day during the first 21-days of complete lockdown, which was declared following the coronavirus outbreak. Under complete lockdown, less than a quarter of India's $2.8 trillion economic movement was functional. Up to 53% of businesses in the country were projected to be significantly affected. Supply chains have been put under stress with the lockdown restrictions in place; initially, there was a lack of clarity in streamlining what an «essential» is and what is not. Those in the informal sectors and daily wage groups have been at the most risk. A large number of farmers around the country who grow perishables also faced uncertainty.

Major companies in India such as Larsen & Toubro, Bharat Forge, Ultra Tech Cement, Grasim Industries, Aditya Birla Group, BHEL and Tata Motors have temporarily suspended or significantly reduced operations. Young startups have been impacted as funding has fallen. Fast-moving consumer goods companies in the country have significantly reduced operations and are focusing on essentials. Stock markets in India posted their worst loses in history on 23 March 2020. However, on 25 March, one day after a complete 21-day lockdown was announced by the Prime Minister, SENSEX and NIFTY posted their biggest gains in 11 year

Atmanirbhar Bharat Abhiyan

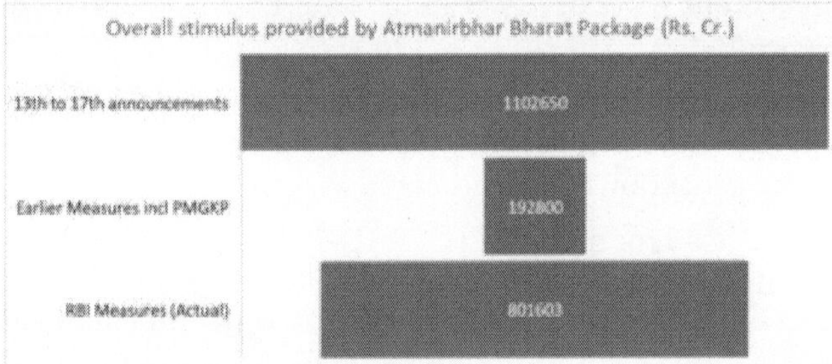

Overall stimulus provided by Atmanirbhar Bharat Package

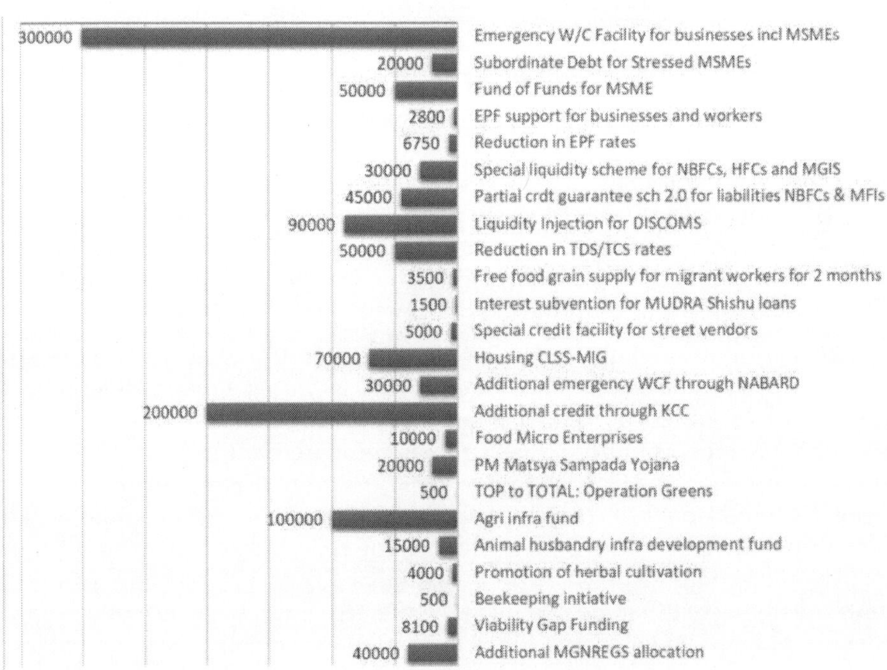

On 12 May, the Prime Minister, in an address to the nation, said that the coronavirus crisis should be seen as an opportunity, laying emphasis on domestic products and "economic self-reliance", a Atmanirbhar Bharat (transl. Self-reliant India) through a Atmanirbhar Bharat Abhiyan (transl. Self-reliant India Mission). The following day the Finance Minister started laying out the details of the Prime Minister's vision which would continue into the next few days. The Finance Minister stated that the aim was to "spur growth" and "self-reliance", adding that, "self-reliant India does not mean cutting off from rest of the world". The law and IT minister, Ravi Shankar Prasad, also said that self-reliance does "not mean isolating away from the world. Foreign direct investment is welcome, technology is welcome self-reliant India... translates to being a bigger and more important part of the global economy."

Economic Situation

In India, up to 53% of businesses have specified a certain amount of impact of shutdowns caused due to COVID-19 on operations (FICCI survey). By 24 April the unemployment rate had increased nearly 19% within a month, reaching 26% unemployment across India, according to the "Centre for Monitoring Indian Economy". Around 140,000,000 (14 crores) Indians lost employment in the lockdown. More than 45% households across the nation have reported an income drop as compared to the previous year. Various business such as hotels and airlines have cut salaries and have laid off employees. Revenue of Ola Cabs went down nearly 95% in March-April resulting in 1400 layoffs.[It is estimated that the loss to the tourism industry will be ₹15,000 crore (US$2.1 billion) for March and April alone. CII, ASSOCHAM and FAITH estimate that a huge chunk of the workforce involved with tourism in the country faces unemployment.[Live events industry has seen an estimated loss of ₹3,000 crore (US$420 million).

A number of young start-ups have been impacted as funding has fallen. A Data Labs report shows a 45% decrease in the total growth-stage funding (Series A round) as compared to Q4 2019. According to a KPMG report venture capital in Indian start-ups has fallen over 50% in Q1 2020 from Q4 2019. Government revenue has severely been affected with taxes collection going down, and as a result the government has been trying to find ways of reducing its own costs.

Agriculture

A study during the first two weeks of May by the Public Health Foundation of India, Harvard T H Chan School of Public Health and the Centre for Sustainable Agriculture found that "10% of farmers could not harvest their crop in the past month and 60% of those who did harvest reported a yield loss" and that a majority of farmers are facing difficulty for the next season. Due to logistical problems following the lockdown tea estates were unable to harvest the first flush. The impact of this on the second flush is not known. The entire Darjeeling tea based tea industry will see significant fall in revenue. Tea exports could drop up to 8% as a result. During the lockdown, food wastage increased due to affected supply chains, affecting small farmers.

From 20 April, under new lockdown guidelines to reopen the economy and relax the lockdown, agricultural businesses such as dairy, tea, coffee, and rubber plantations, as well as associated shops and industries, reopened. By the end of April, ₹17,986 crore (US$2.5 billion) had been transferred to farmers under the PM-KISAN scheme.

Supply chain and logistics

Following the lockdown, certain essential supply chains broke down. Britannia Industries, supporting the lockdown, urged the government to ensure inter-state movement of the raw material for the food processing industry was not hampered. The Managing director of Britannia stated that "if even one link in the supply chain is broken, the country could run out of stocks of packaged food in the next 7-10 days. Although inter-state travel has been banned, it doesn't apply to essentials, and in places like Maharashtra the state police are yet to streamline the process, disrupting supply chains. Vidya Krishnan writes in The Atlantic that due to the lockdown even movement of medical goods were affected.

On 29 March the government allowed the movement of all essential as well as non-essential goods across the country during the lockdown. The milk and newspaper supply chains are also allowed to function.[

Salaries

The Prime Minister on 19 March urged businesses and high income segments of society to take care of the economic needs of all those who provide them services. During the live telecast, he also appealed to families to not cut the pay of domestic help. Following the lockdown, the government circulated advisories[and directives ordering companies to keep paying employees among other things. The Ministry of Finance issued an Office Memorandum on 23 March 2020: wherever such contractual, the casual and outsourced staff of Ministries/Departments and other organization of Government of India is required to stay at home in view of lockdown order regarding COVID-19 prevention they shall be treated as "on duty" during such period of absence and necessary pay/wages would be paid accordingly. These instructions shall apply until April 30, 2020.

A few days later worries grew as to how wages could continue being paid and if the directive was legal or not. There were also concerns raised by migrant workers regarding the implementation of the orders as many daily-wagers have no records of being sacked or salaries being paid or deducted; the concerns also expand to uncertainty in the government's ability to enforce minimum wages under lockdown when it couldn't even do so during normal times. On 15 May, the Supreme Court announced that the government should not take "coercive action" against employees for not paying wages during the lockdown. The court was commenting on 29 March government order.

Migrant workers and labour force

Due to the lockdown, daily-wage workers (the urban poor and migrant laborers) were left with no work. At the same time, the lockdown restrictions put a stop on the movement of buses and trains. Large numbers of migrant workers ended up walking back to there villages.

Soon after a central government directive in late March, state governments set up 21,000 camps to house over 660,000 migrants and stop the exodus. Over 500 hunger relief centres were set up by the Delhi government by the last week of March. By 5 April 75 lakh people were being provided food across the country in food camps run by the government and NGOs.As of 12 April, 37,978 relief camps and 26,225 food camps had been set up. Migrants in such camps in Kerala were provided with medical essentials such as masks, sanitizers, and medicines.

Soon after the nationwide lockdown was announced in late March, FM Sitharaman announced a ₹1.7 lakh crore (US$24 billion) spending plan for the poor. This consisted of cash transfers and steps to ensure food security. To help provide jobs and wages to workers, the average daily wages under the MGNREGA were increased to ₹202 (US$2.80) from the earlier ₹182 (US$2.60), as of 1 April. On 14 May, FM Sitharaman further announced free food grains for the migrant workers, targeting 80 million migrant workers by spending ₹ 35 billion (US$490 million).

Railways transported 48,00,000 migrants back to their homes in the special trains allocated for them between 1 and 27 May. While this service was not initially free, with additional charges over the normal fares,the central government later made the Railways offer an 85% subsidy on the train fares, and the state governments funded the remaining 15%.In the same time period, a total of 91 lakh migrants travelled on both trains and buses.

The governments of Uttar Pradesh,Madhya Pradesh and Gujarat sought to temporarily revise their labour laws in early May with the purpose of attracting industries and investments. Labour unions criticized this as being harmful to the migrant workers while giving more authority to the employers.

Centre and state collaboration

Numerous center versus state tussles have taken place during the COVID-19 pandemic, having a socio-economic impact other that the immediate political impact. Some tussles are not directly related to the pandemic such as the Telangana Chief Minister over the Electricity (Amendment) Bill.Oth Liquor became another source of dispute. Some states have had disputes with the centre related to how the lockdown should be implemented.

The Modi government, in view of the coronavirus pandemic, suspended Members of Parliament Local Area Development Scheme (MPLADS) for two years. This action has been called problematic in many ways, including causing a centralisation of power, being anti-federal in nature, and having an affect on local level development and MP influence at micro levels of the society to handle distress.

There have been calls for halting the ₹ 20,000 crore (US$2.8 billion) re-development of the central vista project in Delhi instead.

During the exit of the lockdown there has been a lack of centre and state collaboration as well as with local authorities. This has been visible in the handling of migrant labour; now that companies are restarting, there is a labour shortage.

Economic recovery

The Technology Information, Forecasting and Assessment Council (TIFAC), Department of Science and Technology, were given the task of preparing a white paper on the revival of the India economy; TIFAC has a "mandate to think for the future". The Punjab government formed a group of experts led by Montek Singh Ahluwalia. Former Prime Minister Dr. Manmohan Singh will provide guidance. DuvvuriSubbarao, a former RBI governor, said that India could look forward to a V-shaped recovery.

There have been numerous suggestions for how India should go about its economic recovery. On 23 April Confederation of Indian Industry (CII), in a paper titled "A plan for economic recovery", outlined three measures that the government should take such as cash transfers to JAM account holders, a credit protection scheme for Micro, Small and Medium Enterprises (MSMEs) and creation of a special purpose vehicle (SPV/SPE) to limit «Government exposure while providing adequate liquidity to industry». On 24 April Bina Agarwal, an Indian development economist, has suggested that the government "create green worker pools, not green zones" as part of lockdown re-calibration efforts.On 28 April Anand Mahindra said a graded re-opening of the lockdown will make economic recovery 'painfully slow'; with the economy being so very interconnected, the lockdown should be lifted comprehensively once the optimal lockdown period as per research has been completed.

Economic re-opening

In the second week of May, companies started preparations for restarting operations. Some companies have opened offices with the maximum permitted strength of 33% while others took a more cautious approach of as low as five per cent.

The struggle against hunger and poverty: the rise of voices

The most sufferer of this COVID lockdown situation are the people who are living on the edge of the society. Among them the migrant labourers are the worst sufferers. After not being heard and not being paid attention to, they have stated to walk on their own paths. They travelled miles after miles to reach their native places. On their path many unexpected incidents happened. For instance, sixteen labourers have died on a railway track while they were sleeping on the tracks and a train suddenly passed smashing their bodies brutally. In another incident of Auraiya, UP twenty four migrant labourers were died in an accident while they were returning home.

These were rather their silent protests against the deaf society and the indifferent government. The privileged has the opportunity to enjoy the lock down, nor these poor. Because sitting idle and doing nothing is the haves' cup of tea, the have nots neither have the fortune nor have the opportunity to enjoy the same.

11 CHAPTER

EPILOGUE

SUMMARY AND CONCLUSION

Summary

Even with the swashbuckling claim on growth and prosperity on the present civilization, the other side of this prosperity is so brick and disastrous that have no match for the past centuries even. Out of around 7 billion population of the world, 1.5 billion are hungry. They don't have adequate access to food, if it is there, the quality doesn't stand any where near to fulfil their calorie requirement. In India 350 million people are living below the poverty line and of them, 200 million people have become victim to moderate to extreme hunger indexes. 42 per cent of the new born babies are under weight. 60 per cent of the children are suffering from moderate to high level of anaemia experiencing stunted growth. Beyond the curtain of hunger, there is another problem that is chronic hunger. Based on hunger index we the nation is occupying 100th position in the world (IFPRI Report,2017). The scenario of chronic hunger is even worse and astoundingly it is worse than African nations as well.

Quoting back of Gunner Myrdal, a German analyst (Wrote' Asian Drama '), the reasons for sluggish growth of Indian Agriculture is her Agricultural work force, who themselves are suffering from malnutrition, hunger and social impoverishment.

The present study has conducted in two separate socio-ecological strata. One study has conducted among the farm families and the other one has conducted on the non farm families. Here we have only selected women at the age group of 15-60 as respondents. The farm locality is Beraberi gram panchayat, Habra II Block, North 24 pgs and the non farm locality is Bilkanda gram panchayat, Barrackpur 2 block, North 24 pgs. For the farm sector 21 and for the off farm sector 19 variables were taken. Pooled analysis was done by selecting 13 common variables from both the sectors. Dependent variables were hunger, poverty, voice and cognitive differential.

1. FARM SECTOR: DESCRIPTIVE STATISTICS RESULTS

It has been found from the study that the mean Age group (x_1) is 38.49 years with standard deviation 10.44 for the total distribution taken for the study. Coefficient of variation of Age (x_1) is 27.13% which shows a very high level of consistency in its distribution. The independent variable Education (x_2) of farmer, has been found from the study that the mean 6.30 with standard deviation, 4.24 for the total distribution taken for the study. Coefficient of variation of Education (x_2) is 67.30% which show a medium level of consistency in its distribution. The independent variable Family size (x_3) of farmer, has been found from the study that the mean 3.90 with standard deviation, 1.19 for the total distribution taken for the study. Coefficient of variation of Family size (x_3) is 30.51% which show a high level of consistency in its distribution. The independent variable Economic motivation (x_4) of farmer, has been found from the study that the mean 1.13 with standard deviation, 0.54 for the total distribution taken for the study. Coefficient of variation of Economic motivation (x_4) is 47.58% which show a high level of consistency in its distribution. The independent variable Stress Risk orientation (x_5) of farmer, has been found from the study that the mean 0.52 with standard deviation, 0.22 for the total distribution taken for the study. Coefficient of variation of Risk orientation (x_5) is 42.48% which show a high level of consistency in its distribution. The independent variable Management orientation (x_6) of farmer, has been found from the study that the mean 5.02 with standard deviation, 2.87 for the total distribution taken for the study. Coefficient of variation of Management orientation (x_6) is 57.16% which show a high level of consistency in its distribution.

The independent variable Stress perception on hunger (x_7) of farmer, has been found from the study that the mean 5.25 with standard deviation, 2.19 for the total distribution taken for the study. Coefficient of variation of Stress perception on hunger (x_7) is 41.74% which show a high level of consistency in its distribution. The independent variable Stress perception on poverty (x_8) of farmer, has been found from the study that the mean 5.37 with standard deviation, 2.77 for the total distribution taken for the study. Coefficient of variation of Stress perception on poverty (x8) is 51.61% which show a high level of consistency in its distribution. The independent variable Stress perception on voice (x_9) of farmer, has been found from the study that the mean 6.07 with standard deviation, 1.32 for the total distribution taken for the study. Coefficient of variation of Stress perception on voice (x_9) is 21.74% which show a very high level of consistency in its distribution.

The independent variable Size of holding (x_{10}) of farmer, has been found from the study that the mean 2.70 with standard deviation, 3.49 for the total distribution taken for the study. Coefficient of variation of Size of holding (x_{10}) is 129.17% which show a very low level of consistency in its distribution. The independent variable Stress Cropping intensity (x_{11}) of farmer, has been found from the study that the mean 233.65 with standard deviation, 85.51 for the total distribution taken for the study. Coefficient of variation of Cropping intensity (x_{11}) is 36.60% which show a high level of consistency in its distribution. The independent variable Livestock count (x_{12}) of farmer, has been found from the study that the mean 1.27 with standard deviation, 1.04 for the total distribution taken for the study. Coefficient of variation of Livestock count (x_{12}) is 82.01% which show a low level of consistency in its distribution. The independent variable Livestock yield (x_{13}) of farmer, has been found from the study that the mean 1.66 with standard deviation, 2.17 for the total distribution taken for the study. Coefficient of variation of Livestock yield (x_{13}) is 130.72% which show a very low level of consistency in its distribution. The independent variable Pond and fish (x_{14}) of farmer, has been found from the study that the mean 0.39 with standard deviation, 0.19 for the total distribution taken for the study. Coefficient of variation of Pond and fish (x_{14}) is 49.18% which show a high level of consistency in its distribution. The independent variable Total crop yield (x_{15}) of farmer, has been found from the study that the mean 6883.98 with standard deviation, 16565.58 for the total distribution taken for the study. Coefficient of variation of Total crop yield (x_{15}) is 240.64% which show a very low level of consistency in its distribution. The independent variable Cost of cultivation (x_{16}) of farmer, has been found from the study that the mean 56664.55 with standard deviation, 116911.83 for the total distribution taken for the study. Coefficient of variation of Cost of cultivation (x_{16}) is 206.32% which show a very low level of consistency in its distribution. The independent variable No. of fragments (x_{17}) of farmer, has been found from the study that the mean 4.54 with standard deviation, 4.25 for the total distribution taken for the study. Coefficient of variation of No. of fragments (x_{17}) is 93.70% which show a low level of consistency in its distribution. The independent variable Communication variables (x_{18}) of farmer, has been found from the study that the mean 1.03 with standard deviation, 1.48 for the total distribution taken for the study. Coefficient of variation of Communication variables (x_{18}) is 143.61% which show a very low level of consistency in its distribution. The independent variable Marketed surplus (x_{19}) of farmer, has been found from the study that the mean 37.71 with standard deviation, 36.46 for the total distribution taken for the study. Coefficient of variation of Marketed surplus (x_{19}) is 96.67% which show a low level of consistency in its distribution. The independent variable Energy consumption (x_{20}) of farmer, has been found from the study that the mean 5.75 with standard deviation, 2.69 for the total distribution taken for the study. Coefficient of variation of Energy consumption (x_{20}) is 46.77% which show a high level of consistency in its distribution. The independent variable BMI (x_{21}) of farmer, has been found from the study that the mean 26.42 with standard deviation, 5.16 for the total distribution taken for the study. Coefficient of variation of BMI (x_{21}) is 19.53% which show a very high level of consistency in its distribution.

CORRELATION RESULTS

- **Hunger vs. 21 independent variables**: Economic motivation helps building up label of compliances or resilience with the risk and jerk. Hungry people has to comply with lots of jerk in the forms of inflicting physiological stress, dejection by the society and accelerating physio-social entropy. But eradication all these negative vibes, the farming community if become motivated, charged, positive and striving, many of their problems would get solution. That is how it stands by the significant correlation with the hunger and vice versa. Cropping intensity and marketable surplus are no doubt, two important factors as far the level of hunger is concerned. Because is cropping intensity would be high, then less resources would be required for more output. Subsequently due to higher production, marketable surplus would be higher, and this automatically would positively affect on the status of hunger. Energy indicates the status of the economy. The more is consumption the better is the status. So the farm ecology with the higher energy consumption indicates the better deterrence to inflicting hunger. Here, BMI is also gets significantly correlated with the level of hunger, rightly because a normal BMI is an indicator of good nutrition, similarly an aberrated BMI is an indicator of bad hunger status.

- **Poverty vs. 21 independent variables:** Age is significantly correlated with poverty. There in our study we have found many of the rural and agrarian old women who are physically unable to do their farming job and some of them are suffering from loneliness as their children has quit agriculture and migrated to the towns and cities. So they are subjected to poverty and living a life of destitute. Hungry people in a family like situation or in a hunger endemic social ecology are through in fear psychosis and constant intimidating surroundings. Chronic hunger and sufferings of the farm women have driven them into a stage of distressing rejection and simmering frustration. So hunger and stress perceptions have been clinically linked with each other.If the income is less rather the people are poor then there would be a psychological stress on the stakeholders about their livelihood and bread earning. The cost of cultivation is negatively correlated with the poverty status because higher cost of cultivation means lower profit and lower profit subsequently creates impact on the poverty status.
- **Voice vs. 21 Independent variables:** Poverty status and stress due to lower income make a person frustrated, depressed, dejected and socially aloof. If a person is poor then he or she would be suffering from lower self esteem. An inferiority complex would be develop within him/her. The same thing is happening in case of these farm women under study. Due to their poor condition, they have developed a kind of withdrawal syndrome with low confidence and morale. They are considering themselves as incapable of giving suggestion, unentitled of being vocal and this condition has push them into the doom of absolute voicelessness. BMI is not directly impactful as far the the level of voice is concerned. It is an important parameter of hunger and poverty status. Therefore as hungry and poor people has got more aberrated BMI, as well as are also suffering from voicelessness, in that way they both(voice and BMI) are linked.

- **Cognitive differential vs. 21 Independent Variables:** Management orientation helps orchestrating time cost and resources and bringing all these on a platform of performance and achievement. The poor farm women mostly are in a fragile reality are not that adept either to manage cost or to save resources. Poor management and poverty are organically linked with each other and sometimes they together build up a resonance of concurrent impoverishment. So, inefficient management of resources and higher cost of cultivation caused lower profit margin which is substantially impactful on poverty status and creates stress. Stress restrain a person from communication and interaction , ultimately ended up with higher cognitive differential. Upscaling of productivity needs optimum investment below which crop will decline and above which there is an incremental impact. Fragmentation has got didactic effect, having both incremental and depletive effect when number of fragments are more, it indicates the cumulative size of holding will be more and basket of choices across the ecosystem within a micro-farming, elasticity would be more. So in the present situation the number of fragments have been a provider for redressing risk out of crop failure or yield losses. The more is the number fragments the more will be the flexibility of redressing the poor and reducing yield losses.Lower the resource like livestock counts, lesser would be their economic potentials. Livestock products like milk and egg additionally can be provided as resources daily or weekly. If these resources are not adequate the rural women would consider themselves as deprived and subject of exploitation as well as deprivation. Higher the resource, higher would be the marketability and this would impact positively on their self confidence and self-esteem. Then they can pull themselves off from the pale, dark and bleak scenario of silence and can cry out the louder for their legitimate rights.

STEP DOWN REGRESSION RESULTS

- **Hunger vs. 21 independent variables:** Cropping intensity is the provider of food and nutrition as well as integrated to the food security. So, the higher is the crop the higher has been the cropping intensity and for the growers, it offers better security for them. The livestock count is an indicator for directing and determining the spill over amount for the livestock raisers' own family.So these three variables in terms of managing hunger can be traced as marker variables which has contributed collectively 30.30 per cent variance in hunger.

- **Poverty vs. 21 independent variables:** The farmers are suffering from poverty because in most cases for the area under study, are vulnerable to low yield, poor communication access to critical inputs and rising cost of cultivation. These three menaces collectively has contributed a deleterious impact on their economy, social well being and livelihood. Thus three variables can be considered as marker variables and together has contributed 28.30 per cent variance in poverty.

- **Voice vs. 21 independent variables:** The on going punitive and its long drawn impact on the bio-physical and psychological status of the farmers has conspicuously being reflected through poor economic motivation, fragile mental ability and feeling of being dejected subsequently enrooted into a variables character. May be livestock in addition to agriculture can help to get over the stressing pursuits of agriculture.So these variables can be traced as strong determinants for characterizing voice of oppressed.The rural people in a response to increasing chaos and stress being inflicted into their life and livelihood, they have become a relentless hunter for options, choices, experiments and moving from an initial combination of crop-crop to crop-horticulture-livestock-fishery in a cascading manner to ultimately move from integration to co integration. In their endless hunching for a better livelihood and earning a decent status in the community, the voices of the subalterns have come up as critical as inputs and machineries, money and motivation.The R2 value being 0.277, it is to infer that 27.70% of variants in the consequent variable has been explained by the combination of these 3 causal variables.
- **Cognitive Differential vs. 21 independent variables**: The cognitive differential or knowledge variance among the respondents can well be predicted by two clusters of independent variable which again are attuned to each other and these are motivational factors and agro managerial factors. When the respondents are exposed to agro managerial factors, their cognitive changes are discernable in terms of package of practices, ecological functions and techno managerial pursuits. Given these are properly done, they can generate a higher level of confidence to manage both stress and stressors. So these are the marker variables and collectively have contributed to 47.70 per cent of variance in cognitive differential.

Factor Analysis results

21 variables can be operationally conceived in five factors which inputs offers a strategic conglomeration of apparently different variables into strategic constellation of five factors. Factor analysis offers a data reduction process wherein the sprawling variables are brought under small homogeneous groups of variables called factors. These conglomeration is based on varimax rotation for reducing the standard error and Eigen values for enhancing the integrations amongst variables. Factor analysis offers immense strategic importance for an efficient management of different variables, dimensions, characters and variance, putting them all into a commonality of interactions.

Path Analysis:

- **Hunger vs. 21 Independent Variables:** The best way in redressing hunger can be possible by augmenting the productivity of the operating farm of the respondents. Whenever farm productivity goes low, the problem of hunger and poverty would turn more complex and untreatable. That is why total crop yield, cropping intensity and risk orientation have recorded the direct and substantive impact on hunger status. Whenever the ower of marginal and fragmented holding keep loosing their confidence to confront against risk, they turn more vulnerable to hunger stress. So, productivity, cropping intensity and risk orientation are moving together interovenly to generate impact on hunger.Cost of cultivation, total crop yield and size of holding, the other set of independent variable,have exerted substantive indirect effect; it is good enough to conclude that these variables are to be treated isochronously whenever we would be considering the direct effect of such three substantive variables(x_5, xx_{11}, x_{15}).The residual effect being 0.575, it is to infer that 57.50 per cent variable in the dependent variable that is hunger could not be explained even though there has been a combination of 21 exogenous variables. This would rather suggest that some collateral or peripheral variables or even some few core variables have not been considered or selected either. That is why the spectrum of residual effect could not be rationalized. If these comes through more core variables and some other natural or peripheral variables would have been considered, then this level of residual effect could have been lesser.

- **Poverty vs. 21 Independent variables:** The path analysis shows that size of holding still comes as a strong socio economic determinant to decide on poverty and quite related to it two other variables, total crop yield and cost of cultivation are also exerting substantive direct effect on poverty and interestingly the set of these three variables have also exerted the substantive indirect effect on the consequent variable poverty. So, in the social ecology of poverty the system variables like size of holding, crop yield and cost of cultivation stand so relevant as well as virulent in terms of dealing with the poverty issues of the respondents. The residual effect being 0.494, it is to infer that 49.40 per cent variable in the dependent variable that is poverty could not be explained even though there has been a combination of 21 exogenous variables were operated in this interaction. This would rather suggest that some collateral or peripheral variables or even some few core variables should be considered or selected in some future study. That is why the spectrum of residual effect could not be rationalized. If these comes through more core variables and some other natural or peripheral variables would have been considered, then this level of residual effect could have been lesser.
- **Voice vs. 21 Independent variables:** A person can raise their voice when he/ she is free from stress and the size of holding is bigger, associated with an ancillary income from the size of holding. The variables livestock yield , no. of fragments and BMI have exerted substantial indirect effect to infer that these variables do retain in associating properties in order to characterize the performance of the consequent variable voice.Interestingly no. of fragment and BMI are impacting conjointly on the voice character of the respondents. The poor and suffering people remain silent when economy goes fragile, getting more and more uncertain and thereby inviting more and more stress on the

psychic domain. So poverty the alleviation programmes , as these are being done usually in a developing economy, should not deal with the economy of a target group in isolation. For example providing critical credit support to the marginal farmers cannot go amply successful unless it being associated with motivational programmes. Any change intervention in any socio ecological transformation process miss technology input credit ,but above all a psychic energy and motivational back up to translate this digitized entity into an euphoria of success.The residual effect being 0.555, it is to infer that 55.50 per cent variable in the dependent variable that is voice could not be explained even though there has been a combination of 21 exogenous variables were operated in this interaction. This would rather suggest that some collateral or peripheral variables or even some few core variables should be considered or selected in some future study. That is why the spectrum of residual effect could not be rationalized. If these comes through more core variables and some other natural or peripheral variables would have been considered, then this level of residual effect could have been lesser.

- **Cognitive differential vs. 21 independent variables:** The table of path analysis rightly suggests that the differential cognitive can well be predicted by the variables age, education and energy consumption. The cognitive differential results from lack of experience, educational level and also the mobility in a given social space which has got an energy equivalence of course. So to deal with these consequent variable that is cognitive differential, these three variables to be considered with a well directed operating strategy. The substantial indirect effect are coming from economic motivation, total crop yield and no. of fragments. So to acquire knowledge and information about different poverty alleviation and other developmental projects, the respondents need both experiential and cognitive leaning. Education is the prime driver for acquiring knowledge and information in a given social ecology. Here energy indicates the consumption of fuel, one of which mobility fuels like diesel, petrol etc. which indicates the mobility and interaction of a respondent in a social ecology. The residual effect being 0.623, it is to infer that 62.30 per cent variable in the dependent variable that is cognitive differential could not be explained even though there has been a combination of 21 exogenous variables were operated in this interaction. This would rather suggest that some collateral or peripheral variables or even some few core variables should be considered or selected in some future study. That is why the spectrum of residual effect could not be rationalized. If these comes through more core variables and some other natural or peripheral variables would have been considered, then this level of residual effect could have been lesser.

- **Canonical covariate Analysis:** Hunger offers both physiological and psychological stress. Combination of which is lethal and deleterious to humanity. That is why the hunger and associating stress are moving together to offer them a propensity of forming a psychological diode and these two are substantially characterized by the contribution made from the x set of variables ($x_4, x_6, x_8, x_9, x_{10}, x_{12}, x_{13}, x_{15}, x_{17}, x_{20}, x_{21}$). Similarly the two dependent variables y_2 and y_4 are moving together to generate an apparent inference that the poorer people are assessing or can assess less information regarding sources of income or livelihoods and even different projects meant for their welfare and well being. So, it is more important to access or build up knowledge about the welfare projects rather than just drive the means of welfare to an unclear destination by the government or NGO or any other institution and these diode of independent variables have got a distinct affinity towards the set of these x variables ($x_1, x_2, x_3, x_5, x_7, x_{11}, x_{12}, x_{14}, x_{16}, x_{18}, x_{19}$).

2. Off Farm sector: Descriptive Statistics Analysis

It has been found from the study that the mean Age group (x_1) is 32.1 years with standard deviation 10.723 for the total distribution taken for the study. Coefficient of variation of Age (x_1) is 33.42% which shows a very high level of consistency in its distribution The independent variable Education (x_2) of farmer, has been found from the study that the mean 8.86 years of schooling with standard deviation, 3.27 for the total distribution taken for the study. Coefficient of variation of Education (x_2) is 36.83% which shows a high level of consistency in its distribution. The independent variable Family size (x_3) of farmer, has been found from the study that the mean 4.93with standard deviation, 1.12 for the total distribution taken for the study. Coefficient of variation of Family size (x_3) is 22.74% which shows a high level of consistency in its distribution. The independent variable economic motivation (x_4) of farmer, has been found from the study that the mean 1.31 bigha with standard deviation, 0.59 for the total distribution taken for the study. Coefficient of variation of size of holding (x_4) is 45.17% which shows a high level of consistency in its distribution. The independent variable Risk orientation (x_5) of farmer, has been found from the study that the mean 0.43 with standard deviation, 0.17 for the total distribution taken for the study. Coefficient of variation of Risk orientation (x_5) is 39.30% which shows a high level of consistency in its distribution. The independent variable Management orientation (x_6) of farmer, has been found from the study that the mean 5.55 with standard deviation, 1.95 for the total distribution taken for the study. Coefficient of variation of Management orientation (x_6) is 35.11% which shows a high level of consistency in its distribution. The independent variable Stress Perception on Hunger (x_7) of farmer, has been found from the study that the mean 3.56 with standard deviation, 1.14 for the total distribution taken for the study. Coefficient of variation of Stress Perception on Hunger (x_7) is 32.15% which shows a high level of consistency in its distribution. The independent variable Stress Perception on Poverty (x_8) of farmer, has been found from the study that the mean 4.46 with standard deviation, 1.63 for the total distribution taken for the study. Coefficient of variation of Stress Perception on Poverty (x_8) is 36.54% which shows a high level of consistency in its distribution. The independent variable Stress Perception on Voice (x_9) of farmer, has been found from the study that the mean 5.36 with standard deviation, 1.07 for the total distribution taken for the study. Coefficient of variation of Stress Perception on Voice (x_9) is 19.92% which shows a high level of consistency in its distribution. The independent variable Homestead Land (x_{10}) of farmer, has been found from the study that the mean 0.14 with standard deviation, 0.11 for the total distribution taken for the study. Coefficient of variation of Homestead Land (x_{10}) is 77.80% which shows a low level of consistency in its distribution. The independent variable Occupation (x_{11}) of farmer, has been found from the study that the mean 14.15 with standard deviation, 8.68 for the total distribution taken for the study. Coefficient of variation of Occupation (x_{11}) is 61.36% which shows a moderately High level of consistency in its distribution. The independent variable Ancillary Income (x_{12}) of farmer, has been found from the study that the mean 83.48 with standard deviation, 99.09 for the total distribution taken for the study. Coefficient of variation of Ancillary Income (x_{12}) is 91.69% which shows Low level of consistency in its distribution. The independent variable Expenditure (x_{13}) of farmer, has been found from the study that the mean 1203.21 with standard deviation, 631.10 for the total distribution taken for the study. Coefficient of variation of Expenditure (x_{13}) is 52.45% which shows a Moderately High level of consistency in its distribution. The independent variable Savings (x_{14}) of farmer, has been found from the study that the mean 143.07 with standard deviation, 178.71 for the total distribution taken for the study. Coefficient of variation of Savings (x_{14}) is 94.90% which shows a Low level of consistency in its distribution. The independent variable Bank Account (x_{15}) of farmer, has been found from the study that the mean 0.38 with standard deviation, 0.14 for the total distribution taken for the study. Coefficient of variation of Bank Account (x_{15}) is 37.57% which shows a High level of consistency in its distribution. The independent variable Communication Variables (x_{16}) of farmer, has been found from the study that the mean 1.44 with standard deviation, 1.46 for the total distribution taken for the study. Coefficient of variation of Communication Variables (x_{16}) is 98.43% which shows a Low level of consistency in its distribution. The independent variable Management Variables (x_{17}) of farmer, has been found from the study that the mean 118.54 with standard deviation, 101 for the total distribution taken for the study. Coefficient of variation of Management Variables (x_{17}) is 85.20% which shows a Low level of consistency in its distribution. The independent variable Energy Consumption (x_{18}) of farmer, has been found from the study that the mean 5.17 with standard deviation, 3.20 for the total distribution taken for the study. Coefficient of variation of Energy Consumption (x_{18}) is 61.90% which shows a Moderately High level of consistency in its distribution. The independent variable BMI (x_{19}) of farmer, has been found from the study that the mean 25.37 with standard deviation, 5.70 for the total distribution taken for the study. Coefficient of variation of BMI (x_{19}) is 22.46% which shows a High level of consistency in its distribution.

CORRELATION ANALYSIS

- **Hunger vs. 19 independent variables:** we can say that in the off farm social ecology, the family size is very much important because higher the family size higher would be the income requirement. But as an old saying goes, that population does not only means one stomach, but it also means a pair of hand, it is also true for family size. Higher the number of members, higher would be the working hands. But if the number of eating members are much higher than the number of earning members, then there would be a gap in between what they need and what they earn. Quite logically the stress per perception is organically linked with hunger. A hungry stomach suffers from stress and a stressed mind is suffered from esurience vis a vis hunger. Similarly, nature of occupation is related to the total volume of income as well as the nature of income. If the family has some ancillary income rather than the main source of income, they are surely getting a chance of having better livelihood and better economic security. Less expenditure and higher savings ignite a person to open a bank account and to save more for future. Homestead is certainly another predictor for the livelihood status of a respondent. Higher the possession lesser would be the economic and social vulnerability. Lastly BMI is directly linked with the hunger and nutritional status but it does not mean that higher BMI will always mean higher nutrition. In our overall study may it be in case of farm sector or off farm sector, the mean BMI is higher than that of the normal range, because they are having more carbohydrate in their diet, that is why they are bulky but not strong as their protein requirement is hardly fulfilled.

- **Poverty vs. 19 independent variables:** Due to poverty and lack of economic motivation, the selected respondents could not exert their entrepreneurial skills and this affects their livelihood improvement negatively. Similarly the management orientation is strongly related to the poverty status. Lack of competency in managing resources would impact on the economic status. The respondents who are managing their resources and skill well, (for example there were some people who are doing some side business like bidi binding, tailoring, livestock management etc.) they are getting some ancillary income as well as can save more and this ultimately improves their economic status by eradicating the poverty to some extent. Stress perception is also an indicator of bad economic status and poverty and status are both way linked. In a poverty ridden economy, the people would always suffer from stress and stress ridden society would suffer from lack of energy and positive vibes that ultimately leads to lower economic development. Stress perception on voice here is an important marker variable. Poor people are hesitant to speak about their rights and needs, they suffer from dejection which culminates into poorer economic status. Homestead land is another marker variable of poverty status because there is a common perception that higher the homestead land higher would be the economic status. Interestingly energy consumption is correlated with poverty. Energy consumption is related to the status of mobility, the power consumption status in households and in business sectors. Energy consumption may be considered as a neo normal variable with the pace of modernization where the energy has got a tremendous role to play. The more would be the fuel and power consumption, the more incensement in social and economic status vis a vis income and security would be seen.

- **Voice vs. 19 independent variables:** Family size has also been figured up in the other variable hunger. There hunger in one hand can be decreased if the number of working hands are more, or the hunger status can be increased if the number of stomachs to feed is more. Same way, if the family size is more in one hand the voices can be loud, but in another hand if the hunger status is increased then stress perception due to hunger and poverty would increase and this would ultimately ends up with lowering of legitimate voices. Ancillary income, homestead land, economic motivation, access to banks and management orientation; these are the variables lack of which economic status is downtrodden and with the poor economy voices are also lower. Now in a non farm economy, those who are having higher entrepreneurial skills, they would have higher voices that is why the transformation from farm economy to off farm economy, may be in some cases signifies the transformation from lower voiced population to a louder one. Last but not the least, communication variables are quite obviously directly impactful on voice because more the communication, more would be the gathering of information, more would be the participation and more would be the voice.
- **Cognitive Differential vs. 19 independent Variables**: Cognitive differential actually means the gap between what are the information available and what are the information accessed. Here, communication variables are playing a pivotal role, as better the communication, narrower would be the cognitive differential. Energy consumption and BMI are the indirect indicator of cognitive differentials. Poorer BMI apparently indicates a bad health. So there would be lesser energy consumption, lesser mobility, lesser communication and lesser gathering of information. The culminating point is broader cognitive differential.

STEP DOWN REGRESSION RESULTS

- **Hunger vs. 19 independent variables**: It has been found that three variables age (x_1), education (x_2) and stress perception on voice (x_9) have been retained in the last step of step wise regression analysis to infer that, these three variables are important determinant in estimating hunger of the respondents dwelling in an off farm social ecology. Now age is a very important determinant of Hunger . Hunger is basically a physiological demand. With the age maturity grows and with the maturity subsequently the consciousness about the hunger status increases. Coming to education, it is quite obvious that those who have poorer education, they have a lesser accessibility of economic opportunities. So, In order to redress the problem of hunger; the basic education, non formal education , informal education or the education as a whole should be improved among the respondents. Hunger is both physiological as well as psychological stress. With a prolong hungry status, there comes a mind set of dejection and silence. So these three variables in terms of managing hunger can be traced as marker variables which has contributed collectively 35.70 per cent variance in hunger.

- **Poverty vs. 19 independent variables:** Savings and communication variables have retained in the last step to imply that both these variables have a strong and determinant impact on predicting the consequent variable poverty. It is quite obvious that the poor people will have a meagre savings and due to their lower income, their communication modes would be constrained . For example, their mobility would be lesser, accessibility of mass media would be tough and also due to their more struggle to earn their bread and butter, they get less chance to interact with their peen groups. Therefore, this constraint of communication creates a barrier to know about the government schemes and probation which they could get benefited from. So these two variables in terms of managing poverty can be traced as marker variables which has contributed collectively 58.50 per cent variance whereas the 19 variables together explained 60.30 per cent variance in the concerned consequent variable poverty. That means these two (x_{14} and x_{16}) together has explained 97.01 per cent{(58.50/60.30)*100) }of the total variance explained by all those 19 variables.
- We can site here the famous Nobel laureate economist Abhijit V. Banerjee when he said " A lot of us have been saying that we need a stimulation package.....Giving money in the hands of everybody, so that they can buy in stores or they buy consumer goods...I think spending is the easiest way to revive the economy"(Source an edited video posted by the national president of congress Rahul Gandhi on twitter).
- **Voice vs. 19 independent variables:** Whenever the age is changing with the level of maturity, the voices will automatically encrypted with the personality. At the same time with the education, self confidence and empowerment level would be up. As far as the economic motivation is concerned, if someone wants to have a change in their life and livelihood altogether to make it a better economic proposition, their voices would be up quite logically. Now coming on to the risk orientation, higher the risk, better would be the gain. But the problem with the respondents concerned is that they don't know how to comply with the risk, therefore scared to take risks in their life and occupations. Due to such timid attitude they are not keen to take credit from banks to do a better exposure. Here, the risk orientation and taking credit from banks can go in compliance with each other. Hence, bank credits can be considered as an important dummy variable.So, with the increasing age, better education and scaling up of both economic motivation and risk orientation, voices would automatically higher. These four variables in terms of managing voice can be traced as marker variables which has contributed collectively 18.90 per cent variance in voice.

- **Cognitive Differential vs. 19 independent variables:** Higher the family size, higher would be the no. of ears to gather information and higher would be the no. of brains to process the information, so the cognitive differential would be more. Quite logically Bank account is a very good estimator to elucidate the cognitive differential. Because going to bank gives a chance to interact with people with cognitive differentials and this subsequently gives them a chance to gather information and come up with a different level of perception. Here, we can clearly see that two causal variables one is bank account and the other is communication variables are also linked together. Lastly, whenever the respondents are having poorer body mass index, they would be suffering from psycho-physiological stresses and their cognitive persuasion and cognitive differential, both will be affected impacted on. So these four variables in terms of managing cognitive differentials can be traced as marker variables which has contributed collectively 46.60 per cent variance in cognitive differentials.
- **Factor Analysis results:** we can say that, out of these 19 variables, only seven dominating factors are extracted by combining the related variables together. That means seven conspicuous variables are actually there. These seven variables, out of this empirical research, are offering tremendous strategic importance at the micro level situation and reality. It is well discernible that these all strategic implications are restricted to the research locale selected for study. To get higher customized and standardized result this research needs to be replicated in different geo spatial economic and ecologic status. Higher the variance across the ecologies, higher would be the reliability. So, it is also one of the recommendations that we need to replicate this study to a higher core of ecological sets to get and extrapolate the micro level variance into a more standardize protocol.

PATH ANALYSIS RESULTS

- **Hunger vs. 19 independent variables:** Higher family size, poorer education at any stage of growth (age) are impacting on hunger. The other variables like economic motivation (x_4), management orientation (x_6) and stress perception on voice (x_9) have clearly recorded associating properties i.e. these variables will ultimately characterize the behaviour of hunger. So, poor motivation, less management orientation and higher stress perception are offering a reticulate interactive behaviour with other variables to ultimately deciding on the severity of hunger and enduring ability of the person being inflicted with the hit of hunger. The residual effect being 52.13 per cent is inferring that even with the combination of nineteen variables, a substantive portion of variance in hunger could not be explained with the participation with these nineteen variables in the entire wave of interaction.
- **Poverty vs. 19 independent variables:** Expenditure is a very important determinant in shaping of the level of poverty. people are poor because

their expenditure is more over their income. If they are compelled to spend more in spite of their low income, then there comes the poverty situation. On the other hand, to fight poverty everyone needs an ancillary income apart from their main source of income. Here in an off farm social ecology, ancillary income plays a very important and determinant role as far as the poverty status of the people is concerned. In today's circumstances, the off farm economy is more stronger than that of the on farm one and sometimes these two become inter changeable. Now coming on to savings part. It is quite obvious that if one needs to get over the brunt of poverty, he/she needs savings. The daily wage earners are the most vulnerable category under any undesirable economic situation. For instance we can take the example of COVID 19 pandemic situation, where we have seen that the most affected class is the migrant labourer category, because they are the daily wage earners with a meagre savings. To fight poverty and to increase the savings of the poor and deprived class, Jandhan accounts were opened by the government of India in 2015 which really a commendable step towards the eradication of poverty by increasing people's savings. Savings also means the possibility of investment. If a person has savings, then he/she can invest and take moderate risk and that is where the opportunity of becoming an entrepreneur lies. Very interestingly we may notice that here bank account is exerting the highest indirect effect on poverty which rightly proves the importance of opening bank accounts of poor people. Another important variable which is having the highest indirect effect is homestead land. Bigger the homestead land higher would be the opportunity to complete in the realm of economy as in today's era of modernization and urbanization, the cost of homestead land is getting higher. The residual effect being 13 per cent is inferring that even with the combination of nineteen variables, a substantive portion of variance in poverty could not be explained with the participation with these nineteen variables in the entire wave of interaction. Though 87 per cent of the variance is explained in this case which shows the less error occurrence.

- **Voice vs.19 independent variables:** Management orientation basically is a psychic, corporeal and operational engagement which is also associated with mobility, communication and interaction. Management orientation has a sizable contribution in raising the voice. When someone is very good in management process, the confidence would go up and when one's confidence is boosted, his/ her voice would automatically start rippling up. Good in management orientation means he/she is good in dealing with the risk, harvesting the opportunity and setting a plan for the future goals. Now, as far as the stress perception is concerned stress perception and voice are reciprocally related. A person with higher stress be there morose and exhausted, so couldn't raise voice. Now as far as the management variables are concerned, better handling of the variables of management gives lesser stress and lesser stress helps in better communication, better communication leads to better information gathering ability, which ultimately leads to a confident raising of voice. Age is also a variable which is exerting a good impact on voice and it is effecting the voice indirectly. With the age maturity to handle stress is growing up and entrepreneual vision becomes keen. So, together with other variables, age exerts a swashbuckling impact on voice.

- **Cognitive Differential vs. 19 independent variables:** Cognitive differential basically means difference in knowledge among the different respondents. Our basic perception in this study was, there would be a difference in cognation in between the stakeholders and the mangers and this would lead to an endemic hunger, poverty and voice situation. Expenditure in this case is a very important indicator. Higher the expenditure, higher would be the numbers of point of interaction. Communication plays an important role in reducing the cognitive differential. Higher the level of communication, lesser would be the cognitive differential and that is why the people should be getting the opportunity to speak. As far as the BMI is concerned, people with better health and nutritional food habit, backed up with better economy, income and livelihood would be more confident and wishful to get information and reduce the ignorance, in other words they would then eager to decrease the cognitive differentials. The impactful indirect effects are age, stress perception on hunger and BMI. Age is directed related to maturity and maturity is closely related to stress. These variables when linked with other variables like income, savings, homestead land etc. then they are exerting a significant effect on the cognitive differential. Here BMI is a variable which is having both significant direct effect and significant indirect effect. BMI solo can effect on cognitive differential and also it is highly impactful in combination with stress, income, expenditure, saving etc.
- **Canonical covariate analysis:** So, the off farm economy and ecology as have been found in the semi urban or peri urban ecosystems suggest that there is income variabilities, expenditure variabilities, multi layered poverty syndromes and differential exposure to sources of information. So, variability in information access, poverty status and voice have formed a complex cob web and they have together presented a selective choices from the right side echelon. Similar things have gone for the other predicted variable hunger (y_1) to suggest that hunger has got a unique character for off farm economy. The off farm occupations are characterized with, while compared with farm based ecology, it can be observed that income is here daily or weekly, less seasonal and less speculative. That is why hunger has moved solitarily here and at the same time has got a clear choice from right side echelons.

RESULTS OF POOLED DATA

- **Artificial Neural Network Analysis :** Artificial neural network helps us to interpret and estimate the non linear pathway of selected input variables on the output variables. Three distinct layers are over here, one is input layer the second is hidden layer wherein the errors are minimized through activation and the output layers or the dependent variables. For the graphical presentation of the first neural network we can find that x11 that is communication variables, x_{12} that is energy consumption variable have rooted the distinct and dominant impact on the output variable y_1 that is hunger. It has passed through the two hidden layers for the rationalization vis a vis the reduction of error. So we can predict hunger with least of errors from these two input variables one is x_{11} that is communication variables and x_{12} that is energy consumption. So hunger has been contributed characteristically and dominantly by the communication variables and at the same time it has been picked up energy consumption pattern as one of the important predictor for the level of hunger of the respondents.

Now we are switching over to the second output variable poverty. Through the same way we can found that the management orientation x_6 variable, stress perception on voice x_9 variable and homestead land x_{10} variable have contributed in a decisive manner on the output variable poverty. Interestingly all these variables have been passed through three hidden layers for the management of their errors vis a vis estimation of their effect on the output variables.

Now we can switch over to the neural third neural network graph, here we can find that the predicted characters of the output variable that is cognitive differential, different level of knowledge and consciousness about different projects which are meant for poverty eradication and redressal of hunger. We can find that two variables that are x_8 (stress perception on poverty) and x_{11}(communication variables) have in a swashbuckling manner characterized the output variable y_3 that is voice. Here only one hidden layer would be considered through which these two variables have passed.

Now we are moving to words wherein all these output variables have been considered together and at the same time all the input variables have been taken care of. Here we can find that five hidden layers are there and through which all these input variables have generated their impact on the consequent variables those are hunger, poverty, voice and cognitive differential. We take one case for example that is x_8. Here we can find that the x_8 variable that is stress perception on poverty has passed through two hidden layers, one is hidden layer 1: 2 and the other is hidden layer 1:5 and after rooted through these two hidden layers as well as being reduced in terms of error these variable x_8 has impacted on the output variable y_2 that is poverty. So we can see that whenever the respondents are under serious stress they are morose and don't show the interest to interact with others. Then they undergo a kind of withdrawal syndrome and then they lack the information about the projects simply because of there lethargy to access the same. So these variables if we see that to be taken care of to characterize the variable poverty. In another case we can find that the variable x_{11} that is communication variable after being passed through the hidden layer 1:1 and 1:5, they have been characterized the output variable y_4 that is cognitive differential. So, it is interesting to note that the communication variables are the main reason why the respondents have developed their different levels of cognitive differentials on the awareness of different government programmes. These four neural network paradigms have been successful in isolating the dominant variables, determining the number of hidden layers and ultimately how they have characterize the respective and corresponding output variables.

Now based on these four categories of responses we have got four quadrants and whenever we are taking care of the communication variables (x_{11}) and it is considered as an integral part of the respondents, we can find that for the quadrant third most of the respondents are conglomerated. So the skewness of these respondents which are basically dominant in the quadrant 3, we can find that the respondents 135, 61, 58, 21 and 26 are so closely, intensely and integrally links with these variable y_1 that is hunger. Hence we can go strategically to identify these respondents to deal with the issues of communication vis a vis hunger. It is also true for the second picture where there is as usual four quadrants are there and for the third quadrant we can find that respondents 28, 41, 129, 140, 148,40, 80 and 81, they are organically linked with the variable homestead land (x_{10}). So for these respondents and in order to deal with the issues of poverty (y_2), homestead land can be considered as an important predictor.So the social ecology wherein the respondents are haling from, they are quite conspicuously having the size of their homestead land in estimating the poverty. Accordingly for the third graphics we can find that the respondents are almost equally distributed for the first two quadrants and for the third quadrant we will find that the communication variables have been integrated for the responds 85, 92, 95, 96, 97, 98, 99 and 100 are characterized for the output variable voice. So in order to deal with the problem of voices why the respondents are so shaky or they are being dominant by withdrawal syndrome, we can take care of these respondents because the communication variables are organically linked with them. For the last graphics we can find that almost all the respondents are equally distributed among the four quadrants (++,+-,-+ and --). So, cognitive differential is a kind of dependent character which is almost universal for all the respondents. Therefore we can deal with this dependent variable for all the respondents in a uniform manner as well.

So these are the analysis of artificial neural network graphs and in a concluding manner we can say that we need not bother about all the variables while selecting. We have to let the input variables passed through the hidden layers and select few ones. So, for all the dependent variables a few selected variables are there. Therefore for any kind of neural network it is interesting to isolate the dominant variables out of the trivial variables and all these interactions and integrated characters again can be distributed based on their beta regression co efficient and the way they are characterizing the y_1, y_2,y_3 and y_4 all these can be divided into four quadrants. For each quadrants, unique respondents are retaining and for them we can go for unique treatments without bothering about all at a time. Therefore ANN is a very useful and effective technique in delineating and describing the micro level strategical interventions.

PARTIAL CORRELATION ANALYSIS

- **Hunger vs. 13 independent variables:** It has been seen that the risk orientation , stress perception and education are creating highest partial impact as far as the hunger status is concerned. Things may so happened that due to poorer education the respondents are having less risk bearing ability and more psycho somatic stresses and thus collectively they effect on hunger in a substantive manner.
- **Poverty vs. 13 independent variables:** It has been seen that the age , homestead land and management orientation are creating highest partial impact as far as the poverty status is concerned. These three variables have the highest substantive effect than all the remaining independent variables on the said dependent variable. with the age maturity comes and with the maturity the sense of status comes. People starts understanding their actual condition. On the other hand homestead land is an indicator of the status of a person whether he/she is from farm or off farm back ground. The third substantially impactful variable is management orientation. Be it in a farm situation or in an off farm situation the well managed resources are always the key factor. The entrepreneurs or the agripreneurs now a days are more successful than a simple shop owner or a farmer. Indian farmers are poor because they are poor in agripreneurship. They are innovatively poor. They are not managing their resources like the Vietnam or the Israel farmers are doing. So, why the indian farmers are poor? is it only because they are monetarily poor or resource- poor? Or because they a poor in ideas or innovation? Hence, from this above discussion one thing can be clearly concluded that if we want to change the status of our poorer population, we have to train them, groom them, motivate them and encapacited them to generate ideas and to utilize their resources the fullest.
- **Voice vs. 13 independent variables:** age , strees perception on poverty and communication variables are creating highest partial impact as far as their capacity to raise voice is concerned. As with the age the sense of status is grown, in the same manner with the age the desire and capacity to break silence is also elevated. Stress perception is also a stimulator in this issue. If the stress is high, people are frustrated, depressed, morose and finally suffer from withdrawal syndrome where they are totally indifferent about their surrounding ambience. Lastly, the communication variables. They are basically the precursors of one's voice. With communication, information come. And with these information the urge and need for voice in generated.
- **Cognitive Differential vs. 13 independent variables:** it has been seen that the stress perception on voice, energy consumption and education are creating highest partial impact as far as the cognitive differential of the respondents are concerned. Like we have discussed earlier, stress causes depression, inactive nature and lastly withdrawal syndrome. So, it is very different for a person in stress to increase their informative behaviour and awareness.

Energy consumption is basically the indicator of a person's overall mobility. energy consumption is becoming a very important and predominant factor in the entire dynamics of the ecological changes in the sector of agriculture. Now, as far as education is concerned, lesser the education more would be cognitive differential. You are less educated(non formal and informal education is also considered) means you are less aware. Sometimes you can be ignored rather. Why you are not allowed to speak, is can simply because you are a drop out.

- **Canonical covariate Analysis:** In statistics, CCA is a way of inferring information from cross-covariance matrices. If we have two vectors, $x = (x_1, \ldots\ldots, x_n)$ and $y = (y_1, \ldots\ldots, y_n)$ of random variables and there are correlations among the variables, then canonical correlation analysis will find linear combinations of the kind yj which have maximum correlation with each other. T.R Knapp notes that" virtually all of the commonly encountered parametric tests of significance can be treated as special cases of canonical correlation analysis, which is the general procedure for investing the relationships between two sets of variables". The method was first introduced by Harold Hotelling in 1936. CCA for this study has been applied to extract the canonical covariates between two sets of variables. The left side variables and the right side variables. It has been observed that the LS variable again has formed two conglomerations further. Here, voice(y_3) and cognitive differential (y_4) both have picked together seven independent variables those are family size (x_3), economic motivation (x_4) , management orientation (x_6), stress perception on hunger (x_7), stress perception on poverty (x_8), stress perception on voice (x_9) and Communication variables (x_{11}). On the other hand hunger (y_1) and poverty (y_2) have picked together the remaining six independent variables those are age (x_1), education (x_2) , risk orientation (x_5), stress perception on hunger (x_7), Homesteadland (x_{10}), Energy consumption (x_{12}), BMI(x_{13}). So, from CCA we have come to know that the groups of y variables have got precise selectivity to ultimately form a splendid strategy as to cater component related interaction to characterize the perception hunger, poverty and voice.

- **Multiple stepwise logistic forward regressions:** It also appears from MDS analysis the underlying factors that respondents used when they created these dissimilarities. In the Perceptual mapping for rural sector identifies that voice and cognitive development are in the negative side of dimension 1 (implies the efforts to reduce the social stigma) where as voice and cognitive development are in positive side of dimension 1 in case of Perceptual mapping for semi-urban areas. Dimension 2 implies the impact of social stigma that causes social entropy. It is clear that (From dimension 2) in case of rural sector, the combined effects of poverty, voice and cognitive development results hunger which is the driving force of social entropy in the region. But in case of semi-urban sector, the effect of poverty and cognitive development in mass scale cause the hunger.

CONCLUSION

The following are the conclusions which have gone in compliance with the present study.

1. Hunger poverty and voice are interrelated to each other and performing isochronously in the domain of entitlement and empowerment of the target respondents in the study.
2. When four consequent variables (hunger, poverty, voice and cognitive differential) have been estimated and elucidated in terms of 21 exogenous variables, it has generated unique and significant interaction and interrelation having huge strategic implications.
3. Hunger can well be estimated in case of farming respondents by Cropping intensity (x_{11}), Livestock count (x_{12}), Total crop yield (x_{15}), poverty can be estimated by Cropping intensity (x_{11}), Cost of cultivation (x_{16}), Communication variables (x_{18}), voice can well be estimated by Economic motivation (x_4), Stress perception on poverty (x_8), Livestock yield (x_{13}) and cognitive differential can well be estimated through Stress perception on hunger (x_7), Stress perception on voice (x_9), Cropping intensity (x_{11}), Livestock yield (13).

Similarly in the case of non farming respondents hunger can well be estimated

Through Age (x_1), Education (x_2), Stress perception on voice (x_9), poverty can well be estimated by Savings (x_{14}), Communication variables (x_{16}), voice can well be estimated by Age (x_1), Education (x_2), Economic motivation (x_4), Risk orientation (x_5) and cognitive differential can be estimated by Family size (x_3), Bank account (x_{15}), Communication variables (x_{16}) and BMI (x_{19}).

However a general inference from the pooled data comprising of both farm and off farm sector can be drawn in the following line-

(a) Hunger poverty and voice are not necessarily dovetailed to either economic or agro ecological variance. It is also related to communication, psychological and techno managerial variables as well.

(b) Similarly to deal with the issues of poverty and voice of the targeted respondents only economic or agricultural variables are not enough to redress it. It can only be redressed with psychological, motivational and sociological variables at the same time.

(c) The factor analysis has been carried out to generate a single marker variable or character rather than dealing with a score of apparently homogeneous variables and to deal with the issues of hunger poverty and voice the new variables vis a vis factors are farm efficiency, family efficiency, livestock economy. stress psychology and psychosomatic communication has been found very effective in redressing whole issues and dimensions together so instead of bothering for 21 variables , we can better take these five factors as strategic variables to engineer entire issues of hunger poverty and voice.

(d) The canonical co variate analysis has helped us to know the direction of the dependent variables, where hunger and voice are moving together by creating a small Holon of two variables and another two dependent variables poverty and cognitive differential are moving together. So there has been a sub conglomeration of predicted variables that is hunger, poverty , voice and cognitive differential. They have recorded a clandestine selection of exogenous variables in compliance with their interactive homogeneity from the right side set of exogenous variables.

(e) Now the artificial neural network analysis has been carried out to reduce their error through the hidden layers and extracting the significant marker variable which are dominantly charactering the dependent or predicted variables. Here voice has been significantly dominated by stress perception on poverty that means this variable has the highest strategic implication for dealing with the issues of voice.

LIMITATION

1. Since The study is unique by concept it needs further replication in different socio economic, socio ecologic and geo spatial strata.
2. The highest intensity of randomized control trial need to be organized to generate severe hard evidence.(poor economics by Abhijit V Banerjee).
3. The R square value for some of the step down regression is not of satisfactory level. It may indicate that some errors may prevail in the interaction of the variables. Which needs to be corrected further.
4. A community level cataloguing and clustering is necessary to generate " inform reality" and "inform comparison" in between the different components of the diaspora under study. These can be gender group, ethnic group, core agricultural group, off farm group and gender group which can undergo both temporal, economic and socio logical comparison.

BIBLIOGRAPHY

- A.Sen, Poverty and Famine: An Eassy on Entitlement and Deprivation.(1981) *Clarendon press Oxford* , page: 56-87.
- Anannya Chakraborty, S K Acharya, Biswajit Goswami, Chethana.K, *The Social Ecology of Migration in Agriculture:The Micro level reality : ACTA Scientific Agriculture; October 2018,Vol 2 issue 10.*
- Anannya Chakraborty, S K Acharya., *Impoverishment and Vulnerability of The Farmers: A Harsh Reality;* Indian Research Journal of Extension Education, April and July 2019.vol 2 &3.
- Anannya Chakraborty, SK Acharya , *Hunger and Voice: Marginalization and Impoverishment towards a Chaotic Social Ecology;* Indian Journal of Agriculture Business, Vol. 15, No. 1, January-April, 2020.
- Camilleri,-G-M; Mejean,-C; Bellisle,-F; Andreeva,-V-A; Sautron,-V; Hercberg,-S; Peneau,-S, *Cross-cultural validity of the Intuitive Eating Scale-2. Psychometric evaluation in a sample of the general French population,* Appetite-. 2015; 84: 34-42
- Kravva,-V, *Politicizing hospitality: the emergency food assistance landscape in Thessaloniki,* Hospitality-and-Society. 2014; 4(3): 249-274
- Benson,-T; Mogues,-T; Woldeyohannes,-S, *Assessing progress made toward shared agricultural transformation objectives in Mozambique,* IFPRI-Discussion-Papers. 2014; (1370): vii + 29 pp
- 5 Dahl,-B, *"Too fat to be an orphan": the moral semiotics of food aid in Botswana,* Cultural- Anthropology. 2014; 29(4): 626-647
- Baski,-B, *Legally entitled to a full stomach,* D-+-C,-Development-and-Cooperation. 2012; 39(3): 117-119
- Oneko,-J, *Hunger in the Horn of Africa,* D-+-C,-Development-and-Cooperation. 2012; 39(3): 110-111

- Bass,-H-H, *Green renaissance, not revolution,* D-+-C,-Development-and-Cooperation. 2012; 39(3): 107-109
- Gaurav-Desai; Niraj-Pandit; Diwakar-Sharma, *Changing role of Anganwadi workers, a study conducted in Vadodara district,* Healthline,-Journal-of-Indian-Association-of-Preventive-and-Social-Medicine. 2012; 3(1): 41-44
- *Nutrition information in crisis situations,* Nutrition-Information-in-Crisis-Situations. 2012; (24): 70 pp
- Nascimento,-V-T-do; Vasconcelos,-M-A-da-S; Maciel,-M-I-S; Albuquerque,-U-P, *Famine foods of Brazil's seasonal dry forests: ethnobotanical and nutritional aspects,* Economic-Botany. 2012; 66(1): 22-34
- Kunyanga,-C; Imungi,-J; Okoth,-M; Vadivel,-V; Biesalski,-H-K, *Development, acceptability, and nutritional characteristics of a low-cost, shelf-stable supplementary food product for vulnerable groups in Kenya,* Food-and-Nutrition-Bulletin. 2012; 33(1): 43-52
- Mitchell,-J-E; Karr,-T-M; Peat,-C; Wonderlich,-S; Crosby,-R-D; Engel,-S; Simonich,-H, *A fine-grained analysis of eating behavior in women with bulimia nervosa,* International-Journal-of-Eating-Disorders. 2012; 45(3): 400-406
- Wiebelt,-M; Breisinger,-C; Ecker,-O; Al-Riffai,-P; Robertson,-R; Thiele,-R, *Climate change and floods in Yemen: impacts on food security and options for adaptation,* IFPRI-Discussion-Papers. 2011; (1139): viii + 48 pp
- Ecker,-O; Mabiso,-A; Kennedy,-A; Diao,-X-S, *Making agriculture pro-nutrition opportunities in Tanzania,* IFPRI-Discussion-Papers. 2011; (1124): viii + 60 pp
- Headey,-D, *Was the global food crisis really a crisis? Simulations versus self-reporting,* IFPRI-Discussion-Papers. 2011; (1087): vi + 65 pp
- Burkhardt,-M-C; Beck,-A-F; Kahn,-R-S; Klein,-M-D, *Are our babies hungry? Food insecurity among infants in urban clinics,* Clinical-Pediatrics. 2012; 51(3): 238-243
- Nuetah,-J-A; Zuo-Ting; Xian-Xin, *Agricultural export subsidies and domestic support reform under the WTO system: what does it mean for welfare in West Africa?,* World-Economy. 2011; 34(12): 2044-2062
- Minaxi,-R-P; Acharya,-K-O; Santosh-Nawale, *Impact of climate change on food security,* International-Journal-of-Agriculture-Environment-and-Biotechnology. 2011; 4(2): 125-127
- Eksi,-A; Isci,-A, *Global hunger phenomenon and seeking for solution,* GIDA-Journal-of-Food. 2012; 37(1): 39-45
- *Special Issue: Frugality and food in contemporary and historical perspective,* Food-and-Foodways. 2012; 20(1): 1-92
- *Special Section: Food security,* D-+-C,-Development-and-Cooperation. 2012; 39(3): 98-119

- Webb-Girard,-A; Cherobon,-A; Mbugua,-S; Kamau-Mbuthia,-E; Amin,-A; Sellen,-D-W, *Food insecurity is associated with attitudes towards exclusive breastfeeding among women in urban Kenya,* Maternal-and-Child-Nutrition. 2012; 8(2): 199-214
- Hartwig,-R; Grimm,-M, *An assessment of the effects of the 2002 food crisis on children's health in Malawi,* Journal-of-African-Economies. 2012; 21(1): 124-165
- Morini,-P, *Social scientific reflections on hunger, appetite, satiety,* Industrie-Alimentari. 2012; 51(520): 34-36
- Nassirou,-T-Y; He-YuQing, *NERICA: a hope for fighting hunger and poverty in Africa,* Molecular-Plant-Breeding. 2011; 2011: No. 11
- Keatinge,-J-D-H; Easdown,-W-J; Hughes,-J-d'A; Yang,-R-Y; Symonds,-R, *Ensuring future food and nutritional security in the context of a warming world with more limited natural resources,* Acta-Horticulturae. 2011; (916): 47-58
- Allison,-E-H, *Aquaculture, fisheries, poverty and food security.* WorldFish-Center-Working-Paper. 2011; (2011-65): 60 pp
- White,-M-M, *D-town farm: African American resistance to food insecurity and the transformation of Detroit,* Environmental-Practice. 2011; 13(4): 406-417
- Rukundo,-P-M; Kikafunda,-J-K; Oshaug,-A, *Roles and capacity of duty bearers in the realization of the human right to adequate food in Uganda,* African-Journal-of-Food,-Agriculture,-Nutrition-and-Development. 2011; 11(7): 5493-5509
- Pfeiffer,-S; Ritter,-T; Hirseland,-A, *Hunger and nutritional poverty in Germany: quantitative and qualitative empirical insights,* Critical-Public-Health. 2011; 21(4): 417-428
- *Special Issue: Food and public health,* Critical-Public-Health. 2011; 21(4): 385-508
- Rochester,-J-S; Nanney,-M-S; Story,-M, *Assessing foodshelves' ability to distribute healthy foods to foodshelf clients,* Journal-of-Hunger-and-Environmental-Nutrition. 2011; 6(1): 10-26
- Wilde,-P-E; Nord,-M; Zager,-R-E, *In longitudinal data from the survey of program dynamics, 16.9% of the U.S. population was exposed to household food insecurity in a 5-year period,* Journal-of-Hunger-and-Environmental-Nutrition. 2010; 5(3): 380-398
- Richards,-R; Smith,-C, *Investigation of the hunger-obesity paradigm among shelter-based homeless women living in Minnesota,* Journal-of-Hunger-and-Environmental-Nutrition. 2010; 5(3): 339-359
- Popielarski,-J-A; Cotugna,-N, *Fighting hunger through innovation: evaluation of a food bank's social enterprise venture,* Journal-of-Hunger-and-Environmental-Nutrition. 2010; 5(1): 56-69
- Fischer,-G, *How can climate change and the development of bioenergy alter the long-term outlook for food and agriculture?* Looking-ahead-in-world-food-and-agriculture:-perspectives-to-2050. 2011; 95-155

- Nunes,-S-M; Carneiro,-A-A, *GM foods and their impact on food security,* Higiene-Alimentar. 2011; 25(200/201): 40-44
- Chiarolla,-C, *Intellectual property, agriculture and global food security: the privatization of crop diversity,* Intellectual-property,-agriculture-and-global-food-security:-the-privatization-of-crop-diversity. 2011; xxi + 232 pp
- Schutter,-O-De; Cordes,-K-Y, *Accounting for hunger: the right to food in the era of globalisation,* Accounting-for-hunger:-the-right-to-food-in-the-era-of-globalisation. 2011; x + 278 pp
- Evans,-A, *Governance for a resilient food system,* Governance-for-a-resilient-food-system. 2011; 30 pp
- Farrington,-J; Christoplos,-I; Kidd,-A; Beckman,-M; Cromwell,-E, *Creating a policy environment for pro-poor agricultural extension: the who? what? and how?* ODI-Natural-Resource-Perspectives. 2002; (80): 4 pp.
- Johnson,-C, *Uncommon ground: the 'poverty of history' in common property discourse,* Development-and-Change. 2004; 35(3): 407-433
- Schofield,-D-J; Callander,-E-J; Shrestha,-R-N; Percival,-R; Kelly,-S-J; Passey,-M-E, *Labour force participation and the influence of having CVD on income poverty of older workers,* International-Journal-of-Cardiology. 2012; 156(1): 80-83
- Musicco,-M; Pulvirenti,-R-A; Gainotti,-S; Petrini,-C; Riccio,-C; Silvestrini,-B, *A study of the Healthy Growth Charter in socially disadvantaged children,* Annali-dell'-Istituto-Superiore-di-Sanita. 2011; 47(4): 417-423
- Israel,-B-T, *Utilization of non-timber forest products (NTFPs) for economic development in Nigeria,* Nature-and-Faune. 2011; 25(2): 68-71
- Glauben,-T; Herzfeld,-T; Rozelle,-S; Wang-XiaoBing, *Persistent poverty in rural China: where, why, and how to escape?* World-Development-Oxford. 2012; 40(4): 784-795
- Ghaffari,-E; Shahi,-A-S; Ozouni,-D-R; Rostami,-R, *Psychological disorders among inhabit residing in poor social district of Gorgan, Iran,* Journal-of-Gorgan-University-of-Medical-Sciences. 2011; 13(3): 87-93
- Lhamsuren,-K; Choijiljav,-T; Budbazar,-E; Vanchinkhuu,-S; Blanc,-D-C; Grundy,-J, *Taking action on the social determinants of health: improving health access for the urban poor in Mongolia,* International-Journal-for-Equity-in-Health. 2012; 11(15): (20 March 2012)
- Sun-Qiu; Cui-Wei; Zhou-PiDong; Zhang-Pei, *SWOT analysis on rural poverty alleviation of Bijie Experimental Zone in the new period,* Guizhou-Agricultural-Sciences. 2011; (12): 262-267
- Okwoche,-V-A; Obinne,-C-P-O; Otijele,-A-E, *An appraisal of urban poverty and measures for its alleviation in Makurdi Metropolis, Benue State, Nigeria,* Agricultural-Journal. 2012; 7(1): 65-69

- Thilepan,-M; Thiruchelvam,-S, *Microfinance and livelihood development in poor coastal communities in Eastern Sri Lanka,* Tropical-Agricultural-Research. 2011; 22(3): 330-336
- Sedgley,-D; Pritchard,-A; Morgan,-N, *'Tourism poverty' in affluent societies: voices from inner-city London,* Tourism-Management. 2012; 33(4): 951-960
- Quisumbing,-A-R; Baulch,-B; Kumar,-N, *Evaluating the long-term impact of antipoverty interventions in Bangladesh: an overview,* IFPRI-Discussion-Papers. 2011; (1077): vi + 23 pp
- Mogues,-T; Petracco,-C; Randriamamonjy,-J, *The wealth and gender distribution of rural services in Ethiopia: a public expenditure benefit incidence analysis,* IFPRI-Discussion-Papers. 2011; (1057): vii + 62 pp
- Groom,-B; Palmer,-C, *Environmental services and poverty alleviation: either, or, or both?* Discussion-Paper-Series-Environmental-Economy-and-Policy-Research,-University-of-Cambridge. 2009; (46.2009): 23 pp
- Groom,-B; Grosjean,-P; Kontoleon,-A; Swanson,-T; Zhang-ShiQiu, *Relaxing rural constraints: a 'win-win' policy for poverty and environment in China?* Discussion-Paper-Series-Environmental-Economy-and-Policy-Research,-University-of-Cambridge. 2008; (30.2008): 55 pp
- Bussolo,-M; Hoyos,-R-de; Medvedev,-D, *Free trade in agriculture and global poverty,* World-Economy. 2011; 34(12): 2019-2043
- Vira,-B; Adams,-B; Chetan-Agarwal; Shrinivas-Badiger; Hope,-R-A; Jagdish-Krishnaswamy; Kumar,-C, *Negotiating trade-offs: choices about ecosystem services for poverty alleviation,* Economic-and-Political-Weekly. 2012; 47(9): 67-75
- Ayoade,-A-R; Ogunwale,-A-B; Adewale,-J-G, *Impact of national special programme for food security on poverty alleviation among women in Oyo State, Nigeria,* Nternational-Journal-of-Agriculture-Environment-and-Biotechnology. 2011; 4(2): 139-151
- Hooft,-K-E-van't; Wollen,-T-S; Bhandari,-B-P, *Livestock-keeping systems and poverty,* Sustainable-livestock-management-for-poverty-alleviation-and-food-security. 2012; 43-52
- Hooft,-K-E-van't; Wollen,-T-S; Bhandari,-B-P, *Sustainable livestock management for poverty alleviation and food security,* Sustainable-livestock-management-for-poverty-alleviation-and-food-security. 2012; x + 194 pp
- Wani,-S-P; Yin-DiXin; Li-Zhong; Dar,-W-D; Girish-Chander, *Enhancing agricultural productivity and rural incomes through sustainable use of natural resources in the Semi Arid Tropics,* Journal-of-the-Science-of-Food-and-Agriculture. 2012; 92(5): 1054-1063
- Brooks,-J, *A strategic framework for strengthening rural incomes in developing countries,* Agricultural-policies-for-poverty-reduction. 2012; 23-59

- Brooks,-J, *Agricultural policies for poverty reduction,* Agricultural-policies-for-poverty-reduction. 2012; 191 pp
- Benedict,-C; Brooks,-S-J; O'Daly,-O-G; Almen,-M-S; Morell,-A; Aberg,-K; Gingnell,-M; Schultes,-B; Hallschmid,-M; Broman,-J-E; Larsson,-E-M; Schioth,-H-B, *Acute sleep deprivation enhances the brain's response to hedonic food stimuli: an fMRI study,* Journal-of-Clinical-Endocrinology-and-Metabolism. 2012; 97(3): E443-E447
- Nanama,-S; Frongillo,-E-A, *Altered social cohesion and adverse psychological experiences with chronic food insecurity in the non-market economy and complex households of Burkina Faso,* Social-Science-and-Medicine. 2012; 74(3): 444-451
- Groesz,-L-M; McCoy,-S; Carl,-J; Saslow,-L; Stewart,-J; Adler,-N; Laraia,-B; Epel,-E, *What is eating you? Stress and the drive to eat,* Appetite-. 2012; 58(2): 717-721
- Feng-GongNeng; Yang-WenPing; Wang-AiMin; Li-Jing; Xu-HangFeng; Han-GuangMing; Xu-Pao, *Effects of starvation stress on body shape, chemical composition and blood physiological of Cyprinus carpio,* Journal-of-Shanghai-Ocean-University. 2011; 20(6): 814-819
- Valdivia,-C; Danda,-M-K; Sheikh,-D; James,-H-S Jr; Gathaara,-V; Mbure,-G; Murithi,-F; Folk,-W, *Using translational research to enhance farmers' voice: a case study of the potential introduction of GM cassava in Kenya's coast,* Agriculture-and-Human-Values. 2014; 31(4): 673-681
- Hendrickson,-M-K; Gilles,-J-L; Meyers,-W-H; Schneeberger,-K-C; Folk,-W-R, *Choice and voice: creating a community of practice in KwaZulu-Natal, South Africa,* Agriculture-and-Human-Values. 2014; 31(4): 665-672
- Carro-Ripalda,-S; Astier,-M, *Silenced voices, vital arguments: smallholder farmers in the Mexican GM maize controversy,* Agriculture-and-Human-Values. 2014; 31(4): 655-663
- James,-H-S Jr; Sulemana,-I, *Special Symposium: Smallholder farmer voice,* Agriculture-and-Human-Values. 2014; 31(4): 637-681
- DeSouza,-S-I; Rashmi,-M-R; Vasanthi,-A-P; Joseph,-S-M; Rashmi-Rodrigues, *Mobile phones: the next step towards healthcare delivery in rural India?* PLoS-ONE. 2014; 9(8): e104895
- Morojele,-P, *Towards gender equitable schooling: insights from rural teachers' voice in the local context,* Studies-of-Tribes-and-Tribals. 2014; 12(1): 103-111
- Barbole,-A-N; More,-S-M, *A study of sustainable development of Self Help Groups: a review of literature,* Review-of-Research-Journal. 2014; 3(6): ROR-666
- Scott,-K; Park,-J; Cocklin,-C, *From 'sustainable rural communities' to 'social sustainability': giving voice to diversity in Mangakahia Valley, New Zealand,* Journal-of-Rural-Studies. 2000; 16(4): 433-446

- Calaguas,-B-U, *A voice in policy and practice: advocacy in the water and sanitation sector,* Waterlines-. 2000; 18(4): 8-11
- Newholm,-T, *Consumer exit, voice, and loyalty: indicative, legitimation, and regulatory role in agricultural and food ethics,* Journal-of-Agricultural-and-Environmental-Ethics. 2000; 12(2): 153-164
- Kerr,-B, *CANEGROWERS - the strong voice of the Australian cane growing industry,* International-Sugar-Journal. 2001; 103(1233): 425-429
- Pingali,-P-L; Rozelle,-S-D; Gerpacio,-R-V, *The farmer's voice in priority setting: a cross-country experiment in eliciting technological preferences,* Economic-Development-and-Cultural-Change. 2001; 49(3): 591-610
- *Engendering development: Through gender equality in rights, resources, and voice,* Engendering-development:-through-gender-equality-in-rights,-resources,-and-voice. 2001; 364 pp.
- Hutchinson,-S-L; Samdahl,-D-M, *Reflections on the "voice of authority" in leisure research and practice,* Loisir-et-Societe. 2000; 23(1): 237-250
- Smith,-C, *Strengthening the voice of supporters,* Soccer-and-Society. 2000; 1(3): 13-16
- Kleef,-E-van; Trijp,-H-C-M-van; Luning,-P; Jongen,-W-M-F, *Consumer-oriented functional food development: how well do functional disciplines reflect the 'voice of the consumer'?* Trends-in-Food-Science-and-Technology. 2002; 13(3): 93-101
- Bennett,-E-L; Cauley,-H-A; Peters,-C-M; Donovan,-R-Z; O'-Connor,-J-M; Putz,-F-E; Romero,-C; Ghazoul,-J, *Timber certification: where is the voice of the biologist?* Conservation-Biology. 2001; 15(2): 308-319
- Cooney,-A, *Voice of the people,* Health-Club-Management. 2002; (August): 36-37
- Santosh-Mehrotra; Jarrett,-S-W, *Improving basic health service delivery in low-income countries: 'voice' to the poor,* Social-Science-and-Medicine. 2002; 54(11): 1685-1690
- Robb,-C-M, *How the poor can have a voice in government policy,* Finance-and-Development. 2000; 37(4): 22-25
- McClelland, John (1859). *Sketch of the Medical Topography or Climate and Soils, of Bengal and the N.W. Provinces. London: John Churchill. OCLC 884189606.*
- Pinnell, L. G. (1944), *ThePinnell Archive on the Bengal Famine: Evidence to the Famine Inquiry Commission 1944. British Library Doc EUR Doc 911.*
- Wavell, Archibald Percival (1973). *Moon, Penderel (ed.). Wavell: The Viceroy's Journal. Oxford University Press. ISBN 0192117238. OCLC 905255837.*

- *Wavell, General Sir Archibald P. (2015) [1948, War Office, London]. "Despatch on operations in Burma 15 December 1941 to 20 May 1942". In Grehan, John; Mace, Martin (eds.). The Fall of Burma 1941–1943. Barnsley, South Yorkshire: Pen and Sword. pp. 1–111. ISBN 978-1-4738-6360-6.*
- Zhang-Bin; Tian-DeRun; Yu-ChunShui; Zhang-Jing; Tian-Xiao; Deneen,-K-M-von; Zang-YuFeng; Walter,-M; Liu,-Y-J, *Altered baseline brain activities before food intake in obese men: a resting state fMRI study, Neuroscience-Letters. 2015; 584: 156-161*

APPENDIX

DEPARTMENT OF AGRICULTURAL EXTENSION

BIDHAN CHANDRA KRISHI VISWAVIDYALAYA

MOHANPUR, NADIA, WEST BNENGAL

INTERVIEW SCHEDULE OF PHD IN AGRICULTURAL EXTENSION

CONDUCTED BY- ANANNYA CHAKRABORTY

1. Geographical Situation:
   (A) Name of the village:
   (B) Name of the block:
   (C) Name of the district:
2. (A) Name of the farmer:

DEPENDENT VARIABLES

Hunger (Y_1)

1. Amount of food accessed (accessed/recommended)
2. Calorie intake
3. Intake of protein
4. Scheduling of food
5. No. of eggs consumed/ week

6. Gms of meat consumed/week
7. Gms of fishes consumed/ week
8. Gms of vegetables consumed/ week
9. Mls of milk consumed / week
10. Gms of fruits consumed/ week
11. Gms of pulses consumed/ week

Poverty(Y_2)

1. Income(rs/month/head)
2. Wage (rs./day/head)
3. Man days (per month)
4. Proportion of farm and off farm income
5. Regularity of income
    (a) Per day
    (b) Per month
    (c) Per season
    (d) Annual
6. Purpose and amount of credit taken
    (a) From banks
    (b) From co-operatives
    (c) From money lenders
7. Per head credit taken

SILENCE(Y_3)

1. Count of information transferred to
2. Count of information received from
3. No. of times acts as a lead speaker in a community meet
4. No. of times acts as a listener
5. No. of occasions shared information about
    (a) Agriculture
    (b) Health
    (c) Food and nutrition
    (d) Others
6. Needs essential but never spoken out(count)

7. Needs essential partially spoken out
8. Needs felt and spoken out totally
8. Hours of speaking (24- rest hours)

ACCESS TO WELFARE OR DEVELOPMENT PROJECTS (Y_4)

(i) What are the development or welfare projects implemented here

(ii) I was communicated/ got to know somehow/I don't know or communicated either/ I tried but wrongly communicated

(iii) I know the objectives(y/n)

(iv) I am a recipient of the benefits

INDEPENDENT VARIABLES

Social personal variables(X_1)

1. Age
2. Education(yr. of schooling)
3. Family size(C+Y+A)
4. Community affiliation(G/SC/ST/OBC)
5. Economic motivation(10 points)
    (i) Don't you want a better choice?
    (ii) Do you want to try a better option?
    (iii) Do you want to migrate elsewhere for a better option?
    (iv) Are you ready to trigger a better option?
6. Risk orientation(what are the risks due to poverty and how)
    (i) Health
    (ii) Education
    (iii) Political
    (iv) Entrepreneurial risk
7. Management orientation(what are the changes needed to manage poverty and how)(10 points)
8. Stress perception on hunger(what are the anxieties related to poverty)(5 point)
9. Stress perception on poverty(what are the anxieties related to hunger)(5 point)
10. Stress perception on silence(why are you silent)

System and resource variables (X_2)

1. Size of holding(bigha)
2. Income(rs/yr.)
3. Cropping intensity(GCA/NCA*100)
4. Livestock count(no. of cattles/goats/poultry)
5. Livestock yield(egg and milk)
6. Pond and fish
7. Crop yield(q/ha)
8. No. of fragments

Communication variables (X_3)

1. Exposure to media(hrs. of watching TV, reading newspaper, listening to radio)
2. Participation in community gatherings
3. Interaction with input dealers (days/week)
4. Interaction with panchayet pradhan(days/month)
5. Interaction with doctors(days/month)
6. Interaction with opinion / political leaders(days/month)
7. Interaction with KVK/KPS(days/ month)
8. Interaction with ICDS(days/month)
9. Interaction with bank(days/month)
10. Communication gadgets

Management variables (X_4)

1. Cost/ unit(rs./bigha)
2. Wage(rs./day/head)
3. Marketable surplus(%)
4. Marketed surplus(%)
5. Energy consumption
   (a) Fuel(kg/head)
   (b) Electricity(rs./head)
   (c) LPG(kg/head)

Signature of the respondent (With date) | Signature of the interviewer (With date)

INDEX